River Republic

RIVER REPUBLIC

The Fall and Rise of America's Rivers

Daniel McCool

COLUMBIA UNIVERSITY PRESS

NEW YORK

Columbia University Press
Publishers Since 1893
New York   Chichester, West Sussex
cup.columbia.edu

Copyright © 2012 Columbia University Press
All rights reserved

Library of Congress Cataloging-in-Publication Data
McCool, Daniel, 1950–
River republic : the fall and rise of America's rivers / Daniel McCool.
        p.   cm.
Includes bibliographical references and index.
ISBN 978-0-231-16130-5 (cloth : alk. paper)—ISBN 978-0-231-50441-6 (ebook)
  1. Rivers—Environmental aspects—United States—History.   2. Water—
Pollution—United States—History.   3. Stream restoration—United States.
    4. Stream conservation—United States.   I. Title.
GB1215.M34   2012
333.91′620973—dc23   2012006043

Columbia University Press books are printed on permanent and durable acid-free paper.
This book is printed on paper with recycled content.

Printed in the United States of America

c 10 9 8 7 6 5 4 3 2 1

References to Internet Web sites (URLs) were accurate at the time of writing. Neither the author nor Columbia University Press is responsible for URLs that may have expired or changed since the manuscript was prepared.

This book is dedicated to
Jan Winniford
My wife, my friend, my refuge

CONTENTS

PREFACE

I began working on this book a few months after the 9/11 tragedy. Racial and religious hatreds were poisoning world dialogue. George W. Bush had lost the popular vote but was president anyway, eager to roll back environmental protection. War loomed on the horizon. Talk of world-wide calamity brought on by anthropogenic climate change was beginning to make its way into the nightly news. The world seemed cloistered in doom, and it was difficult to find a ray of hope. Rather than wallow in hopelessness, I made a conscious effort to focus on what was right with the world. Surely there had to be a hint of brighter days somewhere. My mind went back to a day in 1994, when there was a knock on my office door.

"Hi, I'm Zach Frankel. I just graduated with a B.S. in Biology, and I want to talk to you about Utah rivers." Zach had a long blond ponytail down his back, a scratch of a beard on his boyish face, and was dressed in cargo shorts and sandals. He got right to the point. "I want to save Utah rivers. I love rivers, but they are under attack everywhere in the state. The water buffaloes [high-ranking water officials] don't care about them." His voice began to rise. "But I wanna take on the buffaloes, lobby the legislature, and make them change the way they treat rivers. I want to change the laws that govern rivers." Zach had become quite animated.

"OK," I said, trying to hide my skepticism. I could just see people in the state capitol—a bastion of concretized thinking—rolling their eyes when Zach came to bend their ear. "How can I help you?"

"I want to look at your files on Utah water. I've got to get educated."

I had edited a book on Utah water policy, so I had lots of files. I agreed to loan my files to Zach.

I occasionally heard about Zach over the next year. Someone told me he had started a new river activist group, the Utah Rivers Council. Two years later I got a call from Zach asking me to attend a meeting of Utah water-conservancy-district officials—the water buffaloes. These officials, the most powerful people in the state when it came to water, planned to discuss the Central Utah Project, a massive federal project referred to by its acronym, the CUP.[1] The basic idea behind the project was to suck water out of the Uintah Basin in northeastern Utah, pipe the water over mountains, store it in reservoirs, and allocate it liberally to hay farms and homes in the middle of the state. One component of the CUP was the proposed Monk's Hollow Dam, to be built on the Diamond Fork River in Utah's Wasatch Mountains. CUP proponents argued that the dam was a crucial part of the project, and dire consequences would ensue if it were not constructed. But Diamond Fork just happened to be one of Zach's favorite rivers, and he wanted to stop the dam. He began working with other environmental groups, elected officials, and the news media. He quickly learned how to be an effective agitator.

At the water-district meeting there was a long, officious table with district officials arrayed behind it. It was like the Court of Kings of Utah water. Zach was one of the few public speakers allowed to address the assembled royalty. He had a sharp haircut and a clean shave and was dressed in a dark business suit. The media were there, which was unusual. Water-district board meetings are usually about as exciting as an accountant's diary, but Zach had called every media outlet in the area and promised them a show. They would not be disappointed.

Zach used flip charts full of graphs and data and revealed a critical knowledge of the CUP, water law, regional economics, and, most important, actual demand for water. In effect, he used the district's own data to prove that it was wrong about "the need" for Monk's Hollow Dam. As he spoke, the Court of Kings grew visibly restless; they were unaccustomed to being challenged, and they clearly did not like the way the media were paying rapt attention to Frankel's presentation. When Zach was finished, the reporters were still scribbling notes; it was the only sound in the room. There was a long, awkward silence, and then one of the white-haired water managers spoke up. "Zach, you just don't understand."

In fact, Zach Frankel understood all too well. He was now dangerous, because if he shared his knowledge with elected officials, the media, and the public, the days of free rein for the water buffaloes would be over. Utah water politics crossed a political Rubicon that day. From now on, grandiose plans to develop every river in the state, regardless of the ecological consequences, would be scrutinized very carefully. It was less than a year after that meeting that the water buffaloes announced they would not build Monk's Hollow Dam. A kid with a ponytail had started a movement to do the unthinkable, and succeeded; they stopped a dam.[2]

As I sat in my office on that gloomy day in late 2001, I smiled broadly, thinking back to those days, and I began wondering: How many other Zach Frankels were out there? How many rivers were being protected or restored? I did a little research and was surprised to find river-restoration and protection efforts going on all over the nation. Small, grassroots river organizations were popping up all over, taking on the near-impossible task of challenging the status quo on rivers—and winning. Clearly, I needed to find out more about this phenomenon. Clearly, I needed to write a book.

This book is a journey into hope. At a time when there is a tsunami of bad news about the environment, America's rivers are a beacon of brighter days ahead. The politics of river restoration and preservation is about grassroots democracy. Ordinary people are literally changing the face of the American landscape with their activism. They are proof that you don't have to arrive in a Learjet to get politicians' attention. River activists are establishing what I call a "river republic," meaning that people from all levels and sectors of society are taking control of their rivers and demanding that they be managed for the benefit of all, with policies determined by an open, democratic process.

But to understand the possibilities of the future, we must first know the past. The first part of the book, "The Fall," sets the historical context. It begins with an overview of the river-restoration and protection movement and then offers what are essentially minibiographies of the two largest water-development agencies, the U.S. Army Corps of Engineers and the U.S. Bureau of Reclamation. I describe these as biographies because old, established agencies generate their own organizational culture and personal signature; like people, they develop personalities. Both the Corps and the Bureau have unique folkways, belief systems, and traditions that have had an enormous impact on America's rivers.

The second part, "Dismemberment," describes why and how we "parted out" our rivers, destroyed their essence, and diminished their value as a natural resource. Each chapter focuses on a use that pulls rivers apart and serves special interests rather than the public interest. These chapters explore agriculture, hydropower, barging, flood control, and water pollution.

The third part, "Resurrection," focuses on less invasive, nonextractive uses of rivers, including urban amenity use, habitat and wildlife, and tourism and recreation. There is an enormous demand for healthy, intact rivers so that the public can participate in these activities. The final chapter takes a look at the future and the evolution of a wholly different concept of the way we interact with rivers.

It is important to understand that this book is not about the supply of water, but rather the supply of rivers. There are many excellent books that describe our diminishing supply of water, but this book focuses on the use of water in its natural habitat—a river. To tell that story, I use dialogue and personal anecdotes, as well as technical data and political analysis. I visited numerous restoration projects and interviewed hundreds of people over a span of nearly ten years. I selected projects to visit on the basis of the diversity of their geography, politics, and objectives. I use poetry and music throughout the book to illustrate the cultural relevance of rivers. The book is also partly a memoir. This hybrid methodology is the best way to capture the complexity and depth of our relationship with rivers.

This book is about America's rivers, but by context it is a book about all of us, a synecdoche writ large. The nation's history, culture, and fate are tied to our rivers. The way we interact with our waterways is a microcosm of what we believe, how we live, and what we have learned in over 200 years of national experience. This book also portrays some of my personal experiences with rivers. They are as much a part of my life as family, friends, and home. Like nearly everyone, I have memories of rivers. I have been moved by their beauty, calmed by their serenity, awed by their power, and at times chastised by their unforgiving force. In many ways the nation has had the same experience.

ACKNOWLEDGMENTS

A book project that required ten years of field research generates a lot of thank-yous. My first debt must be paid to the hundreds of people I interviewed. They graciously gave their time and energy so that I might have a better understanding of their perspective. A lot of people helped me along the way, including Beth Blattenberger, Stephanie Duer, David Harpman, James Kay, Mary Marshall, Julie Morgan, Chris Peterson, Andre Potochnik, Bernard Stolp, John Weisheit, and Gerrish Willis. The late Peter Lavigne, Jay Letto and Dawn Stover, Polly Wiessner, and John and Jan Halter all provided a roof over my head at various times while I was traveling. Several people read early versions of the manuscript, including my close friends Art Lang and his wife Becky Rambo, my son Weston McCool, and my wife Jan Winniford. To them I am forever indebted; they caught many embarrassing errors, made numerous useful suggestions, and convinced me to excise some, but not all, of the bad jokes.

At the University of Utah, several people were instrumental in assisting me while I was writing this book, including Megan Anderson, Jason Berry, Jason Hardy, former department chair Ron Hrebenar, Akiko Kurata, emeritus professor Floyd O'Neil, Jennifer Robinson, and the members of the University Research Committee. Special thanks go to Tasha Seegert, the former associate director of the Environmental and Sustainability Studies Program. I am also very appreciative of the staff at Columbia University Press, especially Patrick Fitzgerald and Bridget Flannery-McCoy. Charles Eberline, John Donohue, and Roy Thomas did a masterful job of copyediting. I also want to put in a good word for William Cullen Bryant, whose evocative nineteenth-century poem "Green River" appears as epigraphs at the beginning of each chapter. Bryant was America's first superstar poet and wrote beautifully about the natural world. He was also a leading political voice, one of the first to support an unknown Illinois politician named Abraham Lincoln. (A complete version of "Green River" follows the Acknowledgments.) Finally, I would be remiss if I did not thank Jeanne Nienaber Clarke, my professor in graduate school who pointed me toward rivers and gave me my first chance to write about them; she is an inspiration.

Closer to home, my son has been an inspiration to me throughout the years I worked on this book. During that time he grew from a small boy into a thoughtful, gracious, and insightful young man who engages me in forthright discussions of ideas, beliefs, and actions—both his and mine. I also need to thank all my river-running friends who got the permits, organized the cook crews, and pulled me out of the water when I tested the submersible potential of my raft. Most of all, I thank my wife, Jan. We got married in the midst of my final binge of writing this book. I was writing day and night on the manuscript when I took a break to get married and go on a honeymoon. It was like there were three of us on that trip, the book never far from my mind. She took it all in stride in a show of loving support and encouragement. She counseled me throughout the writing process, and we celebrated joyously upon its completion.

GREEN RIVER

When breezes are soft and skies are fair,
I steal an hour from study and care,
And hie me away to the woodland scene,
Where wanders the stream with waters of
    green,
As if the bright fringe of herbs on its brink
Had given their stain to the waves they drink;
And they, whose meadows it murmurs through,
Have named the stream from its own fair hue.

Yet pure its waters—its shallows are bright
With colored pebbles and sparkles of light,
And clear the depths where its eddies play,
And dimples deepen and whirl away,
And the plane-tree's speckled arms o'ershoot
The swifter current that mines its root,
Through whose shifting leaves, as you walk
    the hill,
The quivering glimmer of sun and rill
With a sudden flash on the eye is thrown,
Like the ray that streams from the diamond-
    stone.
Oh, loveliest there the spring days come,
With blossoms, and birds, and wild-bees' hum;
The flowers of summer are fairest there,
And freshest the breath of the summer air;
And sweetest the golden autumn day
In silence and sunshine glides away.

Yet, fair as thou art, thou shunnest to glide,
Beautiful stream! by the village side;
But windest away from haunts of men,
To quiet valley and shaded glen;
And forest, and meadow, and slope of hill,
Around thee, are lonely, lovely, and still,
Lonely—save when, by thy rippling tides,

From thicket to thicket the angler glides;
Or the simpler comes, with basket and book,
For herbs of power on thy banks to look;
Or haply, some idle dreamer, like me,
To wander, and muse, and gaze on thee.
Still—save the chirp of birds that feed
On the river cherry and seedy reed,
And thy own wild music gushing out
With mellow murmur of fairy shout,
From dawn to the blush of another day,
Like traveller singing along his way.

That fairy music I never hear,
Nor gaze on those waters so green and clear,
And mark them winding away from sight,
Darkened with shade or flashing with light,
While o'er them the vine to its thicket clings,
And the zephyr stoops to freshen his wings,
But I wish that fate had left me free
To wander these quiet haunts with thee,
Till the eating cares of earth should depart,
And the peace of the scene pass into my heart;
And I envy thy stream, as it glides along
Through its beautiful banks in a trance of song.

Though forced to drudge for the dregs of men,
And scrawl strange words with the barbarous
    pen,
And mingle among the jostling crowd,
Where the sons of strife are subtle and loud—
I often come to this quiet place,
To breathe the airs that ruffle thy face,
And gaze upon thee in silent dream,
For in thy lonely and lovely stream
An image of that calm life appears
That won my heart in my greener years.

—*William Cullen Bryant*

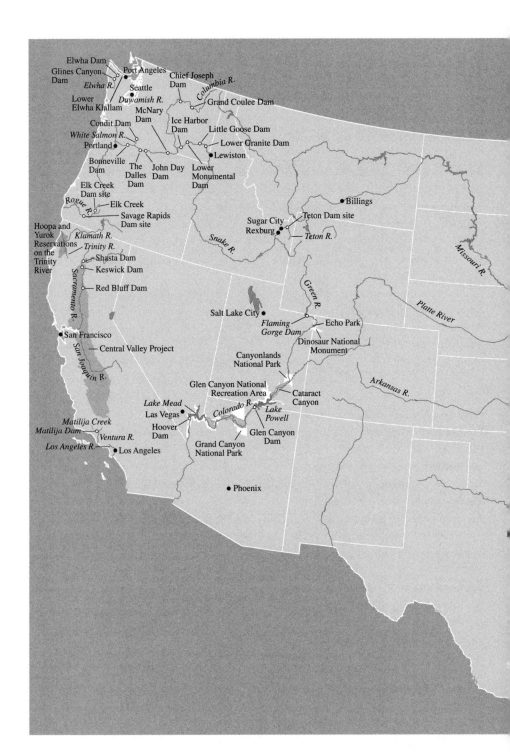

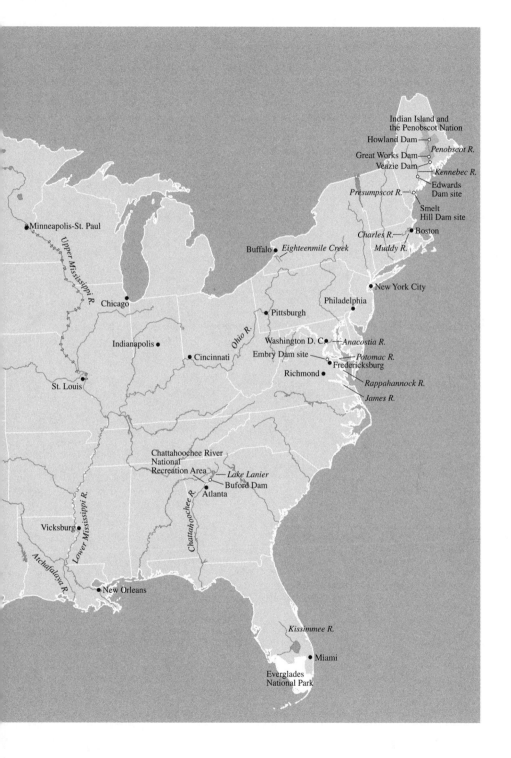

I

THE FALL

America grew up with its rivers. At first wild and beautiful and fearsome, the land and the rivers gradually became peopled and tame and familiar. But there have always been exceptions, places both terrestrial and riverine that defied the hand of settlement, resisted incursion, and remained pristine. For the first 150 years of the nation's existence, these exceptions were viewed as a temporary impediment, a contrarian aberration that needed to be brought under heel. But gradually, we began to see our wild rivers as a special kind of resource, a place of distant memory that offered something that was being lost. That impulse gave rise to the notion that maybe we needed more of these untamed rivers. In this section of the book we will look at the burgeoning river-restoration movement (chapter 1) and explore its premise, its promises, and its challenges. We will then explore the history of American river development through its major proponents, the U.S. Army Corps of Engineers and the U.S. Bureau of Reclamation (chapters 2 and 3). These two agencies remade the rivers of America, often to great service, but sometimes in obeisance to our myopia and greed.

CRUMBLING EDIFICE

But I wish that fate had left me free
To wander these quiet haunts with thee
—FROM WILLIAM CULLEN BRYANT, "GREEN RIVER"

M atilija Dam, sixteen miles from the California coast, sits astride the narrow canyon of the Ventura River amid the velvet green foothills of the Santa Ynez Mountains. I visited the dam in the spring of 2006 to see what the future might look like for many dams.

The road to the abandoned dam was strewn with fallen rocks and debris. The dam had a dark and uninviting look, like an empty house where some crime had been committed. The windows of the operational office were smashed, and rusting cables and slashed wires hung from the abutments. Razor wire conspicuously adorned the top of a fence, just above a sign that said "Danger Razor Wire," as though visitors might not notice that the dam looks more like a concentration camp than a public utility. The face of the dam was a filigree of cracks, and falling boulders had smashed the staircase to the top of the dam. This area has the geologic stability of a stack of greased bowling balls, and one good shake of Mother Nature's mane would turn this dam into beach fill. But this damage was due to neglect, not earthquakes.

Matilija Dam was built by the U.S. Army Corps of Engineers (the Corps) in 1948, during an era when the Corps was an empire unto itself.[1] But, like Matilija Dam, the Corps' edifice has suffered significant damage in the ensuing years. The current lessee of the dam, the Casitas Municipal Water District, would apparently prefer that the public not see this concrete disaster; signs warning TRESPASSING LOITERING PROHIBITED BY LAW were ubiquitous. I sat down beside one of the signs to rest, utterly alone. At my feet were 16,000-pound chunks of concrete that had

Matilija Dam, Ventura River Basin, California, slated for removal.

already been removed from the lip of the dam. I saw a piece of dam about the size of a fist and put it in my pack, a souvenir of the world's largest dam slated for removal. Once the pride of the Corps of Engineers, this dam is now junk.

The actual removal of Matilija won't begin until 2014, but the dam has been notched several times so that the rim of the dam descends through a series of steps from the abutments to the center of the dam. On the spring day when I visited, recent rains had raised the shallow pool of water behind the dam (no one would seriously call it a reservoir), and water spilled over the lowest points on the rim, creating a series of translucent sheets of water that dropped gracefully down the 168-foot face of the dam. The Santa Ana winds were gusting with force, causing the falling water to twist into swirls and drifts. I was briefly tempted to see beauty in these cascades, but the dark, dilapidated face of the dam, marked with painted red circles, quickly dispelled that notion.

The red circles, surrounded by white squares, looked vaguely like targets, as though the dam owners were hoping some nutcase would blow this thing to dam heaven and save them a lot of money and environmental-planning costs. But this is no monkey-wrenching terrorist target; if the dam failed, the meager amount of water stored behind it would create no more than a spring freshet downstream. Most of the reservoir bed is filled with sand, silt, and rock. This renders the dam incapable of storing water or dissipating floods. It generates no electricity, and swimming in the shallow water behind the dam would be hazardous. This dam is utterly, irrevocably useless, other than serving as an eyesore reminder of a time when America built dams willy-nilly without consideration of their long-term impact or utility.

As with all dams, Matilija affects the entire watershed. Because of the dam, the beach at Ventura, downstream, is disappearing. The sand that used to build the beaches is now trapped behind the dam. The beach sand is slowly eroding away, exposing a jumble of rocks and boulders. The lower Ventura River is not much to see. In Spanish its name means "happiness" or "contentment," but the river today evokes less sanguine emotions. Its meager flow quietly edges into the Pacific just north of Surfers Point Beach in the town of Ventura. The river estuary is trapped between two levees. Signs indicate that this area is officially the "Ventura River Sensitive Habitat," although "Insensitive" might be a better term. The river is crowded by a racetrack, a freeway, and a motor-home court. Its course is braided into two or three narrow channels and is choked with dense vegetation, primarily an aggressive invasive species called the Giant Reed. The occasional shopping cart and worn-out tire protrude from the mud. This river clearly has the look of a resource that has been discarded, serving only as a gutter that transports the detritus of society to an outlet on the coast. I saw a lone fisherman, squatting in the mud, watching a pole immersed in a pool of water no larger than a backyard swimming pool; he must surely be hoping for the return of the once-magnificent steelhead runs that brought thousands of anglers to these waters in ages past.

For some people, visions of a massive steelhead run are a spur to action—a dream for a brighter future for the Ventura. To others, a steelhead run is an ominous threat. Both points of view were evident at a meeting to discuss the dam's future, held in the small town of Ojai just downriver from the dam.

The Ojai City Council's chambers look more like an expensive lodge than a government building. The arched verandas, wood-beam ceilings, and white stucco walls lend an air of carefree nonchalance. But that was decidedly not the mood of the crowd the night I sat in on a meeting. The local groundwater management board was holding a public hearing to consider issues related to the recovery of the endangered steelhead trout, and various interests wanted to make sure their perspectives were heard. This was water democracy in action, and it resembled thousands of other contentious water meetings all over America: old users versus new users, farmers versus urbanites, and downstream versus upstream. It was as though the ghost of Mark Twain was sitting in the back row, deftly scribbling the famous phrase, "Whiskey is for drinking, water is for fighting."

Although most people at the meeting agreed that Matilija Dam should be removed, the unanimity quickly broke down when water rights were brought into the picture. The local water district handed out information packets that raised the specter of higher water prices and the possibility that groundwater might someday be used to augment Ventura River flows as part of the effort to restore the steelhead runs.

As an outside observer, my goal was to sit quietly and observe the local political dynamics. It did not take long to discern the perspectives of the people sitting around me. Judging from the guffaws, grumbling, and inappropriate laughter, it was clear that the orange and avocado growers were sitting to my left, and the fish advocates were on my right; they were separated by an aisle and a serious attitude gap.

The "farm people" (their term) were fearful of losing precious water just so fish might spawn. The "fish people" (my term) groaned audibly when one of the farm people spoke of preserving her orange groves and warning that "the government" wanted to take her rights away. It apparently did not occur to her that "the government" built Casitas Reservoir—another dam on the river and the source of her water—at public expense and has delivered that water at greatly subsidized prices for decades to farmers.[2]

On the other hand, her fears are well founded. In a time of scarcity, prices go up, regulation increases, and everyone is asked to get by with less; that is the nature of limited natural resources. The "farm people," despite their apparent oversimplification of the situation, are not naïve; they have been playing the water game longer than most, and they have

6

an innate ability to recognize threats to their water. One of them suggested giving the steelhead "some bottled water and a pick so they can crawl up the dry riverbed without enhancing flows." This comment elicited forced laughs from the farm people and raspy mutterings from the fish people. Another farmer went to the microphone to state unequivocally that "letting water flow down the Ventura River to the sea just for the sake of recreation and fish is just plain ludicrous." To him, oranges are obviously more valuable to society than fish, as though the former are a delicacy and the latter a carcinogen. It is always amazing how self-interest, in stark and naked terms, can color perceptions of reality; one can, as the inimitable Don Quixote put it, "hoodwink thyself."[3]

The fish people spoke with growing impatience. They noted that the steelhead was declared endangered in 1997. A habitat conservation plan must be approved—it is required by law—and resistance only delays the inevitable. They did not seem particularly concerned that some growers might go out of business if the price of water rises appreciably. The growers countered that their farms provide the "ambience" that makes the countryside green and fecund. In essence, they argued that agriculture provides a benefit to the public and therefore should be protected. They hinted ominously that their farms would become urban sprawl if they could not get water at affordable prices. The debate grew more voluble as both sides engaged in what H. L. Mencken called "gaudy oratory."[4]

What is both amazing and prosaic about this public hearing is that it is being repeated all over America, at all levels of government. There is not enough water to meet all the demands, needs, and future visions of everyone. It is that simple, but the resulting politics is never simple. Removing Matilija Dam will solve one small problem in a complex series of problems that plague an overappropriated river. In this case, the river is only about 20 miles long and has, even in the best of times, a modest flow. But never underestimate the capacity of humankind to argue over water. Try going without water for a couple of days, and even the most ardent believer in "let's just all get along" will understand the passions that were expressed at the Ojai City Council's chambers that evening. Of all the dams currently being considered for removal in the United States, Matilija Dam is perhaps the least controversial, but it is a hornet's nest of political intrigue, clashing lifestyles, and righteous indignation. The actual removal of the dam has been stalled by disagreements over how to dispose of the 6 million cubic yards of sediment behind the dam—the future building material of Ventura Beach.[5] As Paul Jenkin, the founder of the Matilija Coalition, put it, "Moving away from traditional flood control to ecosystem management is still a stretch for today's engineers and politicians."[6] Still, everyone believes that the $150-million removal project will be completed and the Ventura River will be whole again.[7] The removal of Matilija Dam is just one small part of an enormous mosaic of river restoration efforts.

A new era in national water policy is approaching. But like all fundamental shifts in policy, it will not be accomplished without rancor or in good time. It may be hard to imagine, but the sad ending to Matilija Dam is the fate of every dam in the world. All dams are temporary, and all dams will one day lose their ability to serve the purposes for which they were built. When I look at Matilija Dam, I see the prospect of a brighter future for some of America's streams and rivers. But to those who still perceive dams as structural marvels that form the foundation of modern society, such a visage has all the appeal of a bad car wreck.

THE DISJUNCTURE

There is a gross disjuncture between what America has done to its rivers and what America wants from its rivers. Nearly all of our rivers are developed, dammed, diverted, dried up, or dirtied. But increasingly Americans want rivers that are clean, free-flowing, teeming with fish and wildlife, and inviting for sports and recreation. People want living rivers, not dead rivers. To achieve that, it will be necessary to restore a lot of river miles and preserve the small fraction of our rivers that are still relatively pristine. This book is about that struggle.

The task may seem insurmountable; nearly all of America's rivers have been altered by 200 years of water development. According to the National Inventory of Dams, there are more than 79,000 dams in the U.S. over 25 feet in height.[8] There are an additional 2.5 million small dams.[9] The lion's share of these dams is privately owned, but the federal government built most of the behemoths that fundamentally altered entire riverine systems. Every major river, with the possible exception of the Yellowstone, has been altered in some significant way.[10] According to the Census Bureau, 85 percent of the inland water surface in the United States is artificially controlled.[11]

The complexity and enormity of the nation's water development are difficult to comprehend. In 1992 the National Research Council succinctly summarized the vast array of human impacts on rivers: "780,000 of these [river] miles are affected by effluents from municipal and industrial treatment plants. Incorporated into our major river systems are close to 12,000 miles of inland waterways [i.e., rivers turned into barge channels]. . . . Along our streams, levees and flood walls traverse an estimated 25,000 miles and enclose more than 30,000 square miles of floodplain."[12] About 600,000 miles of rivers have been directly impacted by the building of dams.[13] Even though there is a great demand for clean, free-running rivers, it is very difficult to find them.

Most of the dams, levees, channels, and other water infrastructure serve a useful purpose and present a net gain to society, but some of these structures are obsolete, dangerous, economically irrational, or ecologically damaging. A few of them are all of the above. An estimated 85 percent of the dams in the U.S. will be at the end of their operational lives by 2020.[14] Much of America's water infrastructure is aging and must be replaced, which creates an opportunity to reevaluate the performance of these projects and look at possible alternatives to those structures that no longer offer the best option. The American Society of Civil Engineers (ASCE), in its 2009 "report card" on America's infrastructure, estimated that, in the next five years, the nation needs to spend $12.5 billion on dams, $50 billion on inland waterways for barges, and $50 billion on levees—and that's just for maintenance. A whopping $255 billion will have to be spent on drinking-water supplies and wastewater treatment to keep them functioning.[15] Projections for expenditures beyond five years are staggering. In short, there is a lot of crumbling concrete in the United States.

Some of this failing infrastructure is a hazard to human life. The federal dam safety program is managed by FEMA (Federal Emergency Management Agency). In a report covering data through 2005, the agency estimated that there were 11,800 dams in the U.S. with high-hazard potential.[16] A more recent count put the number at 15,237. Of that number, 4,095 were declared deficient or unsafe; in other words, they were a threat to human life.[17] The Association of State Dam Safety Officials estimates it will cost $50 billion to repair these dangerous dams.[18]

That's the bad news. The good news, or at least the marginally better news, is that there are streams and rivers that have segments of relatively natural, free-flowing water. The National Park Service is required to inventory rivers to see if they are sufficiently pristine for inclusion into the nation's Wild and Scenic River System (which currently covers less than 1 percent of our rivers and streams). Its most recent inventory, using the verbally effusive criteria of "outstandingly remarkable," included a total of 3,400 river segments, which may sound like a lot of river miles, but it is less than 2 percent of the total of 3.6 million miles of streams in the U.S.[19] Unfortunately, these segments are separated by developed or degraded river sections, so they represent only bits and pieces of rivers and streams. There are only sixty rivers in the contiguous U.S. with natural segments longer than 37 miles.[20] We have also destroyed over half of the nation's wetlands.[21]

Those are sobering statistics, but an extraordinary effort is occurring all across America to restore some of the river segments that have been damaged so that they meet society's increasing demand for clean, living rivers. River restoration can take many forms and may or may not include dam removal. Some rivers can be restored to pristine condition while others are simply improved to meet specific needs. Restoration projects run the gamut from small neighborhood efforts to hugely expensive government-led programs. Most of them begin with a very modest effort, symbolizing Emerson's adage that "the creation of a thousand forests is in one acorn."[22]

Restoration projects involve aspects of engineering, biology, hydrology, and other physical sciences, but I am most interested in the people who are behind these projects. Every river restoration project is a human-interest story, usually with an assortment of heroes, villains, and everyday people coming to the fore to fight for a river they love. Indeed, this book is full of love stories, usually acted out in a political drama that tests the bonds of love and commitment. And, sometimes, there is a happy ending.

INSTIGATORS

It is difficult to grasp the scope of river restoration activities in the United States. Efforts to reclaim rivers are literally popping up everywhere. A recent count came up with a total of 37,099 projects, classified into thirteen categories.[23] No one knows how much money is being spent on these projects; a partial accounting estimated that $14 to $15 billion had been spent since 1990, not including megaprojects such as the Kissimmee River in Florida or the Louisiana coastal restoration.[24] Another study estimated that over $1 billion was being spent on river restoration each year, and forty federal programs were involved.[25] Only a small portion of these restorations involve dam removal.[26] According to the river advocacy group, American Rivers, 888 dams have been removed, including 60 just in 2010.[27] Indeed, river restoration has become something of a cottage industry.[28]

So many rivers are now being restored that a burgeoning scientific literature has developed on how to safely and effectively restore rivers. Some of the research focuses on biological variables.[29] Others look at physical factors such as hydrology or fluvial geomorphology (the interaction between rivers and land).[30] And still others examine the economics, politics, and implementation of river restoration projects.[31] But the literature that has yet to develop sufficiently is engineering. Engineers who express the utmost confidence in how to build big dams must now invent the wheel when asked to take one down.

We have virtually no experience removing big dams or restoring entire riverine ecosystems. These projects represent the new frontier in engineering and biology. They are also pushing the standards for citizen involvement, coalitional politics, and collaborative management. In essence, we are making this up as we go along. Restoration projects typically involve large numbers of people from varied and diverse professional backgrounds and a vast array of expertise. This results in another unique aspect of river restoration that is quite remarkable: the coalitions that form to restore rivers—and engage in the River Republic—are remarkably free of partisanship and ideology. The typical restoration advocacy coalition includes people from all over the political spectrum, including many people who would never characterize themselves as "environmentalists." In an era of venomous hyperpartisanship, this is a refreshing approach to activism.

The typical restoration process usually takes years. But how does a restoration project get started? I interviewed hundreds of people and investigated dozens of projects for this book. Although each restoration is unique, the genesis of most projects appears to center on a singular sort of individual that I have come to call "the instigator." Instigators are typically average Americans, quite often from a background that we would not consider a position of power. They are "housewives," students, small business people, retired persons, or a local politician or government administrator. The one thing they all have in common is a passion for a river, a willingness to risk obloquy and personal attack, and a tolerance for progress so glacial even Job would find it frustrating. It is also helpful if they have a sense of the political art, an ability to smile at twisted logic, and the persistence of a Turkish rug vendor.

Restoration projects appear to follow a fairly predictable pattern. First, instigators may get the wild notion that their favorite river can once again live up to their memories or meet their future hopes. They see the decline or destruction of a river as a fundamental loss to their community—what Elinor Ostrom calls "the remorseless tragedy."[32] In an effort to prevent this, the instigator gathers about him or her a few committed individuals, and they announce to the world that they plan to "save" a river. As John Steinbeck put it, "This is the beginning—from 'I' to 'we'. . . . Need is the stimulus to concept, concept to action."[33] Instant ridicule is the usual response, often from those in positions of prominence and influence. This is not unexpected; if instigators wanted an easy job, they would have chosen a different mission. Instead, their fighting spirit comes from the soul, and opposition merely strengthens their resolve.

The next step is for the river advocates to organize as a group, raise money by selling used gear or grandma's recipes, learn how to decipher the arcane lexicon of government acronyms, and start blundering through the loblolly of politics. In essence, they become the van of the River Republic. At some point, an almost imperceptible change in the public's attitude occurs. People begin to see the wisdom of restoration; they begin to imagine a healthy, intact river and what it could do for their community, their businesses, and their quality of life. The opposition switches from ridicule to producing competing claims and data, but eventually finds a way to make peace, and possibly profit from, the restored river. Only then do the restoration work crews show up.

Instigators must be experts at the tedium and humdrum of street-level politics, but their most important role is to get society to think in new ways—to get an entire culture to look in the mirror and see a slightly different face. In this, instigators are part of a long human tradition of challenging what everyone "knows" to be true. There are innumerable examples of this throughout history, from many fields of endeavor, but music provides an excellent illustration. When Bizet's *Carmen* was first performed in Paris in 1875, it was roundly panned; now it is beloved. When Ravel first conducted *Bolero* in 1928, a woman in the audience screamed, "Ravel, you are insane." Now *Bolero* is perhaps the most performed piece of orchestral music in the world. Leonard Bernstein's *Mass* premiered in 1971 and was called "vulgar trash." Now it is revered. When rock and roll gyrated onto the national stage, it was rabidly attacked as a Soviet plot using "an elaborate calculated scientific technique aimed at rendering a generation of American youth neurotic through nerve-jamming, mental deterioration."[34] Now *that* would make Beethoven roll over. These examples illustrate what happens when novel ideas are unleashed on the unsuspecting masses; initially they are beyond the absorptive capacity of constrained attitudes and must await a gradual crack in the door of acceptance. At such turning points, the public flinches, but then gradually begins to see what was unseen, feel what was unfelt, grasp what was not understood. The renaissance of America's rivers will require the same kind of revelation.

This thumbnail sketch of social upheaval is, of course, an oversimplification, but it does illustrate the challenge. Ultimately, river restoration is a sociopolitical process, and it goes to the very heart of the concept of participatory democracy. The American system of government is a corrupt enterprise, lathered with special-interest money, but it is also an open system.[35] Anyone foolish enough to take on the Big Boys is welcome to try. Sometimes, the political Davids deliver enough nicks and cuts to the establishment Goliath to bring him down, Lilliputian style. Such is the politics of rivers in the U.S. today as the River Republic takes shape.

Many of our rivers are wounded, but they are not dead, even if they have been dammed, drained, or poisoned. For plants and animals, extinction is forever; not so with rivers. The right combination of science, politics, and enthusiasm can revive them. To shamelessly parrot George Santayana, only the dead have seen the end of rivers. Projects such as the Matilija Dam removal are modest efforts that, collectively, promise to remake America's rivers. Not all projects are of such diminutive dimensions. The massive federally funded projects on the Kissimmee River, the coastal Louisiana projects, or the Columbia River's endangered species programs are ambitious to an extent that would have been considered impossible just a few years ago. How far can the zeal for renewed rivers go?

Perhaps the greatest tilt at windmills is occurring in a remote part of the American Southwest. Glen Canyon Dam in Arizona, the epitome of the big dam era, created a reservoir called Lake Powell with a shoreline longer than the West Coast. Until recently no one would have questioned the value of the dam, but now a small but fervent group of river stalwarts has suggested the impossible: maybe Glen Canyon Dam should be decommissioned, and Glen Canyon restored as a national park. It is pure heresy, the unthinkable thought, pushing a big hole in the envelope. But it does illustrate how rapidly our concept of rivers is changing.

THE GLEN

Lake Powell is perhaps the most beautiful "lake" in the world. It is not really a lake but a man-made reservoir, but the Bureau of Reclamation wanted to give it an appealing, natural-sounding name. In a nod to history the Bureau named it after John Wesley Powell, the first man to brave the wild Colorado River through the Grand Canyon.[36] It did so without Powell's approval; whether he would be proud of his eponym, or disgusted by it, will never be known.

The reason Lake Powell is so beautiful is because it did not flood the uppermost 10 percent of Glen Canyon. The higher canyon walls escaped the fate of the lower recesses and remain above water, giving the reservoir a spectacular cradle of amber-hued walls, ice cream–like domes, and gracefully fluted buttresses. The sheer immensity of the reservoir and its surrounding cocoon of beauty have made it an extremely popular destination for water skiers, jet boaters, and those who prefer a placid ride through the red rock in a houseboat. I can certainly understand the appeal. But how many reservoir buffs ever wonder about what's under the water? If the upper 10 percent of this canyon creates a world-class scenic attraction, what does the other 90 percent look like?

Cathedral in the Desert

The year was 2003, and I was sitting below the waterline of Lake Powell, or at least the former waterline. It was spring, when the reservoir should have been brimming with runoff, but the water was at a historic low. My campsite was located on a sandstone bluff overlooking the confluence of the Dirty Devil and Colorado rivers—what used to be arms of the reservoir. A narrow, sinuous channel of moving water made its way down the main-stem canyon amid a wide plain of silt spotted with sprigs of invasive tamarisk. No one had seen this channel for thirty years. Once rich in flora, it was virtually devoid of life. A buoy, lying on its side in the mud, warned ghost boaters in big red letters: MAKE NO WAKE. The buildings and boats of Hite Marina were barely visible in the distance, at least 200 yards from the water's edge.[37]

John Wesley Powell and his men gave the Dirty Devil its name on their first trip down the Colorado. When one of Powell's men asked if it was a good trout stream, his companion replied that no, it was just "a dirty devil."[38] That remark was due to the ruddy color of the silt-laden stream. When the reservoir backed up to this point, the water in the Dirty Devil went still, and the silt fell to the bottom of the reservoir the way snow falls on mountains. It has accumulated since about 1965; I was looking at nearly forty years of annual deposition drying in the sun.

I came to Lake Powell to accompany a group of people from the Glen Canyon Institute on a boat trip. Nearly 2 million people take boat trips every year on Lake Powell, but this one was decidedly different. We came to see features of Glen Canyon that had become exposed to view for the first time in decades. No one under about age 50 had seen these features. The Glen Canyon Institute had gathered together an entourage of reporters, researchers, and Institute members to see how the canyon was renewing itself as the reservoir receded. Richard Ingebretsen, the ebullient, affable president and founder of the Institute, is dedicated to decommissioning Glen Canyon Dam and restoring the canyon. For many people, that's like announcing you want to make Manhattan into a wilderness area and give it back to the Indians. But Ingebretsen, a classic instigator, is unfazed by the enormity of his undertaking or the outrage of his critics. Rich floated through the canyon as a Boy Scout in 1968, just before it was inundated. He spoke of it with reverence, the way one might speak softly of a lost loved one.

Rich wanted us to see a formation called Cathedral in the Desert. It is one of many scenic features in "The Glen," as it is affectionately known by those who love it. The Glen—by its name alone it evokes a yearning to see it, an inviting, cool, green oasis in a sea of blistering desert. Other places in the canyon with lyrical names had yet to emerge from the water—Music Temple, Hidden Passage, The Crossing of the Fathers—but the low water level allowed us to get a glimpse of the Cathedral.

Our first stop was Hite Overlook, a viewing area beside the highway that provides a panoramic view of the upper end of this giant reservoir, which is 186 miles long when full. But on that day the view consisted of a broad field of dried sediment, bisected by a meandering Colorado River. A verdant cloak of nascent tamarisk had sprouted from the exposed silt, prompting someone to mention that it looked a bit like a golf course, plopped down right in the middle of the Southwest's mightiest river.

Standing at the lip of the overlook, Katie Lee looked out over the newly exposed mud flats at the confluence of the Colorado and the Dirty Devil. Katie, in her 80s, is a zealous advocate for the return of Glen Canyon. She pointed to the newly emerged main channel of the Colorado. "That channel did not veer to the north as it does now, but instead hugged the cliffs on the far side of the inner gorge." She was reaching back in time forty years, pulling from her encyclopedic knowledge of the canyon. I had read her book, *All My Rivers Are Gone*, an intensely personal account of her years in the canyon.[39] She got to know herself by getting to know Glen Canyon. Her friend Tad Nichols spent years photographing the canyon before its inundation. The soft elegance and sensuality of The Glen was delicately conveyed in his black-and-white images. As Katie and I stood at the edge of that cliff, I could see that she was lost in memories, reliving her days on the river. The spell was broken by someone asking us to "look at this thing."

An information plaque, placed by the Park Service, explained the geography of the area, but a hand-lettered sign had been taped over the plaque. In bold block letters it announced: "Welcome to the Glen Canyon National Restriction Area. The Future of Lake Powell Overlook." Then, in parentheses: "look upstream—imagine the future . . ." At the bottom, in very small letters, it stated that the "sponsors" of this sign were the Colorado Compact Cartel and the Lake Powell Now and Forever Boosters. For every interest group, there is an opposing, although not necessarily equal, interest group.

At Bull Frog Marina we piled into four speedboats and headed down the main channel to Escalante Canyon, an arm of the reservoir that becomes very crowded in the summertime as boaters vie for a little seclusion and a taste of the red rock narrows.

We first visited Fifty-Mile Canyon in search of Gregory Natural Bridge. Ingebretsen showed us photos taken before the bridge was submerged that showed the rock features in the area. After negotiating a thin ribbon of water that required careful boating, Rich parked his boat below a rock buttress that extended from the canyon wall. "This is Gregory Bridge," he announced with considerable officiousness. It was not obvious to the rest of us; the opening under the bridge was still entirely submerged, but the upper arch of the bridge extended perhaps 60 feet out of the water. At full pool the entire feature is underwater, but a few years of drought could expose the entire bridge to the sun.

Our next objective required that we motor up the narrow confines of Clear Creek, another side canyon of the Escalante. The channel undulated through a series of S-curves. The canyon appeared to abruptly terminate in a huge amphitheater, but a narrow slit of water led up-canyon to yet another amphitheater, so grand and overwhelming that it forced an involuntary inhalation from everyone in the group. "This feature is," announced Rich in hushed tones, "one of the more renowned in the canyon. This is Cathedral in the Desert." Named by early Mormon visitors to the area, the name reflects their spiritual reverence for the area and its extraordinary beauty.

Rich Ingebretsen clearly shares their sense of spiritual awe. He is a man on a mission. To him, one of God's finest creations—The Glen—has been defiled. To him, taking a noisy, smoke-belching motorboat into a place like Cathedral in the Desert is a sacrilege, like riding a Harley into St. Peter's Basilica. Rich's love for this canyon and his visceral sense of awe for its beauty were contagious. It was easy to catch his exuberance

Rich Ingebretsen and friend at Cathedral in the Desert, Glen Canyon National Recreation Area. This feature is currently under the waters of Lake Powell.

and to begin entertaining thoughts of a fully restored Glen Canyon. But at that exact moment, a large powerboat plowed into the Cathedral, full of happy vacationers, and reality intruded upon the dream.

Rich's commitment to the canyon's restoration may be the greatest quixotic crusade since that errant knight jousted with windmills. But water policy is full of surprises. Dismissed as an unrealistic gadfly just a few years ago, Rich Ingebretsen now has an audience. People are curious and are no longer immediately dismissive. Respected water experts and mainline environmentalists are now reevaluating the costs and benefits of maintaining a full reservoir behind Glen Canyon Dam. I don't think politics will ever drain Lake Powell, but upstream diversions and climate change may lead to that eventually.[40] Ingebretsen is a sterling example of how one person, tirelessly working on a dream in his spare time, can have an impact on the fate of a river.[41]

After visiting the Cathedral, we sped "downriver" to Escalante Canyon. This deep, red rock fissure would, by itself, be a national park in any other location in the United States. But it has the misfortune of flowing into the Colorado River in Glen Canyon, and thus the most impressive part of the canyon is submerged under Powell Reservoir. Escalante Canyon, like most of the ninety-six side canyons of The Glen, is marked with buoys, like street signs at busy intersections. We took a hard right turn in our speedboat and entered a maze of side streams, dead-end cirques, and spiraling rock buttresses. A GPS salesman could get lost in here if it were not for the street signs, so we paused for traffic to clear, turned left into Davis Gulch, and slowed the motor to a crawl. Still, our wake bounced off the narrow rock walls and rolled back and forth across the channel. We followed this avenue of water, its surface like green polished marble, around innumerable switchbacks and meanders. All the while the water channel became more confining while the surrounding walls remained high and often overhanging. It was as though we were descending a cleft in the crust of the earth itself. After a mile or two (distances were nearly impossible to judge), we came to a sandy beach that stretched across the canyon—the end of this arm of the reservoir. Perhaps 100 feet above us a distinct change in the coloration of the rock walls indicated the previous high-water mark. We beached the boat and began walking on ground that had not been trod since John F. Kennedy was president. The area closest to the current waterline was practically

The Escalante Arm of Lake Powell, Utah. This location in the canyon is currently submerged under Lake Powell.

devoid of vegetation, but emergent sprigs of grass, tamarisk, and cheat grass had started to appear a little farther upstream. As we hiked up-canyon, the typical array of canyon flora began to appear in profusion. After half a mile, we were in a part of the canyon that had been free of inundation for almost two years. A meadow had formed on both sides of the stream; evening primrose, ephedra, and bunch grass competed for space. Across the creek, a dense copse of young cottonwood trees crowded the bank, some of them 6 feet tall. We knew this area had been free of the reservoir for less than two years because boaters had camped on a sandbar just above the trees and spray-painted their names and the year they visited on the cliff above their fire ring. The fecundity and re-generative power of the canyon were spectacular, especially compared with the dead zone that immediately surrounded the reservoir's edge. The great push and rush of spring floods had fed a portly load of seeds, soil, and moisture into these previously inundated areas, bringing them to life in just a season or two. Canyon country is fragile, but it does have a habit of aggressively reclaiming its own. Davis Gulch was coming alive, recovering its natural green velour and its complex web of desert wildlife.

In 1934 an idealistic young man named Everett Ruess disappeared and was last seen in this canyon. He had devoted his life to wandering these stupendous canyons, many of them unexplored, when some ill-fated turn of events, some encounter with either hostile humans or unforgiving elements, snuffed out his life. If he were to return today, he would not recognize the lower reaches of this canyon, but the portion that was freed from static waters had a rough resemblance to the wild country Everett Ruess knew.[42]

Ruess loved the canyon country because of its natural beauty. But the reservoir that now covers this canyon has a starkly surreal aspect to it; there is nothing natural about cold clear water pressed up against varnished desert sandstone. It is a wholly contrived environment, designed to entertain those who are, it is assumed, incapable of enjoying unaccented nature. Glen Canyon was one of the most awe-inspiring canyons in the world, but today it is but a backdrop to motorboating, like a stage prop behind a prima donna. The standing water of the reservoir has created a compelling distraction that pulls visitors out of a natural setting and places them in a TV-like environment. Visitors need only to sit and be entertained, without expending effort or becoming deeply enmeshed, physically or spiritually, in their surroundings. The lake is deep, but the experience is not. At Lake Powell the crowds literally skim across the top of nature's work, never developing a sense of place because most of "the place" is invisible below the waterline.

As our journey into Glen Canyon drew to a close, I could not help but feel an unstinting, discomfiting sense of envy. Rich, Katie, and a handful of others who had floated the canyon before inundation had experienced something exquisite, primeval, and sublime—something now utterly unattainable to me. I felt like I had been robbed, my son had been robbed, and his son after that. The whole world had been robbed.[43]

Not everyone shares that sense of loss.[44] The Friends of Lake Powell, a group formed to protect the reservoir, lists "25 good reasons not to drain" the reservoir. The dam generates 1,320 megawatts of hydropower. It is enormously popular for flat-water recreation; nearly 2 million people visited Lake Powell in 2009 to cruise in powerboats, scream across the water on Jet Skis, or sit on houseboats in inebriated bliss and watch the scenery go by. The backers of Lake Powell are just as passionate as those who would restore the canyon, reminding us how much people love to recreate in, on, or by water, although the form of that enterprise varies greatly. Quite simply, our species loves rivers, in many different forms.

WATER HUBRIS

This book will tell the story of how America's rivers are experiencing a renaissance. We have come a long way since the days when rivers caught fire and water that flowed to the sea was considered wasted. This is a story about politics in the largest sense because a new era in river policy will require a fundamental shift in values and attitudes. During two centuries of building water-control structures, the nation developed an attitude that I call "water hubris." There are several aspects to this belief.

The first is the notion that water development can produce benefits without incurring costs or trade-offs. During the first two centuries of this country's existence, water development agencies such as the Corps and the Bureau of Reclamation had a peculiar blind spot. Their mission was to build dramatic symphonies of concrete, but there was little or no realization that such man-made contrivances were also destructive. They acted as though only benefits, free of costs, could be obtained by replacing nature with edifice. The professionals in these agencies simply reflected the values and preferences of the society from which they sprang, but in this case it was an attitude that was institutionalized in the form of great government agencies, ordered to "develop the nation's rivers" without apparent awareness of attending loss. This mentality can be seen in a motto for the Bureau of Reclamation: "Our rivers: total use for greater wealth."[45]

Another component of water hubris is the assumption that human works are inherently superior to nature; that we can always "improve" upon a natural river through "conquest." This attitude is exemplified in a remark by William J. McGee, the vice-chair of President Theodore Roosevelt's Inland Waterways Commission, who intoned that controlling the nation's rivers "was the single step remaining to be taken before Man becomes master over Nature."[46] Apparently McGee never saw a category-5 hurricane. Alexander Powell, writing for *Sunset* magazine in 1913 about new irrigation projects, proclaimed that "nowhere has the white man fought a more courageous fight or won a more brilliant victory than in Arizona. . . . The enemy which he has conquered has been the most stubborn of all foes—the hostile forces of Nature."[47] When Hoover Dam was dedicated, Interior Secretary Harold Ickes proclaimed that "pridefully, man acclaims his conquest of nature."[48] In the most extreme version of this, nature was viewed as downright evil. Secretary of Interior Frank Lane averred in 1916 that "the mountains are our enemies. . . . The sinful rivers we must curb."[49]

Another part of the water hubris mentality is the belief that society has a moral right, or at least deserves, to conquer rivers—sort of the water version of manifest destiny. This belief is often couched in religious terms—that God gave us the power to divert water for our use, so we must do so. Sometimes it is overtly jingoistic or racist. William Smythe, an early propagandist for irrigation, wrote that "irrigation . . . is a religious rite. Such a prayer for rain is intelligent, scientific, and worthy of man's divinity."[50] Floyd Dominy, the controversial Commissioner of Reclamation (described in detail in chapter 3), claimed that "man serves God, but nature serves man."[51] A speaker at the 1907 National Irrigation Congress justified huge irrigation projects by declaring that the West "was destined by the Almighty for a white man's country."[52] A. E. Powell, quoted above, wrote glowingly that irrigation in the Southwest was "one of the epics of civilization . . . and its heroes, thank God, are Americans."

Water hubris can be seen in the design and construction of Glen Canyon Dam. Its builder proudly proclaimed that it would stand for 700 years, never thinking to ask if society would want a dam there for 700 years. The Bureau was so sure of its success that it built a depth gage—a giant yardstick on the upstream side of the dam—to a point well below what it considered "normal" reservoir level. In 2004, after five years of drought, the depth gage was 30 feet above the water level. Apparently it never occurred to the Bureau that Mother Nature might have less respect for its achievements than it did.

Water hubris is slowly giving way to a new water ethic. Evidence of this ethic can be found throughout the following pages, but at its most basic level it simply means living like we care about the planet, care about the future, and care about one another. It means treating rivers as the common property of all, cared for by all, managed to serve the nation as a whole for innumerable generations. The political corollary to that is the River Republic, where citizens of all stripes take part in a newfound participatory democracy of rivers.

On my way home from the Glen Canyon boat trip, I stopped at Hanksville, Utah, a town that seems to consist solely of strip motels, minimart gas stations, and abandoned trailers. While I was waiting for a cup of coffee at one of the gas stations, I struck up a conversation with the young man behind the counter. "I guess this isn't the busy season yet," I said. I was the only customer, and he had to brew of pot of coffee just for me.

He tilted back his gimme cap. "Yeah, last month I could count the customers on two hands, but in the summer, with all the people going to Lake Powell, we have to stay open twenty-four hours a day."

"I don't suppose you've heard about the idea to drain Lake Powell?" I had no idea what kind of answer I would get.

"Oh yeah, I've heard about it, but it will never happen. Why, they'd have to practically shut down southern California. They get half their electricity from that dam—it makes electricity, you know. And there's a pipe from the dam that goes to California; that's how they get most of their water. They'd have to cut down all their orchards if that dam wasn't there." He smiled while looking at the floor, clearly amused at the absurdity of the idea. "That's insane," he added.

None of these "facts" are true, and I began to wonder how such fabrications get started. As I walked out the door, the young man looked at me, as though he was wondering, "Is he one of them . . . ?"

"You might be right," I said with a weak smile. I was tempted to ask him if he liked *Bolero*.

2

PLANTERS, SAWYERS, AND SNAGS

The U.S. Army Corps of Engineers

And mingle among the jostling crowd,
Where the sons of strife are subtle and loud
—FROM WILLIAM CULLEN BRYANT, "GREEN RIVER"

Wayne Stroupe knows how to make people feel welcome at the Corps of Engineers' premier laboratory, the Engineering Research and Development Center (ERDC). Formerly called the Waterways Experiment Station, the Center's sprawling campus is just outside Vicksburg, Mississippi. Wayne is a Mississippi native and earned a degree in journalism at Southern Mississippi University by participating in the Corps' cooperative student program. "If it hadn't been for the Corps, I couldn't have finished college," he says.[1] That was twenty-five years ago. He is now the director of the Center's public relations, and he gives substance to the phrase "Southern hospitality." Mr. Stroupe has hosted a great variety of prominent people, from Jacques-Yves Cousteau to Jack Palance, all of whom came to the Center to see how the Corps studies America's rivers. Wayne does not look like the stereotypical Corps bureaucrat. His dress is casual, and he has a close-cropped beard, wears a modest gold earring, and is unflinchingly amiable. He has an upbeat, relaxed attitude that makes it clear he appreciates the Corps and loves his work. Twenty-five years of dealing with both the public and a great variety of Corps personnel have given him a unique insight into the organization—he understands where it has been and where it is going. He knows the Corps' foibles as well as its victories. He is proud of the Corps' accomplishments ("We provide answers and solutions to complex problems"), but he is willing to admit that some projects have been pure pork barrel.[2] And he is eager to face the new challenges of river restoration and environmental mitigation. In many ways, Wayne Stroupe is the personification of the new Corps of Engineers, the Corps of the future, the Corps that must save America's rivers from, among other things, the damage done by past Corps projects.

Mr. Stroupe set up numerous interviews for me at ERDC, ranging from a fish biologist to a soon-to-be retired veteran who specializes in navigation channels and to people who develop methods to restore rivers and estuaries. The diversity among the people I interviewed speaks volumes about the past and the future of this agency. The old-timer with whom I met had recently decided to retire early because, as he put it, "the Corps just isn't what it used to be." He described himself as "an ole Corps boy," but instead of the imperious panache one might expect from a man who once controlled rivers like a stern overseer, he was self-effacing and modest about his accomplishments—which were many. He explained he would not retire early if the Corps was the same agency it had been when he joined twenty-five years ago. "Back then, the Corps would get it right up to a 90 percent probability. Then they started to change; they'd say, get it right up to 75 percent, and we'll fix later what does not work. I guess I'm the kind of guy that sees the glass half-empty."

But his attitude, and his decision to retire early, may have more to do with mission change and less to do with "getting it right." His expertise is in building big-river navigation channels, but there is not much demand these days for new ways to build big navigation channels. His brand of Corps work was once the pride of the agency; now the agency must deal with the damage done by navigation structures. Time and a changing political climate have left him behind. It is surely difficult for him to watch the Corps, after a lifetime devoted to navigation, expand into contradictory missions. Many of the old-timers prefer to retire rather than adjust to the new reality of the Corps. His glory days, when the Corps tried to manhandle giant rivers without regard to the environment, are over, and he is willing to concede that and move on. He is the personification of the old guard in the Corps, a group of men (I have yet to meet a woman in that group) who learned engineering on a slide rule and equated the conquest of rivers with the betterment of mankind. They were the crème de la crème of a society that was vainly confident in man's ability to "improve" the natural world, from redirecting rivers to the mix of species in a reservoir. Words such as *subdue, capture, control,* and *overcome* had an innate appeal to them. And being a military organization, the

Corps may have been guilty of what John Hay (Lincoln's personal secretary) called the "insolence of epaulettes."[3]

James Eads, who built the first bridge over the Mississippi and the first jetties in the Mississippi River delta, once declared: "I believe [man] capable of curbing, controlling, and directing the Mississippi, according to his pleasure."[4] Such water hubris is not unknown today—even after the great floods of 2011—but there is a greater awareness of both the limits of human control and the long-term environmental consequences. We no longer live in a world of engineered certitude and single-minded goals—a reality reflected in the contemporary Corps of Engineers' complex and sometimes contradictory agenda.

"BUILDING STRONG"

A neophyte to the politics of water might logically ask: Why is the army involved in river policy? It is a fair question and has been asked repeatedly for the last 200 years. The Corps of Engineers is one of the oldest agencies in the federal government, created when Washington, D.C., was literally, not just figuratively, a swamp. At that time, the nascent republic was vulnerable, unsure of its borders, and surrounded by hostile forces. The most effective way to defend the nation was to move troops via water, either from sea harbors or on rivers. Thomas Jefferson, who had big plans for western expansion, finagled the Louisiana Purchase, sent Lewis and Clark to search for a western river passage, and signed the authorizing act for the Corps—all at about the same time. The Corps was authorized to hire 20 people (it now has 26,000).

The new agency built coastal forts and harbors and cleared rivers for the passage of boats. By 1824 the principal transportation corridor linking together the growing nation was the Mississippi River and its largest tributary, the Ohio. Congress ordered the Corps to rid those rivers of dangerous "planters, sawyers or snags."[5] This was the beginning of a long era of expansion for the Corps—an era that has yet to end. It did not take long for both the agency and the Congress to realize that some form of Corps project, paid for by the taxpayers of America, could generate a lot of votes and contributions for a legislator's next campaign. The great tradition of water pork barrel was born. Delivered directly to the folks

Riverboats in a lock, Upper Mississippi River.

back home, water projects would help a lot of legislators get elected—again and again and again. They would also keep the Corps busy for a long, long time. Projects became a kind of political currency, to be traded in the halls of Congress for favors and votes.

That, in a nutshell, is why we have so many dams, levees, channels, and waterways. The projects were sometimes in the national interest, occasionally in accord with sound economic principles, but rarely built in an environmentally sensitive manner, and sometimes a gross waste of money. But they made for good logrolling, and the Corps, allied with Congress and beneficiary groups into an "iron triangle" of special-interest politics, has literally changed the face of riverine America by building 11,750 miles of levees, 12,000 miles of navigation channels, 276 locks, 350 hydropower facilities, 926 harbors, and 692 dams.[6] The Corps adopted the motto "Building Strong," intended as a reference to the durability of its construction works, but also a telling metaphor for the political relationships the Corps has nurtured over two centuries.

The Corps' work was virtually unopposed in Congress because it had bipartisan support. Democrats and Republicans who could not agree on the color of a stoplight would stand shoulder-to-shoulder in line for their favorite water project. Occasionally, presidents would challenge the Corps' suzerainty. Franklin D. Roosevelt challenged the Corps' autonomy, but failed utterly.[7] Richard M. Nixon tried to move its river duties into a new Department of Natural Resources; Congress ignored him. Then came the peanut farmer from Georgia, and things began to heat up.[8]

NOT JUST PEANUTS

Jimmy Carter came to Washington with a naïveté that was both refreshing and dangerous. He assumed that people wanted their tax money spent wisely, so upon his arrival at the White House he began to search for ways to cut unnecessary spending. He immediately learned a basic truth about government: "wasteful spending" is an evil in the abstract, but a well-deserved boon to those who receive it. Only a month in office, he issued a directive to cut nineteen projects from the fiscal year (FY) 1978 budget, eleven of which were Corps projects. Carter was out to save the American taxpayer $5.1 billion. He ended his directive by saying, "We must work together."[9] The press immediately dubbed Carter's cuts "the hit list."

Congress was not the least bit interested in working together. Legislators from all over the country went ballistic. Many claimed they were angry because they weren't consulted, as though there was any question as to what they would have said if they had been. Throughout the following year, Carter and Congress played a numbers game, with Carter adding, and then subtracting, projects from his list. By April he had a new list of eighteen projects to be terminated (twelve of them Corps), and five that needed to be significantly modified (two of them Corps). But even more important, Carter issued what was basically a manifesto of the new era in water policy; he outlined five reforms.

First, he challenged the cost-benefit analyses of water agencies that evaluated their own projects. That is a lot like having students grade their own exams. Not surprisingly, the rankings reflect a high level of self-satisfaction. Inevitably, the costs were underestimated and the benefits were inflated.

Second, he wanted to focus attention on dam safety. Teton Dam had just collapsed (described in the next chapter), and many dams in the United States were aging and presented a threat to downstream areas. Third, Carter wanted to introduce cost-sharing, the idea that beneficiaries should pay "their fair share of the enormous capital and operating costs" of water projects. This was a direct challenge to the pork barrel, something-for-nothing approach that had always been the hallmark of water projects. The fourth reform focused on the need for water conservation, noting that "water is not free—it is a precious commodity." And finally, Carter wanted to fundamentally redirect national water policy away from big make-work construction projects that "simply shift economic development [from one region to another] for no apparent policy reason."[10]

Jimmy Carter was a true instigator, even though he operated from a position of power. The reforms he proposed in 1977 seem obvious today; indeed, most of his ideas are now policy. But in 1977 he was thinking way outside the box, and the box didn't like it. Congress screamed, lobbyists fumed, states and localities demanded their handouts, and beneficiaries thundered in righteous indignation and "spluttering wrath."[11] By the end of his term Carter had stopped very few projects, and it appeared that water politics as usual would continue. But even though Carter failed to impose his will on the Corps, he changed the tenor of the debate. He had touched the sacred cow, the public works third rail, and lived to tell about it (albeit as a *former* president). Carter's "hit list" was the starting point of a new dialogue on water policy, a watershed moment that was the beginning of the end for the old way of thinking about rivers. The "fall" was about to turn into the "rise."

When Ronald Reagan became president, he promised to abandon Carter's environmental agenda and initially expressed support for federally funded water development. But Reagan gradually began to pressure Congress to accept Carter's concept of cost-sharing; he was much more interested in spending federal money on weapons than on water development.[12] At the same time, environmental groups and fiscal conservatives were pressing hard for water policy reform. The Corps, which had always had support from both the left and the right, was now taking heat from both sides. Something had to give, and the Corps, always with a fine nose for the political winds, assiduously worked with its allies to accept a modicum of reforms as long as its traditional concrete-centered missions—navigation, flood control, and hydropower—would remain its bread and butter. The Corps, in its official history, described it this way: "With increasing demands on the federal budget and growing doubts about the wisdom of some expensive water projects, a way had to be found to eliminate doubtful projects while responding to legitimate water resource needs in an equitable and efficient manner. The situation required innovation and a willingness to challenge and, if necessary, change old ways of doing business."[13]

The result was the 1986 Water Resources Development Act (WRDA).[14] The only way proponents of change could get some of their reforms through Congress was to accept a massive slew of new projects—377, to be exact. But in exchange for that large dose of pork barrel, Congress agreed to a significant cost-sharing proviso—making special interests pay at least something for their benefits, and requiring the Corps to do fish and wildlife mitigation simultaneously with project construction.

The 1986 Act was a big step forward, but the Corps of Engineers did not suddenly morph into the Wild River Restoration Service. Quite the contrary, while the Corps accepted a new environmental mission, it retained its strong commitment to the very activities that had destroyed so many rivers. Unwilling to give up its traditions, but forced to look to new missions, the agency was trapped, as T. S. Eliot put it, in the "enchainment of past and future."[15] In essence, the Corps began a new life as a schizophrenic agency, tearing up rivers with one hand while restoring and preserving rivers with the other.

This split personality can easily be observed by comparing two projects, one built without regard to common sense or salmon, the other an effort to make up for past indiscretions.

RUSTING REBAR

Imagine a dam so pointless, so obviously irrational, that even the Corps of Engineers did not want to build it, and when forced to do so, this Dams-R-Us agency completed one-third of it and quit. That's Elk Creek

Dam. It sat astride a beautiful mountain stream in south-central Oregon. Up- and downstream from the dam this little river is a verdant natural paradise, coursing briskly through dark volcanic rock and dense stands of pine and Douglas fir. But when I visited this area in 2003, the dam site looked like an industrial nightmare. Giant piles of crushed rock and gravel surrounded the dam, placed there by the Corps for use in the completion of the last two-thirds of the dam. Rusting rebar jutted from the top of the dam, and rolled concrete facing slowly crumbled on the downstream escarpment. The story of Elk Creek Dam should be a short one, but unfortunately it is not. It is a saga that played out over forty-four years from authorization to final demise.

Bob Hunter works for the Oregon river advocacy group WaterWatch. When I interviewed him in 2003, he spoke with the quiet confidence born of having won a spate of victories. He is an attorney by trade, but in his job he must play the multiple roles of lobbyist, diplomat, negotiator, and yes, the lawsuit-filing environmentalist. He has been involved in numerous water conflicts over the years, sometimes under difficult circumstances. At one local meeting, the man sitting immediately behind him leaned forward and said in a low voice, "You know, we shoot people like you."[16] Bob, bearded and bespectacled, at ease in sandals and cargo shorts, looked like he had just stepped off a rafting trip. In a room full of irrigators it would not be difficult to pick out "the river guy."

Bob is a critic of many dams, but he reserved his greatest ire for Elk Creek. "Other dams provide more than enough water for the area. We don't need the flood-control benefits; they are marginal. There is no hydro at Elk Creek. We don't need it for water supply. It's expensive too. It never made fiscal sense. The cost-benefit ratio is 20 cents back on each dollar spent."

Given stats like that, why did it take so long to get rid of Elk Creek Dam? The answer lies in the long tradition of Corps projects and local politics. Elk Creek Dam was the perfect symbol of pork barrel water politics, born in an age when politicians equated dam building with re-election. In this case, the first standard-bearer for the dam was Oregon's Senator Mark Hatfield. He had an obsession for the dam. According to some sources, Hatfield had made a secret commitment to build the dam after floods devastated the area in 1964. Elk Creek was part of a package of dams designed to reduce flooding. The other dams were built, but Elk Creek remained in limbo; clearly it was the weakest of all the proposed dams. Plans to build the dam lingered for twenty years.

In 1982 the General Accounting Office blasted holes in the Corps' cost-benefit analysis, concluding that 76 percent of the benefits were dubious, and costs were significantly underestimated. In short, the Corps was cooking the data in a transparent effort to justify an economically irrational project.[17] Elk Creek Dam was becoming a major embarrassment to the Corps at a time when the agency was facing increasing criticism. When the Corps recommended against building the dam in 1984 because of the abysmal cost-benefit ratio, Hatfield put funding for it in an appropriations bill anyway. The Corps, a military unit, dutifully followed orders and began construction of the 250-foot-high dam in 1986. It started digging keyways and pouring concrete. When representatives from an environmental group met with Corps personnel in Washington, D.C., to discuss Elk Creek, a Corps staff member expressed his frustration with the politics of Elk Creek, asking the environmentalists, "Isn't there anything you can do to get Hatfield off our backs?"[18]

He soon got his answer when environmental groups sued the Corps for failure to complete an adequate Environmental Impact Statement (EIS). They were unsuccessful in the district court but prevailed at the appellate level. That gave the Corps the excuse it needed; it quit work and walked away from the construction site, leaving building materials lying around as though it planned to resume work the next day. That was in 1987. For twenty-one years the unfinished dam just sat there rusting and crumbling, serving no useful purpose. But it did block fish migration up the river, so the government began an expensive trap-and-haul system, which consisted of catching fish in the river, dumping them into trucks, and driving them around the dam—a scheme that had Rube Goldberg written all over it, only Rube Goldberg's contraption did not kill fish.

By the mid-1980s, some elected officials realized that building the dam might cost them more votes than it earned them. The coho salmon was officially listed as threatened in 1997; the unfinished dam blocked 25 miles of salmon habitat. Local newspapers, always a good thermometer of the local mood, began writing dramatic headlines:

"Cutting Federal Deficit Can Start with Elk Creek Dam," *Medford (OR) Mail Tribune* (Dec. 8, 1985)

"Send Back Rancid Pork," *Oregonian* (Nov. 17, 1985)

"Bulldoze Elk Creek Dam," *Sunday Oregonian* (Jan. 5, 1992)

"Demolish Elk Creek Dam," *Oregonian* (Nov. 12, 1992)

"Put an End to Massive Elk Creek Boondoggle," *Medford Mail Tribune* (Feb. 14, 1994)

"Dam's Demise: Scrap Elk Creek Dam as Cheaply as Possible," *Medford Mail Tribune* (Nov. 8, 1995)

"Dam Damage," *Medford Mail Tribune* (Oct. 9, 1997)

"Elk Creek Dam: Good Money After Bad," *Medford Mail Tribune* (Sept. 15, 2002)

Elk Creek Dam had became a metaphor for the schizophrenic mission and conflicted history of the Corps. The argument in favor of Elk Creek was simple: someday the local residents might need the water, and the feds were paying the bill. Local congressmen, county commissioners, and some irrigators supported the dam because of a bedrock belief that smart water users never turn down a handout from Washington.

To a local congressman, a water project offers a visible, tangible, and immediate symbol of legislative prowess, "bringing home the bacon," as it is called. When a water project is completed, there is always a ribbon-cutting ceremony with speeches from elected officials taking credit for "this magnificent new whatever." It does not matter if the project makes economic sense as long as it makes political sense. In the case of Elk Creek Dam, it was as economically logical as a Bernie Madoff investment, but it won votes. Eventually, what did matter was that the dam turned into a vote loser; the tide had gone out on fish-killing dams in the Rogue River Basin.

In 1997 the Corps officially announced plans to notch the dam to allow for fish passage, in response to the listing of the coho and a series of court cases that required extensive environmental review.[19] But local congressman Greg Walden saw it as a betrayal and began a campaign to finish the dam and continue the trap-and-haul operation indefinitely. He inserted a rider into an FY 2003 water appropriations bill that forbade the notching of the dam and instead authorized an $8-million permanent trap-and-haul system, which would cost over $150,000 a year to operate.[20] The National Marine Fisheries Service had issued a biological opinion the year before warning that any option other than dam removal would jeopardize the threatened coho salmon runs in the Rogue River.[21] In 2002, fifty commercial and recreational businesses and organizations publicly condemned the Walden trapping scheme.[22] But the Republican congressman managed to stall dam removal until Democrats took over Congress in 2006. The Corps immediately applied for funds to notch the dam.

The 25-year battle over Elk Creek Dam came to a dramatic close on July 15, 2008, when a carefully engineered blast ripped through the center of the dam. Sixteen more blasts, using 2,777 pounds of explosives, followed, creating an open passage for fish and announcing in a dramatic way that a new era in river policy had arrived on the Rogue. Bob Hunter was there, recording the blasts on a camcorder.[23] In his video, cheers can be heard from the crowd. This was a different kind of "ribbon-cutting ceremony."

Clear across the country, a starkly different Corps project was unfolding. Like Elk Creek Dam, this one was a bad idea that eventually morphed into a good idea, reflecting the curious life and times of the Corps of Engineers.

THE BIG BOO-BOO

In 1962 the Corps of Engineers produced a film to promote its latest big project. *Waters of Destiny* was a promotional documentary designed to justify the Kissimmee River Project, a 100-mile-long channelization project that replaced a thriving wetland river with a sterile channel. The project drained two-thirds of the floodplain in central Florida, interrupted the flow of water into the Everglades, and destroyed an entire ecosystem. The film is pure propaganda, a cross between *Reefer Madness* and *Triumph of the Will*, so outrageously overdone that today it appears as pure camp, a parody of the extremes of that era. But it clearly displays the man-conquers-nature water hubris of the Corps at that time.

According to *Waters of Destiny*, the Corps was determined to quell the "crazed antics of the elements" and the "maddened forces of nature." Central Florida at that time suffered from the "fierce eye of the unseasoned sun." Lake Okeechobee—into which the Kissimmee River flows—was a "monster [that] had to be controlled." The narrator sputters darkly that "progress usually finds anger in its path" while a rattlesnake coils on the screen. After informing viewers that the Kissimmee Project is one of the largest earth-moving projects since the Panama Canal, the narrator explains that "we've got to control the water, make it do our bidding." Once that is accomplished, "water that once ran wild, took lives, and destroyed the land now waits there, ready to do the bidding of man and his machines." The film acknowledged that the massive project would affect wildlife, but assumed that the impact would be positive: the project would "draw the wildlife to its still and quiet surfaces," and "wildlife displaced by Florida's burgeoning population will find a home here." As a result, central Florida would "no longer be nature's fool—a stooge to the impractical jokes of the elements" because "water, once the fierce, uncompromising enemy of this land," would be tamed. The film ends with the confident assertion that the Kissimmee Project would have a 4-to-1 cost-benefit ratio. The project drained 30,000 acres of wetlands, created huge spoil piles, and dug a canal 300 feet wide and 30 feet deep. It was completed in 1971 at a cost of $28 million.

Fast-forward thirty years to a different film, this one titled *River of Dreams*, produced by the South Florida Water Management District in conjunction with—yes, believe it—the Corps of Engineers. The film explains that the Kissimmee River's fish and wildlife were "decimated" by the channelization. With the slow realization that the original Kissimmee Project was an enormous mistake, the local water district and citizen groups began lobbying to undo what the Corps had done. Seeing an opportunity for another big, expensive project, the Corps joined the chorus of voices supporting a second Kissimmee Project to put the river back in its meanders and restore the adjoining wetlands. In 1992 Congress approved the project as part of that year's WRDA, eventually authorizing $578 million for "one of the largest ecosystem restoration projects in the world."[24] According to *River of Dreams*, "The project is viewed worldwide as a model of river restoration efforts."

It is not easy to return a channelized river to a fairly natural state. The project entailed buying over 100,000 acres of land in the floodplain, blowing up a couple of dams and dikes, and converting those huge spoil piles back into river meanders. The Corps says the project will restore "over 40 square miles of river and floodplain ecosystem including 43 miles of meandering river channel and 27,000 acres of wetlands. Restoration efforts will re-establish an environment conducive to the fauna and flora that existed there prior to the channeling efforts in the 1960s."[25] Is the Corps up to such a task? Can it reject the man-conquers-nature mind-set and replace it with a more eco-friendly mission?

The sign on the bridge on Highway 98, just south of Lorida, says "Kissimmee River," but that's a misnomer; it's really a man-made canal as wide as a football field. All that is left of the Kissimmee River at this point is a faint imprint of past meanders in the scrub and grass that now cover the former riverbed. These faint meanders are visible in aerial photographs and are helping the Corps restore the river to its original contours.

The ultimate goal of the Kissimmee Project is to restore a river to its natural watercourse. Ironically, that will require building several dozen new structures—bridges, box culverts, flood gates, and, yes, enlarged channels. How is it that the seemingly simple process of returning a river to its original course could be such an engineering challenge? The answer is that all water policy is determined by politics—a process of compromise in an imperfect world with limited resources. The Kissimmee River floodplain cannot be completely restored; the hand of humankind is much too heavy for that. Houses, highways, strip malls, and an occasional golf course will remain as developers envisioned them—an intrusion in the natural environment. The restoration process must be engineered around these islands of disturbance.

The Kissimmee Project, called the "Big Boo-Boo" by some of the engineers working on the restoration, drained water from an enormous area and put it in a deep, narrow gash cut through the wetlands. Much of this area became grasslands to support the region's cattle industry. Some of it became new subdivisions and other types of structural development. It is neither economically nor politically feasible to remove these structures. Thus, the Kissimmee River restoration process must take place around these structures, an unprecedented engineering challenge. So the Corps uprooted one of its lead engineers, Chuck Wilburn, from his office in Tampa and moved him to Sebring, just a few miles from where he supervises construction of the new water conveyances that will allow water to once again spread across the floodplain, but still protect certain areas from rising waters. Chuck Wilburn's job is to connect the design engineers to the construction teams—to translate the plans of partial restoration into a set of contracts telling concrete companies, electricians, and heavy-equipment operators what to do.[26]

Chuck is an engineer's engineer—efficient, with a keen sense of the relationship between the small, insignificant tasks and the ultimate goal. He talked constantly as he drove me to the various construction sites, displaying an encyclopedic knowledge of not only structural data but also the personalities of the many people who work for him.

"No one foresaw the consequences of channelizing the Kissimmee. Now we are trying to restore it, but it's tricky because there is all this development. We're building box culverts under the highway, we're expanding some channels to handle additional flood runoff, and we have to construct new floodgates. Look at what we're doing around Lake Istokpoga."

Along a narrow country road, near the quaint-sounding Lake Istokpoga (the name means "people have died there" in the native tongue of the region), the soil is the color of tar. Its rich, dark hue is the result of millions of years of episodic flooding when Istokpoga Creek would rise above its banks and spread out like a blanket in a breeze. The soil and the warm climate are ideal for growing flowers, and this area bills itself as the caladium capital of the world. To increase agricultural production in this area, the Corps and state agencies channelized Istokpoga Creek and other local waterways. This allows farmers to both drain off excess water and irrigate in dry periods; they can switch from one to the other nearly instantly, providing the perfect level of water flow. But the rivers no longer flood; they no longer lay down a fecund layer of organic material and silt. Eventually, the soil around the lake will lose its dark hue, and the unique growing qualities of the land will be lost. Sometimes, in an attempt to use more of a resource, we end up destroying it.

But the Lake Istokpoga area is now too developed to be returned to nature, so the Corps is enlarging the two channels that connect the lake to the Kissimmee River and adding more floodgates. This is necessary because the restored Kissimmee River will have much less capacity than the man-made channel and will absorb less water from the Lake Istokpoga area. To make up for this loss of capacity, the Corps must increase the carrying capacity of the side channels. Thus, the larger "de-channeling" project must include digging bigger channels in some areas even while other channels are being back-filled.

If the Corps of Engineers is an unlikely partisan of river restoration, the South Florida Water Management District may be even more so. For most of its operational life, the District, like nearly all water districts, had a myopic view that focused narrowly on development without regard to long-term consequences. But the South Florida District has displayed an uncanny ability to change with the times and adapt to new circumstances, and it is now the major partner with the Corps on the Kissimmee River restoration project. I met with Joe Koebel, the District's senior environmental scientist. He explained how the Kissimmee Project fits into a larger effort to restore an entire ecosystem.[27] "The Everglades has its own restoration project, as does Okeechobee Lake, so there are three restoration projects in South Florida, but they are all connected hydrologically. The Kissimmee River flows into Okeechobee and is the headwaters of the Everglades. We completed the first phase of our project in 2001, and we were able to reestablish continuous flows."

I asked how the District came around to the idea of reversing the damage done by the original project.

In the water district the idea of restoring the river began even before the channelization was complete. There was an organized movement to restore the river; that core group of people fought for the 1976 act in the state legislature [which began the process of investigating restoration as an option]. You have to remember that one of the missions of the District is to protect natural resources, so we had an impetus to push for restoration from the beginning. The water district back in 1994 partnered with the Corps to do this project. One of the key things is that the whole effort is based on good science; that's why the District supports it.

In other words, the instigators of this project worked from within the water establishment to promote restoration. They used their expertise and the broad outlines of the District's mission to build support for a fundamental change in policy. But, as with all major changes in river policy, there was opposition.

Koebel explained. "There was a small group that actively resisted the restoration, called ROAR—Realists Opposing Alleged Restoration. They had a website and they were very outspoken."

ROAR consisted primarily of local county commissioners, and they participated in several project reviews. But ROAR is no longer active; apparently the "realists" discovered that reality was not cooperating with them.

In contrast to ROAR, a host of organized interest groups support the restoration, including the Florida Audubon Society, the Fresh Water Fish Commission, and the Sierra Club of West Palm Beach. They work closely with the Corps and the District to monitor the project and lobby for continued funding. All the recent governors of Florida, including Lawton Chiles, Bob Graham, and Jeb Bush, have been very supportive.

Despite the widespread support for restoration, Koebel made the same point as Chuck Wilburn: they must work within practical limitations. "The law specifies that we have to have at least as good flood protection as the pre-project; this is a significant limitation. We can't restore the whole river; we have to leave areas at the north and south ends of the project still channelized and developed, for that reason."

After meeting with Chuck Wilburn and Joe Koebel, I was eager to see the section of the river that had already been restored, so I headed to a place called the Riverwoods Field Laboratory. There I met with Loisa Kerwin, the director of the laboratory, who offered to take me on a tour of the restored area. Riverwoods is a research unit established as a partnership between the District and the Florida Center for Environmental Studies. I expected to be led out to a waiting boat, but first she wanted to educate me about how things have changed in South Florida; she showed me the films described previously, *Waters of Destiny* and *River of Dreams*. After that, I was quite prepared to see those maddened forces of nature.

The boat, christened the *Kissimmee Explorer*, was something like a houseboat, but designed as a moving open-air observation deck. There were about ten people onboard that day, ranging from students to tourists to scientists. We shoved off and soon passed an enormous alligator basking in the sun—one of the beneficiaries of the restoration. The river itself was a marvel of natural regeneration. Along stretches of the river that had been re-meandered only a year earlier, vegetation crowded along the water's edge, as dense as any jungle. It looked more like the Amazon than

Florida, and it was difficult to believe that such a place could exist surrounded by the huge sprawl of subdivisions and agribusiness.

As we quietly motored along the labyrinthine channels, Ms. Kerwin explained how the staff at Riverwoods had improved their relationship with
the people in the area, who initially resisted restoration: "Our relationship
with locals used to be quite contentious, but now we work a lot with locals
and we have a better relationship, in part because the results of restoration
are clearly visible: improved habitats, better fishing and birding."[28]

I met with one of the locals after our river tour. River restoration
efforts seldom succeed without the support of local people. Indeed, it
is usually the locals who initiate the demand for restoration, and they
rarely fit the stereotype of the green activist. Patrick Luna worked on the
construction crew that built one of the structural components of the
project. He has long roots in this soil.[29]

> I was born in the hard-pine house where we still live. My grand-
> mother was a midwife, and she delivered me. This was a great place to
> grow up. I had a dog and an old square-nosed pole boat. I spent my
> boyhood out on the wetlands. Even today I could go out there at night
> without a light and I'd know where every fishing hole, gator hole, and
> spring is, from Lake Kissimmee to Okeechobee.

After the Kissimmee channelization was completed, Patrick watched
as the world he knew and loved disappeared.

> I roamed this river as a boy, and I saw it change. It used to be wetlands
> with a lot of waterfowl and a lot of different vegetation. After they
> dredged it the birds and waterfowl just left. To me, being a kid, it was
> exciting to see it get dredged, but I later realized it was a disaster. As I
> was growing up, we had a lot of deer, wild hogs [javelina], and turkeys;
> the game was plentiful. Before they dredged, the river was fantastic. It
> wasn't anything to catch fifteen to twenty bass a day; and that was just
> a normal day. Same with catfish and bluegill and shellcracker.

Mr. Luna drove me to a part of the project that involved placing box
culverts under the highway so the road would not flood with the increased
flows coming down the restored Kissimmee. When we discussed the

past, his voice had a tinge of bitterness, but his demeanor changed to hope as he talked about the restoration.

> This valley went from some of the prettiest country in the world to just a desert. The fishing holes went to nothing, the creeks dried up, the water got stagnant. All our creeks would just turn to stagnant water, with no fish. We couldn't swim in them any more. It took years before the water quality started to clear up. I moved away and started farming at Pompano, for twelve years, then came back here in 1980. My dad got sick so I started running the family ranch. But now I'm a superintendent on this project, and I've been working on it for seven years.

Patrick could clearly see the long-term benefits of restoring the Kissimmee River, and he is proud to be part of it. For him, it is a matter of restoring a lifestyle he knew in his younger years, when hunting and fishing were a central part of his life.

> The restoration project will improve the hunting; it's already 90 percent better than just ten years ago. We've seen a few Florida panthers around the house.[30] When I was a kid, we'd see the panthers all the time, but they didn't bother us, they'd run. My grandmother always said just carry a stick with you to scare 'em off. A lot of game is coming back. We're seeing more eagles and more waterfowl. The river restoration is probably one of the greatest things to ever happen in this country to clean up the quality of the water for people in this area, from Orlando all the way down to the 'Glades.

The Kissimmee River restoration project is a marvel of engineering, but it is an even greater testament to the "crazed antics of the elements" that, if given a chance, will re-create what was once thought to be lost forever. It is scheduled to be completed in 2014.[31]

UNDER SIEGE

The Corps deserves kudos for its cutting-edge work on the Kissimmee River restoration. That effort is already becoming a model for how to repair a river. But the hopeful visions conjured up by a revived Kissim-mee must be tempered by other Corps projects that have not changed much since Elvis Presley was at the top of the charts. The schizophrenia of the Corps is very much in evidence today. Indeed, it hardly makes sense to talk of one agency; the various districts have become minifiefdoms, each with a different organizational culture. Some districts are respon-sive to change, while others retreat into a bunker mentality and try to withstand the barrage. In subsequent chapters we will encounter the Corps' dated steamboat mind-set on the Upper Mississippi River, but also see its changing role along the coast of Louisiana. We will experi-ence the agency's obduracy on the Lower Snake River, but see evidence of a growing sensitivity in other western watersheds. We will see how some of the Corps' experts are assisting restoration and taking out dams in some locations while other Corps personnel are fighting restoration.

These contradictions are due to the basic fact that the Corps has never abandoned its tradition of building and maintaining big, pork-laden water projects that serve a narrow but well-organized clientele. For that reason, there has been a relentless effort in recent years to reform the Corps and bring the entire agency—not just parts of it—into the modern era. For over thirty years the Corps has been under siege. The siege be-gan with Jimmy Carter, reached a milestone with the 1986 WRDA, and continues today.

The passage of the 1986 WRDA made it clear that the Corps could be challenged successfully. Some people in the agency thought that the Act gave them a new lease on life and satisfied their critics. But in fact it gal-vanized the agency's detractors and inspired them to push even harder. Critics of the Corps in Congress inserted provisions in the 1990 WRDA that mandated environmental protection as a primary Corps mission and established a "no net loss of wetlands policy."[32] The 1996 WRDA authorized the Corps to engage directly in aquatic ecosystem restoration projects.[33] In 1994 a major review of the Corps' flood-control program was published, called the Galloway Report, which recommended major changes to the Corps' approach to floods (more about this in chapter 7). In short, in the 1990s there was a lot of pressure on the agency to change.

The Corps responded in classic fashion by pushing in two contradic-tory directions. First, it created the *Challenge 21* program, also known by the more verbose title of the *Riverine Ecosystem Restoration and Flood*

Hazard Mitigation Initiative, which required the Corps to "emphasize nonstructural approaches to preventing or reducing flood damages."[34] The Corps publicly acknowledged that river restoration was a "priority project output."[35] By the end of the 1990s, environmental protection and restoration made up about 22 percent of the agency's budget. As one Corps official put it, "The Corps recognizes [that] doing this kind of work is a good opportunity to fix past sins."[36]

But this 200-year-old agency was not about to abandon its longtime clientele or traditional missions. Internally, the Corps began a concerted effort, called the *Program Growth Initiative*, to push hard for big, traditional projects that emphasized structural approaches (i.e., pouring large amounts of concrete and delivering big construction contracts to the districts and states of favored members of Congress). Equally important, the Corps refused to abandon its antiquated planning rules, called the "Principles and Guidelines," and continued to use a cost-benefit formula that would make even the scandal-scarred accounting firm Arthur Andersen blush. These procedures virtually guaranteed a favorable decision for big construction projects while undervaluing nonstructural alternatives.

At the end of the 1990s, the Corps got caught red-handed cooking the data to justify a massive new expansion of the lock-and-dam system on the Upper Mississippi. A Corps economist, Dr. Donald Sweeney, had to file for whistle-blower status when he refused to falsify his cost-benefit analysis (this bizarre story is covered in detail in chapter 6). This was just the latest in a series of embarrassments over the agency's fantasyland economic justifications (see Elk Creek, above). The agency's resistance to change and its willingness to overtly bias its studies and analyses guaranteed that the battle for Corps reform would continue at a fever pitch into the new century.

The attack on the Corps came from many different directions. A 1999 report by the prestigious National Research Council called for a "comprehensive revision" of the Corps' "Principles and Guidelines" and basically accused the agency of being completely out of whack with national goals and contemporary values. The Council suggested diplomatically that this "may be the result of skewed benefit calculations" and "an institutional bias against nonstructural projects."[37] In other words, this emperor was stark naked.

In 2000 the National Wildlife Federation teamed with Taxpayers for Common Sense to produce a hard-hitting report titled *Troubled Waters: Congress, the Corps of Engineers, and Wasteful Water Projects* that listed what they viewed as the twenty-five most wasteful projects. Stopping these projects, they claimed, would save $6 billion. It was like Jimmy Carter all over again. Of course, those who were on the receiving end of the $6 billion were not about to let go of their pork-barrel pocket lining without a fight.

At the same time that *Troubled Waters* was generating headlines, Michael Grunwald of the *Washington Post* began a series of articles on the Corps. Grunwald was an old hand at writing about water issues, and he used the cutting edge of his expertise to flay the agency. Grunwald broke the story on the Corps' self-inflating Program Growth Initiative and wrote extensively about Dr. Sweeney's whistle-blowing activities. Most important, he exposed a dramatic sequence of policy changes at the Corps that clearly revealed who was pulling the strings. In the aftermath of the Sweeney whistle-blowing episode, which was a public relations disaster for the Corps, the Secretary of the Army, with the full support of the Clinton administration, announced major reforms in Corps policy. Then, just one week later, the Secretary reversed himself and put the reforms on hold at the request of three powerful Republican senators. The Senate at that time was under Republican control, and these three senators had enough clout to stop the reforms.[38] One of them was Pete Domenici of New Mexico, who had funneled $14 million in Corps funding to his state the year before.[39] A spokesman for Senator Domenici argued that the senators simply "want[ed] the Army Corps to stay the way it is."[40] It should be noted that many Democratic legislators have been just as cozy with the Corps.

Toward the end of the Clinton administration, the Corps saw an opportunity to roll back wetlands protection—a move that was supported by the incoming Bush administration.[41]

Not everyone in Congress was enamored with the Corps, however. Even though the usual litany of congressmen and senators automatically lined up in defense of the agency's traditional big-ticket projects, a growing number of legislators opposed the environmental degradation and spendthrift ways of the Corps. In 1990 Congressman Ron Kind and Senator Russ Feingold, both Democrats, introduced the Army Corps Reform

Act to require external review of proposed projects and force the agency to perform concurrent environmental mitigation. A similar bill was introduced by Congressman Tom Tancredo, a Republican.[42] Those bills did not pass, but some elements of the environmental mitigation requirements were included in the WRDA that year. But the Corps once again escaped calls for an independent external review of its proposed projects. Another bill was introduced in 2000 to require external cost-benefit analysis of Corps projects, but it did not pass.[43] The students were still grading their own exams.

Critics of the Corps continued to focus on the need for external review, environmental mitigation, more cost-sharing, and the de-authorization of old projects. A bipartisan Army Corps Reform Caucus was created in the House in 2002. At about the same time, a coalition of over 100 interest groups formed the Corps Reform Network (the number of member organizations had grown to 192 by 2011). That same year the National Research Council recommended that all Corps reviews should be overseen by an independent review board.[44] Also that year, Senators Russ Feingold (Wisconsin), Robert Smith (New Hampshire), and John McCain (Arizona) challenged the Corps head-on when they introduced the Corps of Engineers Modernization and Improvement Act of 2002. This bill would have fundamentally changed the way the Corps evaluated projects and would have led to the de-authorization of hundreds of projects. The three senators even had the temerity to include a requirement that the National Academy of Sciences study the feasibility of decommissioning barging waterways that were not economically justifiable.[45] This was not just an incremental change or a benign addition to the Corps' roster; it was a direct threat to the heart and soul of the agency. Any economist this side of Marx could see that some of the nation's waterways do not meet even the most imaginative test of economic rationality, so applying such a standard would require the Corps of Engineers to abandon its archaic tradition of clearing sawyers and snags and find something else to do (such as repairing the damage done by past projects).

A panoply of interest groups lined up in favor of the bill, ranging from the National Taxpayers Union to Environmental Defense. Even elements of the construction industry supported the bill, with one op-ed headlined "Pork: Rotten to the Corps."[46] President George W. Bush got in on the act and chastened the Corps in his budget message for FY 2002: "Serious questions have been raised about the quality, objectivity and credibility of Corps reports on the economic and environmental feasibility of proposed water projects."[47] Finally, it looked like the Corps would be dragged kicking and screaming into the twenty-first century.

The bill never even made it out of the Senate Environment and Public Works Committee. It was reintroduced again in 2004, with requirements to increase the cost-benefit ratio to 1.5-to-1, require independent review, and increase the local cost-share on flood-control projects to 50 percent. Senator Feingold's statement in introducing the bill is worth quoting because it explains the long history of Corps pork barrel and efforts to contain it:

> In 1836 a House Ways and Means Committee report discovered that at least 25 Corps projects were over budget. In its report, the Committee noted that Congress must ensure that the Corps institutes "actual reform, in the further prosecution of public works." In 1902, Congress created a review board to determine whether Corps projects were justified. The review board was dismantled over a decade ago, and the Corps is still linked to wasteful spending. Here we are, more than 100 years later, talking about the same issue.[48]

Again, the bill did not get out of committee.[49] With a show of uphill tenacity, the Corps reform bill was reintroduced again in 2005, and again it never made it beyond the grumbling senators on the committee.[50] The fact was that, despite all the criticisms and exposés and scientific studies, and all the damage done to America's rivers, the Corps still held a trump card in the U.S. Congress; it handed out the goodies, and the recipients of those goodies were entrenched, well organized, and connected. In addition, the water authorizing committees in Congress tend to attract those senators and representatives who stand to gain the most from the Corps' porcine tendencies, so they work hard to stop legislation that would curtail the flow of dollars. On those committees, bringing home the bacon trumps rational economics and the national interest. By 2005 over 500 Corps projects had been authorized but never funded, creating an enormous backlog. Yet this didn't stop the Corps and its allies in Congress from seeking new authorizations.

Still, the critics persisted. The National Wildlife Federation and Tax-payers for Common Sense teamed up again to write another critique of the Corps, detailing the numerous studies that blasted the Corps and pointing out what they viewed as the twenty-nine worst projects, which would cost $12 billion and do enormous environmental destruction.[51] The Government Accountability Office (GAO) issued five critiques of the Corps' cost-benefit methodology, all of them harshly critical.[52]

Reformers in Congress sensed that the combination of bad press for the agency, support from the Bush administration to cut federal spending on water projects, and a ballooning federal debt that was approaching $8 trillion just might get reform over the top. Their opportunity was the next authorization bill, the 2005 WRDA. It was obvious by now that the only way the reformers could get sufficient political support in Congress was to insert reforms into a massive authorization bill that approved hundreds of new projects. A WRDA had not been passed since 2000, so there was a lot of pent-up desire to spend money on new projects. In effect, the reformers were using current pork barrel to help curtail future pork barrel. That is a little like drinking to get sober, but water politics has always been full of perversities.

The fight in the Congress over the reforms and new projects in the 2005 WRDA was fierce. Stunningly, the first thing that Corps proponents added to the 2005 WRDA was authorization for the enlarged locks on the Upper Mississippi that had been the focus of Donald Sweeney's whistle-blowing. Numerous studies had made it clear that the enlargement project made no sense, but Senator Christopher Bond from Missouri—a state that moves vast amounts of grain down the river for sale in Asia—used his senior clout on the Senate Environment and Public Works Committee to insert the Upper Mississippi project into the bill. Arguments over various provisions in the bill stalled it until 2006, when negotiations continued without resolution.

Finally, Congress passed the 2007 Water Resources Development Act. The vote in the House was a lopsided 394 to 25; in the Senate it was 91 to 4. It had taken seven years for a new WRDA to pass. What led to success in 2007, especially after so much conflict and calls for change? The answer involves something known in the halls of Congress as a "Christmas tree," a bill so larded with goodies and ornaments that it looks like a heavily decorated Yule tree. The 2007 Act gave the reformers much of what they wanted, but also authorized hundreds of projects at an estimated cost of $23 billion.[53] There was something for virtually everyone in the bill—except, of course, the American taxpayer. Funds were authorized for the Matilija Dam removal and Kissimmee River restoration (and several other restoration projects that are covered in subsequent chapters). And there was funding authorized for literally hundreds of water development projects, ranging from Alaska's "bridge to nowhere" (Sec. 4007) to stream-bank protection for "St. Johns Bluff Training Wall" in Florida (Sec. 1003) and virtually every wet spot in between. The authorization of the Upper Mississippi River lock enlargements proved beyond doubt that old pork barrel develops a life of its own—in this case nine lives.

In return for accepting all these projects, the reformers achieved several significant changes in the Corps' mission. They forced the Corps to revise its antiquated "Principles and Guidelines" so that they focus more on "protecting and restoring the functioning of natural systems," stop unwise floodplain development, and maximize sustainable development. Moreover, the guidelines must be based on the "best available economic principles and analytical techniques" and must provide equal consideration for nonstructural alternatives (Sec. 2031). This was a direct result of the barrage of criticism directed at the agency's skewed cost-benefit analyses.[54] The new law also established a process of independent review of all projects that cost more than $45 million or were deemed controversial (Sec. 2034). Finally, the students were no longer grading their own exams—at least the big ones. Proponents of reform effused that the 2007 law "enacted a new national water policy that requires a fundamentally different approach to water resources project planning."[55]

The mountain of special projects in the bill provoked a howl of protests from antitax groups. The Heritage Foundation called it a "pork fest" that "benefits[s] the rich and influential who can afford a lobbyist with access to Members of Congress and committee staff."[56] Responding to this criticism, President Bush vetoed the 2007 WRDA, saying lamely that "this bill lacks fiscal discipline." That's like saying Nero had an anger-management problem. Bush urged Congress to send him a "fiscally responsible" bill.[57] It ignored him and easily overrode the veto. Even the Corps' harshest critics supported the Act. For the reformers, pork was simply the price of progress.

The 2007 WRDA did not convert the Corps into a modern river-restoration agency, but it did force it to go in a different direction. Excitement over this new mission must be tempered by the environmental and economic impact of hundreds of existing projects. In FY 2010 the Corps' civil works program received $5.7 billion in funding for a huge array of projects.[58] Some of them serve the American people admirably and deserve continued funding and political support. And some of them should be relegated to that proverbial dustbin of history. Which ones?

THE INVITATION

Charles Eliot, president of Harvard in the early twentieth century, was presented with a dilemma: he had to bestow an honorary degree on a senator, but Eliot was less than enamored with the senator's political activities. So Eliot, the consummate diplomat, described the senator as "[a] man with great opportunities for public service still inviting him."[59] The same thing could be said about the Corps of Engineers; it has the potential to assist America's entry into a new era, but only if the agency can abandon the parochialism and pork barrel of the past and turn its $5.5 billion annual budget toward a new destiny. In this endeavor the Corps must be joined by a Congress that is willing to make some hard choices and begin to think in terms of the national interest rather than individual members' narrow political agenda. To a great extent, the Corps wastes money because Congress wants it to waste money. To achieve meaningful reform on America's rivers, the change will have to occur on two levels.

The first level of change involves the relationship between the Corps and the Congress. The old iron triangle must be replaced by a matrix that centers on the long-term best interests of the nation as a whole; rivers must be treated as a commons, not as the private resource of a small minority. One way to accomplish this is to take a lesson from the Defense Base Closure and Realignment Commission (the BRAC), established by Congress.[60] Since this country's founding it has been plagued by unnecessary military bases scattered throughout various states and congressional districts. Legislators relied on these bases to funnel federal money to the home district, so they fought against closing the hometown base—even while acknowledging that many of the bases were unnecessary.

Finally, recognizing its own lack of backbone, the Congress created the BRAC to identify the bases that were no longer needed. Beginning in 1988 the commission developed lists of bases to be closed, and Congress then had to vote on each list as a whole; no legislator was allowed to pull his or her pet base off the list. This gave legislators the cover they needed; they could lament with great consternation the base closing in their district or state, but also claim to be vigilant stewards of the public's money by closing many bases. This system worked well, and hundreds of bases were closed.[61]

It is time for a BRAC-type commission to evaluate existing water projects and determine which ones still serve the public interest, and which projects need to be retired, removed, or abandoned. A nation burdened with a $14.5-trillion national debt no longer has the luxury of building and maintaining expensive make-work projects. Eliminating projects that do more harm than good would help direct water agencies such as the Corps toward a more useful and productive role while assisting the nation in restoring its rivers and maintaining those projects that actually make sense.

The Corps of Engineers, like the nation as a whole, is a work in progress. It is still learning, evolving, maturing. There are thousands of excellent civil servants who work for the Corps; all they need is a modern, future-directed mission that will propel them to the front lines of the movement to remake America's rivers. Perhaps then people will stop asking why the army is involved in river policy.

3

THE MANLESS LAND

The Bureau of Reclamation

Though forced to drudge for the dregs of men,
And scrawl strange words with the barbarous pen
—FROM WILLIAM CULLEN BRYANT, "GREEN RIVER"

For Americans who lived through the Great Depression, the date of September 11 carries a meaning quite different from what usually comes to mind today. On that date in 1936, Hoover Dam began generating electricity. At the dam's dedication ceremony, held nearly a year earlier, President Franklin D. Roosevelt called Boulder Dam (it was not named Hoover until 1947) a "great feat of mankind."[1] He looked out across the crowd at the new reservoir and the "fiord-like vista."[2] "Gee," he exclaimed, "this is magnificent."[3] The world's biggest dam at that time, it was "the Great Pyramid of the American Desert, the ninth symphony of our day."[4]

Clearly this was no ordinary ribbon-cutting ceremony. Many "experts" had predicted that a dam of such monstrous dimensions could not be built. Some thought it would collapse of its own immense weight. Others predicted that the concrete, poured in such vast quantities, would never harden, and would instead form a huge gelatinous glob that would gradually ooze down-canyon. And a few thought the weight of the dam and reservoir would throw the earth out of orbit and the planet would go hurtling off into the sun like a curveball gone berserk. Instead, American ingenuity, courtesy of the Bureau of Reclamation, proved to the world that we could do the impossible and rise to our highest aspirations. In the midst of the Great Depression, this—an impassioned lifting of the national psyche—was perhaps the greatest benefit of building Hoover Dam.

Today, three-quarters of a century later, the dam looks much as it did on that September day in 1936 when it was dedicated. The great swath of white concrete still sweeps across the river in a massive arc. The art deco motif on the rim provides an expression of artistic zeal to an otherwise strictly utilitarian structure. The shiny brass doors on the elevators, the gracefully sculpted intake towers, and the symmetrical curve of the dam itself combine to create not only an architectural statement but also a testament to the sense of beauty and pride of workmanship that typified an America of another era.

But the artistic motif has a broken link; in 1995 the Bureau of Reclamation, with substantial cost overruns, built a new museum and parking structure at the edge of the dam. The parking garage is a modern multi-storied box, and the museum building is metal and glass, tinted to a garish copper color. They fit in like a commode in a flower shop. They also speak vociferously about how the Bureau of Reclamation has changed since it completed its Ninth Symphony, its engineering dénouement.

When Secretary of Commerce Herbert Hoover first conceived the idea of building a great hydropower and flood-control structure in Boulder Canyon (the dam was actually built in Black Canyon), he had in mind a project that would be financially self-supporting; his deep-seated parsimony would not allow him to propose a project that would be a perpetual drain on the federal treasury and the honest citizens who support it. Hoover was quite adroit at cost-benefit calculus, but he never understood politics. It probably did not occur to him that, if Hoover Dam was wildly successful—and it has been—then virtually every congressman in the nation would demand that one be built in his (and later, her) district. Indeed, this has occurred; building such dams became an article of faith and a visual testament of a legislator's concern for his constituents and his ability to bend Washington to his will.[5] Dams were built without reason or purpose other than to win votes, please powerful interests, and conquer the elements. Not all dams were of such character; some of them were effective responses to the discernible values of their day. But others were built simply to pour concrete and get someone else to pay for it, like constructing a post office in a town with no literate inhabitants.

WATERY SCIENCE

At the beginning of the last century, America was in an expansive mood. We had whipped the Spaniards in war, stolen Hawaii from its natives, bloodily suppressed a quest for freedom in the Philippines, and reduced the American Indian population from several million to 250,000.[6] We had invented so many new contraptions that the commissioner of the Office of Patents thought it was no longer necessary to issue patents because "everything that can be invented has been invented."[7] For good measure we built the Panama Canal, sent our battleships steaming around the world, and shook our big stick at anyone who dared question us. At home, there was only one more thing left to conquer: the desert.

Science would be an important weapon in the war against aridity. By the beginning of the new century, the newfangled concept had taken hold in a big way. It was assumed that science would lead to a new era of prosperity and growth by allowing mankind to control nature. Of course, science then, as now, had a large potential for abuse and was often used as a tool against racial and ethnic groups that stood in the way of "progress." The term *junk science*, as it is used in political parlance today, pales into inadequacy compared with the way social Darwinist science was used 110 years ago to justify everything from genocide to the conquest of nature.[8] But some saw science as a kind of amorphous threat to tradition. The famous poet Oliver Wendell Holmes lamented how this new technique was supplanting art and culture:

> No more with laughter at Thalia's frolics
> Our eyes shall twinkle till the tears run down,
> But in her place the lecturer on hydraulics
> Spouts forth his watery science to the town.[9]

In regard to public policy, there was a strong move afoot to use this "watery science" to irrigate vast stretches of land in the American West. It was thought that, with properly applied scientific technique, farmers could "subjugate" arid lands.[10] The assumption was that desert lands were useless unless they were farmed. The first chief engineer of the Reclamation Service (the forerunner of the Bureau of Reclamation) liked to make reference to the "dead and profitless deserts."[11] In effect, irrigation would make the arid West look like the humid East and blanket the region with salubrious farms, wholesome villages, and the occasional little red schoolhouse.

But the science was only part of the irrigation phenomenon in 1900. The other part was a vast scheme of social engineering that was so grandiose and far-reaching that it would have made Lenin smile. Eastern cities were teeming with new immigrants, primarily from "less desirable" countries in southern and eastern Europe. They mixed roughly with the even earthier hoi polloi, the Irish. Their crowded, fetid tenements were a breeding ground for radical new political ideas such as socialism and anarchism. Some of these unlettered subversives even advocated for labor unions.

The nation's elite feared these people and believed that the salutary lifestyle of the "yeoman farmer," as Jefferson termed it, would cure them of their wayward ideologies. The solution to America's urban political blight, then, was to lure people out of the cities and into an agrarian cornucopia carved out of worthless wilderness. In effect, the government would use dams to settle the West—and solve some pressing political problems. Public land was available for the taking, thanks to the Homestead Act of 1862. If any Indian land was needed, the government could appropriate it under the Allotment Act of 1887 by labeling it "surplus" and making it available for purchase at fire-sale prices. Overlying all of this was a strong sense of ecclesiastical warrant: God wants us to irrigate! This was the sociopolitical milieu in which the U.S. Bureau of Reclamation—then known as the Reclamation Service—was born, in a "blaze of mystical fervor."[12]

In short, the social planners needed three things to expand the nation and disarm the radicals: men willing to move west and farm, land watered by the new science of irrigation to make fecund the landscape of the yeomen, and a government agency with "absolute control of all water in one strong central authority" to build the engineering marvels to make it all happen.[13] A lobbying organization, the National Irrigation Congress, was formed to push the agenda. Boosters needed a catchy slogan and came up with the puerile "Put the Landless Man on the Manless Land."[14] Apparently it takes corn to grow corn. This slogan and others equally pedestrian were often uttered in conjunction with the biblical admonition in Isaiah to make the "desert blossom as [not 'like'] the rose."[15]

The Irrigation Congress, with vocal support from President Theodore Roosevelt, proposed to create a new federal Reclamation Service to build western irrigation projects. The political disadvantage of the enterprise immediately became apparent; operating only in western states (which eventually numbered seventeen), it could not muster political support from the states in other parts of the nation. Eastern legislators simply saw it as a transfer of wealth from their constituents to the West—which it was. The easterners went on the attack. One of the most vocal was Congressman William Hepburn of Iowa. In languorously tepid language, the good congressman opined that the reclamation program was "the most insolent and impudent larceny that I have ever seen embodied in a legislative filching."[16] Not to be verbally outclassed, Congressman George Ray of New York proclaimed reclamation to be "robbery or looting of the Treasury."[17] Prospects looked dim for reclamation.

In a desperate maneuver, western senators threatened to cut off funding for the Corps of Engineers, which at that time focused primarily on the East and South.[18] That was enough to convince legislators from those regions that a little logrolling could go a long way; I'll support your water welfare if you'll support mine. However, in exchange for their votes, the easterners extracted a concession: all reclamation projects must be reimbursed through a revolving fund that consisted of monies paid back from previous projects. In effect, all reclamation projects would be funded by short-term loans, and the program would be self-sufficient—not like the Corps' tax-supported work.

The ink on the Reclamation Act of 1902 was barely dry before the Reclamation Service began digging canals and building dams. The Newlands Project in Nevada was the first, followed quickly by the Yuma Project in Arizona and Oregon's Klamath Project. It took only a few years for it to become obvious that a bunch of novice farmers, eking out a living in some of the harshest climates of the country, could not pay back the costs of their projects. Congress responded by passing a series of laws that rendered the payback provision meaningless and converted the reclamation program into a nationally funded pork barrel much like the Corps of Engineers.[19] In 1923 the Reclamation Service was renamed the Bureau of Reclamation. The agency wasted no time in promoting and building huge, expensive dams and water-diversion projects. Hoover Dam was authorized by the Boulder Canyon Project Act in 1928.

Forty more projects were authorized during the Depression, despite the fact that by then the reclamation program "was so manifest a failure that, had there not been powerful groups and strong cultural imperatives supporting it, federal reclamation would have died an ignominious death."[20] One of the biggest of those projects was Grand Coulee Dam on the Columbia River in central Washington.

THE INLAND EMPIRE AND THE FIFTY PRINCESSES

Grand Coulee Dam lacks the soaring height of Hoover Dam and the stunning setting of Glen Canyon Dam, but it makes up for it in sheer mass. Dam builders are fond of citing big numbers accompanied by a panoply of superlatives, but Grand Coulee sets the record. References to the "mighty giant," the "huge structure," "enormous flow," and "vast project" abound in Bureau promotional materials. A recorded voice at the visitor center compares the dam to the pyramids of Egypt. All of this is accompanied by the stark data of engineering metaphor: the dam is composed of enough concrete to "build a standard six-foot sidewalk around the world at the equator."[21]

None of this prepares the visitor for the actual experience of seeing the dam firsthand. In a land of arid, rolling grasslands, streaked with occasional stands of pine, the dam sits athwart a mighty river and stops it cold. It is nearly a mile in width; its downstream face is so huge the Bureau has nightly laser shows that splash multicolored lights across the dam to entertain tourists. On a quiet night, one can hear the steady hum of the 80-foot-wide hydroelectric generators, churning out nearly 7,000 megawatts of electricity for the Pacific Northwest.

But in some ways the most impressive visual at the dam is not the dam itself, or its great bank of generators, or the hokey laser show; it is, rather, the modest but moving bust of Franklin D. Roosevelt that stands near the eastern abutment of the dam, staring out across the great expanse of water and concrete. The bust is mounted on a slanting pedestal, as though Roosevelt is thrusting forward into a new era. It speaks of determination, commitment, and the promise of a better future. It was Roosevelt and Secretary of the Interior Harold Ickes, who had a vision of transforming the Depression-era Pacific Northwest into a "planned promised land," an "Inland Empire" of small farmers, hydropowered industry, and thriving communities.[22] It was social engineering on a grand scale, what Timothy Egan characterized as "the last gasp of agrarian idealism."[23] For his vision and willingness to commit federal dollars, Roosevelt was rewarded with an opportunity to speak at the dam's inauguration, and the reservoir was named after him.

In the years of the Great Depression, building massive public works projects was a good idea; it got the nation moving, working, thinking, envisioning. It redistributed wealth to people who were suffering from problems not of their own making. But the downside of such efforts is that, once a handout is offered, it becomes a birthright in the minds of the recipients. The lid to the public cookie jar is more easily opened than closed.

From the beginning, Grand Coulee was to be a "multipurpose" project. In the Bureau's parlance, that means that hydropower customers, mainly city folk, would pay extra for electricity to subsidize irrigated agriculture. That subsidy—what the Bureau euphemistically calls the "paying partner"—was essential to the Grand Coulee Project's irrigation component, which was called the Columbia Basin Project. The engineering of the irrigation project is even more complex than the funding schemes; the Bureau built six massive pumps, each with 65,000 horsepower, to pump water uphill 280 feet through 12-foot-diameter pipes to an old riverbed (the true "Grand Coulee"). The resulting impoundment was later named Banks Lake. This required the construction of four more dams, 300 miles of canals, 2,000 miles of laterals, and 3,500 miles of drains and wasteways. The irrigation component cost more money than Grand Coulee Dam itself. The costs were staggering, but the local farmers wanted the water. As Paul Pitzer describes it in his book, *Grand Coulee: Harnessing a Dream*, the farmers demanded that their dry farms be irrigated, but "they wanted someone else to pay the bills."[24] They got their wish; a succession of bills allowed the farmers to take over the project, avoid paying most of the costs, and increase their control over water allocation. In other words, Grand Coulee Dam and its accompanying irrigation project was fairly typical of Bureau projects: early rosy projections of self-sufficient farmers making the deserts bloom as the rose gave way to gritty, high-conflict politics over who was going to pay for an economically irrational endeavor with high risks, enormous capital costs, and harsh market realities.

Political opposition to the Columbia Basin Project was led primarily by budget-conscious Republicans, who complained about spending a great deal of the nation's money to benefit a few people in one remote locality. The PR people in the Bureau wanted to make it look like these Republicans were opposed to mom and apple pie, so they dreamt up a publicity stunt to emphasize that the project created benefits for the nation as a whole. They invited each state and the territories of Alaska and Hawaii to send a gallon jug of water to the dam. Then, during the "Water of All States Ceremony," fifty princesses from the Washington Apple Blossom Festival, decked out in what looked like prom dresses, gathered on the dam with much fanfare and simultaneously emptied the states' jugs. This sounds like a hoary cliché by today's standards, but it does illustrate how much the country and its social values have changed since Grand Coulee Dam was completed.

Past conflicts over funding are not at all apparent when one visits the dam. The Bureau offers tours of the dam's third power plant, which was added to the dam in 1983. Our guide was a middle-aged woman named Sherry, who had been with the Bureau for eleven years and displayed an obvious pride in the big dam. She had a matter-of-fact delivery style that was clearly founded on conviction—and repetition. She was especially fond of the irrigation component. "Before the irrigation project, this land was only good for growing sagebrush and rattlesnakes. Now we grow eighty different crops." She recited a long litany of impressive benefits but did not mention what was lost in constructing the dam.

The Grand Coulee Dam Visitor Center, built in the 1970s, extols the virtues of the project while downplaying its costs, which included a dozen communities that were drowned by the project. And in a scene that is all too familiar, Indian tribes paid most dearly. The Spokane Reservation lost 3,000 acres to the reservoir; the Colville Reservation lost 18,000 acres. But the biggest loss of all is not even mentioned in the numerous Bureau handouts and brochures; only a single sentence, in a display at the visitor center about the Colville Confederated Tribes, mentions it. It says that, with the completion of the dam in 1942, the Indians of the region could no longer fish for salmon on the upper Columbia River, including the traditional fishing site at Kettle Falls—now under water.

At the time of construction of the dam, people still thought of dam building as a form of "conservation." It was, as Woody Guthrie sang, a way to tame a "wild and wasted stream." The builders of Grand Coulee never gave a thought to anadromous fish; the dam blocked the entire upper river to fish runs. Before the dam, the upper Columbia was home to a unique kind of salmon called the seven-year summer hogs—enormous chinooks that spent seven years at sea before running upriver for the spawn. After the dam was built, these giant fish appeared each summer for seven years, thrashing against the wall of concrete that stood between them and their predetermined destiny. This stunning lack of foresight at Grand Coulee and other dams of that era sowed the seeds of future conflicts that are now playing out in a big way in the Columbia River Basin and other rivers of the West.

DINOSAUR

The construction of Hoover Dam, Grand Coulee Dam, and other projects launched the Bureau into a period of major expansion and construction. In 1946 it proposed a massive project to basically replumb the Colorado River and prevent it from flowing "wasted" into the sea. The planning report was titled *The Colorado River: A Natural Menace Becomes a National Resource.* Clearly, a free-flowing river could not be considered a resource. It is worth quoting from the report at length because it so accurately describes the fervor of those times and the pivotal role played by the Bureau:

> Tomorrow the Colorado River will be utilized to the very last drop. Its water will convert thousands of additional acres of sagebrush desert to flourishing farms and beautiful homes for servicemen, industrial workers, and native farmers who seek to build permanently in the West. Its terrifying energy will be harnessed completely to do an even bigger job in building bulwarks for peace. Here is a job so great in its possibilities that only a nation of free people have the vision to know that it can be done and that it must be done.[25]

There it is in one paragraph: postwar/Cold War/American Dream/ nationalistic can-do/conquer-the-desert water hubris. Congress bought into it completely and passed a string of authorizing laws that put the

Bureau to work. Two of the largest multidam projects were the Central Valley Project in California and the Colorado River Project in the Southwest.

The Central Valley Project (CVP) began with the 560-foot-high Shasta Dam on the upper Sacramento River and grew to include Friant and Keswick Dams, and eventually seventeen additional dams. Shasta was built without fish passage, blocking anadromous fish from many of their spawning streams. Much of the water from the project was diverted to flood-irrigate rice paddies in the dry Central Valley.

The 1956 Colorado River Storage Project Act authorized four units in that river basin: Glen Canyon Dam, Navajo Dam on the San Juan River, Flaming Gorge on the Green River, and the Curecanti Unit on the Gunnison River.[26] With time, an additional twenty-one projects were added as "partners" on the Colorado River system. The 1968 Colorado River Basin Project Act authorized the Central Arizona Project (CAP), an enormously expensive canal with siphons and pumps that would divert water out of the Colorado River and move it 300 miles through the desert to Phoenix and then to Tucson. That act also authorized the Delores Project in Colorado, the Central Utah Project, and perhaps the most illogical project of them all, the Animas–La Plata Project, designed to suck water out of the Animas River, pump it uphill to a reservoir in the mountains, and then pipe it down down the other side to the La Plata River so farmers growing hay at 7,000 feet could get an additional cutting.[27]

The Bureau's apex was characterized by a level of water hubris that is scarcely imaginable today. Marc Reisner, in *Cadillac Desert*, described the attitude of Bureau engineers: "They tended to view themselves as a godlike class performing hydrologic miracles for grateful simpletons."[28] With a surfeit of confidence and a paucity of humility, the dreams of water bureaucrats and private developers grew to monstrous proportions. There was talk of diverting Mississippi River water to the high plains of Texas and New Mexico.[29] Some proposed damming the Yukon in Alaska and creating reservoirs the size of European countries and then pumping the water all over the American West.[30] There were serious discussions of sucking water out of the Great Lakes, or the Missouri River, or the Columbia River, and spreading it out on farms in the Southwest and the Great Plains.[31] Not to be outdone by the West, some easterners proposed damming both ends of Long Island Sound and turning it into a reservoir.[32] It seemed that no idea was too outrageous to conjure up visions of blueprints and contracts.

In time the agency would build 476 dams and 348 reservoirs, 58 hydropower facilities, and 56,000 miles of conveyance systems and irrigate 10 million acres of land.[33] It went after the "natural menace" in a big way. The Bureau's heyday lasted for sixty years, even though the problem of "settling the West" had long since passed into memory and TV westerns, and all the good dam sites had been taken. But the Bureau, like the Corps, had become such a convenient vehicle for dispensing federal favors that western legislators were loath to give up the money conduit, so the projects kept coming, regardless of their economic viability or environmental impact. Indeed, the Bureau added millions of acres of new crop land at a time when agricultural surplus was the biggest problem in farm country. In response, the government began to provide price supports for crops that were overproduced. That meant many western farmers were growing subsidized crops with subsidized water delivered from subsidized Bureau projects. Some also received subsidized electricity and fuel.[34] This would all make sense if tax dollars fell like manna from heaven—or you were a western congressman trying to get reelected. Some of the Bureau's projects were worthwhile, but many were simply makework pork barrel.[35] And they devastated the West's riverine landscapes.

To justify the projects, the agency resorted to the same kind of distorted cost-benefit analysis that characterized Corps studies. The Bureau continued to claim that the payback provisions in early legislation still covered most costs, but in fact the "payback" consisted of an interest-free loan with a 10-year grace period and up to 100 years to pay it back, and beneficiaries only had to pay back according to their "ability to pay." That kind of financing makes Freddie Mac look reasonable. It also opened the door to blistering critiques.[36]

The first attempt to challenge the mighty Bureau came soon after it announced in 1949 that it wanted to build a dam on the Green River at Echo Park, which just happened to be in Dinosaur National Monument. The Sierra Club had bitter memories of dams in national parks—it was the construction of O'Shaughnessy Dam in Hetch Hetchy Valley that broke and ultimately killed Club founder John Muir. The Club decided to fight the Echo Park Dam with all its resources. The first full-time director of the Club, who had just come onboard, was David Brower. He would become a legend in environmental circles. Brower and others organized a very effective media blitz that was unprecedented in scope and imagination. With relentless commitment, they pressed Congress to exclude Echo Park from the proposed legislation for the upper Colorado River. It was an unprecedented victory for river advocates when, in 1956, Congress specifically excluded a dam in Dinosaur National Monument from the Colorado River Storage Project Act. Both supporters and detractors of the Echo Park Dam predicted that the fight in Dinosaur would become a rallying cry for future attempts to stop dams. In exchange for their victory at Echo Park, river advocates agreed to support another dam in the Colorado River Basin, one that was not in a national park or monument, at a little-known place called Glen Canyon.[37]

The next great fight over Bureau plans began when the agency announced it wanted to build two dams in the Grand Canyon. One dam would be built at Marble Canyon at the upper end of the main canyon and just outside the park's boundaries—which at that time included only part of the actual Grand Canyon. The other dam was proposed for Bridge Canyon (also called Hualapai), which was at the other end of Grand Canyon. The purpose of these dams, referred to as "cash register dams," was to generate electric power and money.[38] The power would be used to run the pumps needed to move water through the Central Arizona Project. The money would be used to subsidize irrigation.

Brower and the Sierra Club went into full-scale attack. He devised a publicity campaign that became a model of grassroots lobbying that is still emulated today. He generated thousands of letters to congressmen, he arranged to have a large-format book published and sent to each member of the Congress, and he ran full-page ads in major newspapers with provocative leads such as "Now only you can save Grand Canyon from being flooded for profit" and "Should we flood the Sistine Chapel so tourists can get nearer the ceiling?"[39]

The proposed dams generated a howling protest. Even Senator Barry Goldwater expressed opposition to the dam at Bridge Canyon because it would back water up into the park. Under tremendous pressure, Secretary of the Interior Stewart Udall eliminated the dams from the proposed Colorado River Project; he assumed massive coal-fired power plants along the river could provide the electricity necessary to run the pumps on the CAP. As Russell Martin put it in his book, *A Story That*

Green River at Echo Park, Dinosaur National Monument, Utah–Colorado. This is just upstream of the dam site proposed by the Bureau of Reclamation in the 1950s.

Stands like a Dam, "Floyd Dominy and the boys at the Bureau of Reclamation finally had taken their empire-building too far."[40]

The fight over the Grand Canyon dams was iconic, pitting two great powerhouses against each other—David Brower, representing the growing environmental movement, and Commissioner Floyd Dominy, an unapologetic champion of federal reclamation.

DOMINY

"Would you like a glass of wine or some bourbon?" Dominy asked as he greeted me at his door. It was 11 a.m. I declined, but he poured himself a tall glass of Chablis with ice, sat down in a rocking chair below a large painting of Hoover Dam, and began to talk while looking out the window of his farmhouse in the Shenandoah Valley. I was concerned that a 95-year-old man, with a belly full of wine, might be less than coherent. I was mistaken; I underestimated Floyd Dominy, as many men had before me.[41]

Approaching a century of life, Dominy was as combative and self-assured as when he had run the Bureau the way George Patton fought wars. His mind was sharp, his memory was as clear as a Wyoming morning, and he still had an ego the size of the Bureau's biggest dams. He often referred to himself in the third person, as though he was discussing an important historical figure that he had read about.

When Dominy became commissioner in 1959, the Bureau of Reclamation was at its apex. By the time he left ten years later, Reclamation had become the focal point of a grand argument over the future of western rivers. He had become a cult hero to some and a despised villain to others. When he stepped down as commissioner, it was the end of an era. According to Dominy, he was not fired by James Watt, as Marc Reisner claimed in *Cadillac Desert*. "James Watt did not fire me; nobody fired me. I quit on my own accord. I wrote Nixon a letter in May 1969, and said I was going to retire on the first of December 1969. Nixon replied, thanking me for long dedicated service. James Watt was a political flunky; he was in my office every five minutes trying to get me to help him with his career."

Nixon did not fire Dominy. But surely Dominy sensed that his time, and the Bureau's, had come and gone; that America had changed, but the agency and its commissioner had not. In a three-hour interview, I asked him numerous questions; only one question caused him to pause before answering. I asked if any of the Bureau's dams should be removed. Dominy studied the ceiling, studied his hands. He could still run through the list of projects—hundreds of them—like he did in the days when he testified before Congress without notes, impressing them with his command of the facts.

"No."

Were any of the Bureau's projects a mistake? I was searching for perhaps just a narrow slice of repentance or regret.

"No. Mistakes were made, sometimes big mistakes, but every project we built was worth the effort because every project served real people."

To Dominy, "real people" meant people like him, who grew up in the rural West and depended on irrigated farms to keep them one step ahead of starving. To him, building reclamation projects was not a job, it was a holy war, and he was the messiah. The Bureau's official biographical sketch of Dominy describes him as the "most colorful commissioner." I asked what that meant to him.

"I was a surprising choice for commissioner, being a farmer and not an engineer. Also, I was never cowed by the stature of the senators and congressmen, and I wasn't afraid to straighten out the record."

Part of Dominy's record-straightening concerned the payback formula for reclamation projects. He could see right away that the loans, even though interest-free, were beyond the financial capabilities of farmers barely scratching out a living, often on projects that were marginal at best. It was an invitation to financial collapse. In his first hearing before a congressional subcommittee, he explained that the Bureau built some dubious projects because congressmen ordered them to. It took considerable *cojones* to go before Congress and blame it for bad projects, but the congressmen appreciated his candor and bravado. They also sensed that Dominy, with his mastery of facts and no-nonsense style, knew how to rescue the Bureau's troubled projects and get them on a solid footing. He was the man who could save their pet projects. In return, they would give him steadfast political support so that he could make himself a water potentate and bully everyone around him until he got what he wanted. The word "colorful" is an artful massage of his brusque character.

Dominy's Bureau took on everyone. You either gave him your undivided loyalty or you were on his enemy list. When asked to name his most effective political rival, he did not hesitate for a moment. "The Corps of Engineers. They don't have the 160-acre limit, and they could work in all fifty states; I had to depend on just seventeen states. They were my biggest political headache. I did not consider the environmentalists to be a headache; I was perfectly willing to work with them. Even Dave Brower agreed that I was a pretty reasonable guy to work with."[42]

David Brower, the Sierra Club's "most colorful" director, is no longer alive to support or refute that statement. But most environmentalists would be surprised to learn that Dominy did not consider them an enemy; they certainly perceived him in such stark terms. The fight over dams in the Grand Canyon, the effort to protect Rainbow Bridge National Monument, and the near-complete destruction of the salmon run on the upper Trinity River all occurred while Dominy was commissioner.

As our interview drew to a close, I looked around the former commissioner's den. It was practically a museum—to Dominy. There was a large painting of a king salmon leaping through the surf, and it caught my eye as I mused about what Dominy's dams had done to anadromous fish runs.

"I caught that fish in Alaska. Senator Ted Stevens wanted me to come up and survey the proposed route for the Alaskan pipeline. I told him I'd only do it if he gave me four good days of fishing."

In a bookcase I spotted a copy of Edward Abbey's *Monkey Wrench Gang*. I expressed surprise.

"I met that skunk once. I also met Wallace Stegner—I had to help him launch his boat on his last trip through Glen Canyon before we flooded it." He chuckled at the thought.

On the floor, leaning against a desk, was a collection of core samples mounted on a plaque. He had not bothered to hang it on the wall.

"Those are soil samples, a gift from the Bureau upon my retirement. And that big painting of Hoover Dam, behind my chair, so was that. It hung on the wall behind my desk when I was commissioner. I liked it, so I just took it. I suppose it's government property."

Also adorning the wall was the famous painting by Norman Rockwell of Glen Canyon Dam with a Navajo family in the foreground. I asked if it was the original.

"No, but I was instrumental in getting him to paint that. He told me he didn't paint structures, he painted people, so I arranged to get some Navajos to pose for him. We also hired Georgia O'Keefe to paint for us."

Amid Dominy's memorabilia from his contentious public life were dozens of family pictures of his long-deceased wife, three children, eight grandchildren, and twenty great-grandchildren. In the annals of water history Dominy is a larger-than-life figure, but these photographs rendered him on a human scale; there was warmth, and perhaps a little vulnerability, somewhere beneath that iron will.

The most imposing item in the room was a bookshelf sagging from the weight of a long row of leather-bound oversized books. These are Dominy's "diaries," as he calls them—scrapbooks filled with photographs, newspaper articles, and personal correspondence from his Bureau years, twenty-three volumes in all. He invited me to peruse them, so I pulled a volume and randomly opened it to a letter from Senator Lee Metcalf. It made reference to him and Dominy having a drink together in a restaurant.

"Ole Lee was a great guy, an alcoholic. He was drunk half the time he was speaking on the floor of the Senate." Dominy took a swill of wine and told me to pull another volume. I grabbed the final volume and opened it to the last page. I thought perhaps I would see the letter from President Nixon, thanking him for his service. Instead, it was a photograph taken from space. In between the clouds, Glen Canyon Dam was clearly visible.

"That was when Lake Powell was filling. You could already see the lake from outer space." Dominy grinned hugely. A reservoir so enormous it is visible from outer space—a fitting finale to Floyd Dominy's tenure as commissioner of the Bureau of Reclamation.

But even in his heyday Dominy came close to becoming a caricature, an anachronism from an age when white males with slicked-back hair governed by pounding their fist on a desk, and pouring concrete was the only way they could relate to rivers. Dominy was an instigator in his own, unique way, not by starting something new but by taking an old idea to new extremes. Like all instigators, Dominy was just an individual citizen, committed passionately to a cause. He was a man who felt no need to question the righteousness of his mission, an attitude that prevailed in the Bureau of that era. The Bureau was so entrapped in its own conceptual box that it was unable to perceive any other method of approaching water problems; Dominy personified that myopia. Perversely, his aggressiveness, unwillingness to compromise, and self-righteousness may have hastened the downfall of the Bureau's empire and helped convince the nation that a new approach was badly needed.

The national psyche that developed in the late 1960s was weary of chest-thumping politicians and in search of new ways to live in harmony with one another and the planet. But the Bureau, even after Dominy stepped down, didn't get the message. Together with its allies in Congress and the hundreds of irrigation districts that blanketed the West, the agency was prepared to fight change at every step. The firm iron triangle of western water development seemed impervious to external challenges. But two starkly different events occurred that would change the agency forever. One of them was Jimmy Carter's "hit list" (described in chapter 2), and the other happened in a distant corner of Idaho.

SUGAR CITY

"Can you direct me to the flood museum?" I asked the motel manager.

Virtually every locality in the world has experienced floods; they are as old as water. But in Rexburg, Idaho, the word carries a special meaning, heavy with memory and angst. The motel manager explained that the flood museum was only three blocks away, in the old Mormon tabernacle. And then, with just a slight prompt, he told me of his experience that fateful day in June 1976, when Teton Dam crumbled into mud and an entire nation began to have second thoughts about the Bureau and its engineering. People in Rexburg and the nearby towns of Sugar City and Wilford know where they were that morning the way the rest of us remember the moment we heard about the 9/11 attacks or President Kennedy's assassination. The motel manager looked out the window and began a stream of consciousness:

We had bought a farm, just north of town, a modest frame house. On that morning, June 5th, we heard on the radio that Teton Dam had collapsed and a flood was heading our way. We were told to go to the high ground on the east side of Rexburg. I got the wife and kids together and we got ready to leave. Before we left the house we collected a pile of valuables and family heirlooms, but we didn't have time to load them in the car, so we put them on the bed. We didn't really think anything would happen to the house. We started driving toward Rexburg, on the road that took us down into the valley. As we came up over a rise, we saw a wall of water sweeping through town in front of us. It was filled with trees, cars, dead cows, and houses—houses floating right down Main Street. If we'd left a few minutes earlier we would have been right in the middle of it. I turned that car around and drove 80 miles an hour toward higher ground.

At that point the motel manager paused and then stared at the floor. He was clearly viewing a troubling tableau in his head. "I'm sorry, you just wanted directions," he said.

I told him I was very interested in dams, and I appreciated hearing his story.

"I knew one of the lead engineers on that project. He was in my ward [a local jurisdiction of the Mormon Church]. It got so bad after the collapse that he had to move away. He just couldn't live here any more."

"What happened to your house?"

"It was a tough frame house, but it had two feet of mud in it, and was damaged beyond repair. It had to be demolished. But you know that stuff I told you we left on the bed—the family heirlooms? It was just above the mud line, and it was all still sitting there when we went back."

Following the motel manager's directions, I made my way to the "flood museum," which is something of a misnomer. A flood is something natural, like an unusually heavy spring runoff that splays out across a flood plain. This was not a flood. Rather, it was a man-made disaster. But no one wanted to call the museum the "dam failure museum" because to do so would be to admit that building an earthen dam in that location on the Teton River was a colossal mistake. Local politicians had been clamoring for years to put a dam there. The Bureau, already out of good dam sites, was only too willing to oblige. The "flood museum" tells the story of what happens when politicians and bureaucrats fail to heed warnings, fail to think in terms of the big picture, and get a bad case of water hubris.

Pete Thompson greeted me warmly at the museum. He is a volunteer docent, and I was his only customer that morning. I asked him where he was on that fateful morning.

"I was a mailman back then [a letter carrier in today's parlance], doing my rounds, when one of my patrons mentioned that a dam had failed and we were supposed to go to high ground. I didn't take it that seriously. I called the postmaster and he said to just keep delivering the mail. But not long after that the police came by and told me to go to high ground immediately. Good thing they did. After the flood, the cleanup was a terrible job. Ricks College (now Brigham Young University–Idaho) fed us for nearly a month. And volunteers came from all over Idaho and Utah by the busloads to help us clean up."

At that point Mr. Thompson ushered me into a room to view a short video about the flood. The color had faded to a pinkish hue, and the soundtrack crackled like an old LP. It began with peaceful, bucolic images of farmhouses and children playing in the yard to the tune of "Long After Saturday Is Gone." The people of the Teton Valley were described as "plucky," never suspecting that a disaster was impending. These scenes of tranquil security were then interspersed with images of the river violently gushing past the remnants of the dam. The music changed to the theme song from the movie *The Longest Day*. There was no explanation of why the dam failed, but the narrator mentioned that "environmentalists and other groups" had tried to stop the dam, but the dam's promoters had ultimately prevailed.

Teton Dam was an earth-filled structure, 305 feet high and 3,050 feet long. It was designed to create a reservoir 17 miles long. During the three-year construction of the dam the Bureau poured millions of gallons of concrete grout into the fractured rock embankment that was supposed to anchor the dam in place. Ignoring these signs of serious structural weakness, the Bureau began filling the dam rapidly. When leaks developed, it sent two men in D-9 Caterpillars to the top of the dam to try to stop the leaks by dumping vast quantities of rock down the face. The dam nearly collapsed beneath them. As the two men scurried to safety, their giant machines were swallowed up by a rapidly growing chasm of churning water and soil—soil that had previously been the core material of the dam. At that moment, 80 billon gallons of water came rushing violently down the Teton Valley. The people of Wilford, Sugar City, and Rexburg had been warned of the impending disaster, and most had already fled to higher ground by the time the floodwaters hit. As a result, the death toll was remarkably low, totaling about a dozen people.

The collapse of Teton Dam was truly a disaster for the "plucky" people of Teton Valley. But the ramifications were felt across the nation. The once invincible Bureau of Reclamation had overplayed its hand and had been sharply rebuked by nature. Already under attack from environmentalists and fiscal conservatives, Teton Dam gave the nation yet another reason to call into question the Bureau's modus operandi.

TURNING THE CORNER

President Carter's "hit list" directive made specific mention of the Teton Dam disaster, but that was only a small part of Carter's larger ambitions. He gamely took on some of the biggest Bureau projects: the Central Utah Project (CUP) and the Central Arizona Project (CAP), the huge Garrison Diversion in North Dakota, and two of the five "Aspinall" projects in Colorado.[43] He even tried to stop the proposed Auburn Dam, which river expert Tim Palmer characterized as "one dam too far."[44] This was like playing with matches in a dynamite factory. But Carter made some progress. Eventually Auburn Dam was abandoned, as were two of the Aspinall projects. The Garrison Diversion was significantly cut back in 1986. The CUP and the CAP would be built, but in different form and with a different set of beneficiaries and with cost-sharing.

By the end of the 1970s the reclamation program had altered every major watershed in the West. In some basins, entire rivers downstream of Bureau diversions were de-watered.[45] When critics of the Bureau started pointing out the environmental damage caused by these large projects, the Bureau fought back fiercely. Unlike the Corps, an agency that was willing to adapt if it led to an expanded mission, the Bureau drew a line in the sand and declared war on change. Wendy Espeland, in her book *The Struggle for Water*, attributed this to "an insularity that allowed [Bureau] leaders not to notice or acknowledge that the world had become less enamored of its big dams. . . . It was just too easy to dismiss dissenters as radical, irrelevant and unrealistic."[46]

This attitude was abundantly apparent in the defensive posture of the water development community at that time. The Upper Colorado River Commission, which worked closely with the Bureau to develop the CUP, passed a resolution in 1973 lamenting the criticism of the CUP, which was "being subjected to unmerciful, unreasonable, and unconscionable attacks by ecology and environmental extremists." The resolution then made a threat that sounds more like Malcolm X than a public agency: "The staff and appropriate committees . . . are hereby authorized . . . to take any measures necessary and appropriate . . . to expedite construction of the Central Utah Project."[47] The "extremist" critics had pointed out that the CUP's irrigation component was economically irrational—a point that Carter made when he placed the project on his hit list.[48]

Saddled by an obsolete mission and a Paleolithic mind-set, the Bureau tried to stop the New West from happening. The Bureau still bragged about making the deserts bloom, but perhaps another phrase from Isaiah was more accurate in describing the work of the Bureau: "I will lay waste the mountains and hills and dry up all their herbage; I will turn rivers into islands and dry up all the pools."[49] But even mighty pyramids eventually crumble. The excesses of the Dominy period, the disaster at Teton Dam, and the Carter hit list all took their toll. The agency had become an anachronism, living in the past, unprepared for a future that had nothing to do with putting landless men on manless lands. Things had to change. The election of Bill Clinton was the next step in the forced evolution of the agency. His Secretary of the Interior, Bruce Babbitt, was instrumental in forcing a public dialogue on dams and their impact on American rivers. He even went on several "dam-busting" trips.[50] But his most effective move was to put Daniel Beard in charge of the Bureau of Reclamation.

BLUEPRINT FOR REFORM

"I'm a change agent." For Dan Beard, that is something of an understatement. He is the guy who hit the Bureau of Reclamation on the head with a two-by-four, trying to, as he put it, update a "New Deal 1930s bureaucracy in an email age." The dam-building era was over—everyone seemed to know that except for the Bureau. It was Beard's job to tell them.[51]

For decades a host of interest groups and legislators had tried to reform the Bureau, without much luck. All attempts to change the Bureau had come from outside the agency. Dan Beard had a whole new approach: the best way to change the agency's mission, and its culture, was from within. Dan Beard knew the Bureau well, knew its obdurate traditions and hidebound policies. He had served in the Interior Department with the Carter administration, and had been staff director for the House Interior Subcommittee on Water and Power when it was chaired by Congressman George Miller—an outspoken critic of pork-laden, environmentally destructive water policies. During those years, Beard and others had written "elegant policy statements" that immediately disappeared into the ether; the Bureau just ignored them. The only real progress was made when George Miller forced through a compromise authorization bill in 1992 that imposed some changes in reclamation policy. These changes were fiercely resisted by Bureau employees, who found ways to blunt their effectiveness. It did not take long for Beard to realize that the only way to make lasting changes in the Bureau was to change the agency internally, using the agency's own employees. When the Clinton administration came to Washington, Dan Beard lobbied hard to be appointed Commissioner of Reclamation; his plan was to act fast.

"I wanted to be a change agent for the organization, and then I wanted to leave. I did not want to 'drive a bus' and do a regular routine."

When a bold innovator is imposed on an agency, a great deal of resistance is the typical reaction. Beard utilized a number of management strategies to minimize opposition. First, he recognized and employed the Bureau's best attribute. "The Bureau was an organization filled with people who are problem-solvers, but they had been given the wrong instructions. They just needed instructions that were more contemporary so that they didn't continue solving the wrong problems."

One of Beard's first acts was to put together a team of Bureau employees to write a new "blueprint" for the agency. He outlined the general direction he wanted the Bureau to take; working within those parameters, they came up with a "Blueprint for Reform" with a mission statement that was a revolutionary departure from previous Bureau policies. It completely reorganized the agency from the ground up and converted it from a construction agency to a water management agency.[52] It even recognized American Indians, who had suffered grievously at the hands of the Bureau, as part of its clientele.[53]

Dan Beard and his blueprint drove home a critical message to the Bureau: either change or become irrelevant. Some of the old rank and file went about the new mission with all the alacrity of a Barry Manilow box set. Some of them quit rather than adapt, which made the whole process easier. During the two years in which Beard was commissioner, agency personnel went from nearly 8,000 to 6,500. The agency's budget was reduced from nearly $1 billion to $800 million—nearly all of it for operation and maintenance. Authority was delegated to field personnel, and an 8-foot-high stack of regulations was reduced to a 6-inch notebook. The changes were so striking that Harvard's Kennedy School of Government awarded the Bureau its "Innovations in Government" award.

Beard also recognized the importance of symbolism. Anyone who has studied water policy quickly learns that water projects often become instruments of power, prestige, and political gamesmanship. Beard understood this. His first act as commissioner was to open the commissioner's office. For as long as anyone could remember, the commissioner's office had been sealed off from the hallway; the only way to access the commissioner had been to go through two other offices. Beard removed a partition so that his office opened directly onto an open hallway. Almost immediately people started dropping in. Outside his office, in that hallway, was a long row of pictures of past commissioners. Beard knew about the famous artists that Floyd Dominy—another man who understood symbolism—had commissioned for the Bureau. So he had the pictures of past commissioners removed and replaced them with vibrant photographs of scenery and river landscapes. Suddenly, the hallway to the commissioner's office was devoid of reminders of the past and had become a visual cue to the future.

After just two years with the Bureau, Dan Beard moved on. Toward the end of his tenure as commissioner, he overheard two engineers talking at a water conference. "See that guy over there? He's the guy who turned the Bureau of Reclamation on a dime."

THE PENDULUM AND THE ROGUE

In response to these enormous pressures, the Bureau began to change. Some of its projects were experiencing hardships, and some were simply no longer viable. The agency needed a new niche. In 1995 the Bureau found itself in a unique situation in southern Oregon when it was called in to assist the Grants Pass Irrigation District with problems associated with Savage Rapids Dam on the Rogue River.[54] The Bureau had not built the dam, but it had provided long-term, interest-free loans to the irrigation district to install fish screens and radial gates back in the 1950s. By the time the district was encountering fish problems in the 1990s, it still owed the Bureau money.

When the District built the dam in 1921, no one cared about fish. As a local put it, "In the old days, if someone said something about fish, they were ostracized out to eastern Oregon!" But over the years, as salmon stocks dwindled, it became apparent that the dam was a major source of mortality for both smolts and adult anadromous fish. It created a substantial barrier to over 500 miles of spawning habitat. The Rogue River is home to five runs of salmon and steelhead, including the threatened coho salmon. According to a 1995 Bureau of Reclamation study, Savage Rapids Dam killed about 22 percent of those runs.[55] In the "old days," that was not a problem. But when the district applied to the state for an increase in its water right in 1987, it encountered a wall of opposition.

Some irrigation districts have had it all handed to them on a silver platter; the Bureau built their dam, canals, and laterals, and the taxpayers of America covered nearly all the costs. That is definitely not the history of the Grants Pass Irrigation District. It built its own dam, canals, and laterals with its own money raised by the sale of bonds. The dam was 39 feet high and 465 feet long, using the latest in 1920s high-tech dam design, with removable stoplogs, water-powered pumps, and the newfangled idea of constructing thin concrete arches to hold back tons of water. It was an enormous undertaking for such a small group of farmers. But such was their faith that a dam on the Rogue River would secure their futures and their families and make Grants Pass, Oregon, a great place to farm and raise a family. No one worried about the fish, which were literally too numerous to count. There would never be a shortage of fish; that was for certain in 1921. But there was a demonstrable shortage of

Savage Rapids Dam on the Rogue River, Oregon, just prior to its removal.

water for crops that by midsummer wilted in the drying sun. The rain in this region does not come until snow can be seen on the mountains in early autumn; the dam would give them water in midsummer.

Savage Rapids Dam was built at a time when deals were sealed with a handshake, and no one needed a lawyer or an Environmental Impact Statement to get a share of water. Things in Oregon were so laid back that the District did not complete its 1921 water rights certification process until 1982; before then, certification was viewed as a bureaucratic triviality. By then, many of the farms in the District had been converted into hobby farms and suburbs, which resulted in a reduction in water use and hence a reduction in water rights. To reverse that reduction, the irrigation district filed for an increase in its water right with the Oregon Water Resources Department. District officials were more than a little stunned when the Department pointed out that its dam, the Savage Rapids Dam, was perhaps the biggest fish killer on the Rogue River, and maybe it did not deserve to have its water right increased and certified. In exchange for a temporary water right, the state told the District it would have to do something about the fish mortality caused by the dam. The task of solving that problem fell to Dan Shepard, the manager of Grants Pass Irrigation District.

Shepard, self-described as a "redneck," has a likable, relaxed style and a fondness for folksy aphorisms.[56] He is somewhat unusual among western water district managers; early on he saw the handwriting on the wall, and it said "dam removal." As he phrased it, "We could see the pendulum swinging toward the environmentalists. I told the board that if they try to fix the dam, everybody in the world will block them. But there's lots of money if we climb on the backs of the salmon."

As we talked, I got the sense that his transformation did not come easy. He talked like someone who had been hit on both sides of the head with a club. The club in this case was the politics of the New West. On one side there were the environmentalists "who were bird-dogging the fish issue," as he put it. When the District applied for an increased water right, WaterWatch, an advocacy group for Oregon rivers, objected. The battle was on.

The people hitting Shepard from the other side were members of his own irrigation district. Shepard had no choice but to come up with a plan to solve the fish-passage problem, so under his leadership the District did a study in 1994. It clearly indicated that the cheapest alternative was to take out the dam and replace it with electric pumps. Some members of the District went ballistic when they realized the dam had to come out.

Clearly, some people in the Grants Pass Irrigation District did not expect a prolonged attack on their dam. To them, the dam was a symbol of prosperity; those who had water did well in the West, and those who didn't suffered. Mr. Shepard had lived in Grants Pass all his life, except for a stint in the service in Vietnam. He recognized at an early age that access to water divided people into classes: "When I grew up here, the people by the river had something, but the inbreds and the hillbillies lived out by Merlin [away from the river] because they could not afford to irrigate." He laughed as he said that, using the pejorative terminology for effect rather than literal meaning. But his message was simple: if you could divert water, you were much better off than those who could not. To people who grew up with that as a core belief, the idea of removing a perfectly good dam made about as much sense as filing the cutting edge off a knife.

But Dan Shepard saw the future. As we talked, he mentioned the pendulum again and described a situation that could apply to hundreds of rivers all over America. "[The battle over fish] was a sign of the times, that things had changed. Some people said just wait, the pendulum will swing back toward dams and away from the environmentalists, but you look in the papers and you see third graders doing things for fish. This is not going away."

Still, some members of the District's governing board were unwilling to see that, and they decided to fight for the dam. It got ugly. Bitter accusations, often of a personal nature, marred the debate. Dan told of "backyard quarterbacks who did not understand the law." Bob Hunter of WaterWatch described meeting a new member of the district board who introduced himself by saying, "So, you are the enemy."

The controversy over the dam wrought acrimonious divisions among district patrons. It was literally neighbor against neighbor; some thought removing the dam was an obvious solution, while others wanted to fight it out to the last man standing. There were death threats against people who wanted to remove the dam. When some of the patrons began talking about selling their water rights and washing their hands of the whole situation, a district board member sued them. When one patron was elected to the board on a pro-dam platform and later changed his position when the coho salmon was listed as threatened, a nasty recall vote removed him from office. The proponents of the recall demanded to see district phone records; they were convinced the board was secretly in cahoots with environmentalists and Big Brother bureaucrats. The level of animosity was further heightened when "wise use" property rights advocate James Buchal showed up in Grants Pass to ignite local resentment against a nefarious conspiracy consisting of the New World Order and environmentalists. To Buchal and his fellow travelers, Bob Hunter was Satan in sandals.[57]

In 2000 pro-dam elements on the board confidently scheduled a vote on dam removal; they were sure the patrons would vote to fight it out. They did not; instead, they voted to accept a settlement. That settlement, recognized in a 2001 court decree, committed the District to "the Pumping/Dam Removal Plan as the best and only permanent solution to solving the fish passage problems at the Dam."[58] All the parties that signed the decree, including the District, the state of Oregon, WaterWatch, and eleven other fishing and environmental organizations, were then partners in the effort to get the decree implemented. All they needed was congressional authorization and an appropriation of $15 million for the pumps and $5 million for dam removal.

For several years the Oregon congressional delegation introduced federal legislation to authorize the Bureau of Reclamation to build the pumps and remove the dam. That law finally passed in 2004. Funding to start building the pumps was provided in the Bureau's FY 2005 budget. That same year the Bureau provided a draft environmental assessment on installing the pumps and removing the dam.[59] Actual dam removal began in 2007 and was completed in 2009.[60] Yes, that's right; the Bureau of Reclamation helped remove a dam—although it was not one of theirs.

None of this would have happened if it were not for the local instigators. Bob Hunter of WaterWatch championed dam removal when it was dangerous to do so. And Dan Shepard proved to be a wily and astute negotiator for his district, instigating his own set of unprecedented solutions. After interviewing both Shepard and Hunter, I was struck by how each described his opponent. Both men were gracious in their depiction of the other, despite coming from very different perspectives. Each was a visual representation of his viewpoint. Dan wore the blue jeans and snap-button cowboy shirt that is a virtual uniform for western irrigators. Bob looked like a river guide. They are living symbols of the Old West and the New West. For many years they had been on opposite sides of the debate over whether to remove Savage Rapids Dam. But they finally found common ground through the arduous, frustrating process we call democracy. I saw the fruits of that effort when I visited the dam site in 2010—seven years after I stood at the same location and took photos of the problematic dam. There was a spanking new pumphouse on the south bank, and a brisk free-flowing river ran past it. There were dam abutments—but no dam.[61]

If one looks back at the gritty, often antagonistic process of deciding what to do with Savage Rapids Dam, it is easy to see it as both a bitter, divisive fight and, finally, a realization that times have changed and everyone is better off working together. Dam removal threatens iconic symbols, like putting antireligious graffiti on a famous church. The politics of dam removal hovers over a political landscape of sacred cows. But if done with sufficient sensitivity to all stakeholders, and with a recognition of the needs of those who lose when dams are removed, the decision-making process can bring a community together. And in the case of Savage Rapids Dam, it introduced the Bureau of Reclamation to a new role; perhaps there is a place for the Bureau in the River Republic. Recent Bureau announcements have revealed just how the agency thinks it can fit in.

WATER 2025

In 1994 Dan Beard gave a speech to the International Commission on Large Dams at which he said: "For many years, the Bureau of Reclamation largely served the needs of a few agricultural interests. We generally did not serve the needs of an expanding urban population. The result was that the base of support for our program declined."[62] Since then the Bureau has been searching for ways to stem the hemorrhage of support and find a new mission that is relevant to a twenty-first-century West.

In 1997, two years after Dan Beard stepped down as commissioner, the Bureau was a different animal. Its annual report for that year stated flatly, "In recent years, Reclamation has moved from development to management." The new commissioner, Eluid Martinez, wrote hopefully that the agency could undertake "exciting new initiatives such as watershed management investigations and agreements and protecting water quality."[63] The agency's annual budget hovered around $1 billion. Bureau staff was just one-fifth of what it had been in 1993.

The election of George W. Bush brought hopes that the glory days of the past might come back. For commissioner, Bush chose John Keys, a firm believer in the Bureau's past. Keys promised no big changes in mission, which at that point was a relief to the agency's beleaguered defenders. He did, however, hold out the possibility that new dams might be constructed, and he made it clear that he was not in favor of dam removal (except for Savage Rapids—he did favor that).[64] But his first budget reflected the new reality of the Bureau—one that fit around the agency like a straitjacket, forcing it to face in a new direction. Most of the money was for fish and wildlife, river restoration, or simply maintenance on existing projects. The big news that year (2002) was the

100-year anniversary of the agency, so the Bureau threw itself a birthday party. That provoked a howl from Taxpayers for Common Sense, which derisively accused the agency of "celebrating a century of subsidies."[65]

In its continuing search for relevance, the Bureau produced a study in 2003 called *Water 2025: Preventing Crises and Conflict in the West*. The study identified areas of high conflict where there was insufficient water to meet all demands.[66] Ironically, many of those areas are facing a water "shortage" only because Bureau projects have diverted so much water to agriculture that there is not enough left over for other users. The alleged crises are also caused by hopelessly out-of-date western water laws, institutional impediments to change, and limits on water marketing and other forms of efficient water transfer. The report, of course, does not mention those.

Another report, this one with the theme of "Managing for Excellence," followed in 2006. It was in response to a critique of the agency by the National Research Council and notes in the first paragraph that the agency went through many changes over the years, and now "it is time for Reclamation to change again."[67] The Bureau's "WaterSMART" (Sustain and Manage America's Resources for Tomorrow) program, which focuses on improved management and efficiency, is part of the agency's effort to do just that. Michael Connor, President Obama's Commissioner of Reclamation, summarized the new mission in his proposed budget for 2012: "Key areas of focus for FY 2012 include Water Conservation, Landscape Conservation Cooperatives and Renewable Energy, Ecosystem Restoration, Youth Employment, supporting Tribal Nations and maintaining infrastructure."[68] It's as though John Wesley Powell, Jimmy Carter, and Dan Beard spoke with one voice.

HERITAGE OF CONFLICT

John Wesley Powell is best known for an outrageous stunt: running the Colorado River through the Grand Canyon while perched on a chair strapped to a wooden boat—with one arm! This grand adventurer, who benefited from what Wallace Stegner called "the brashness of the half educated," was also a man of prophetic vision; he once warned an 1893 Irrigation Congress that it was "piling up a heritage of conflict" with its plans to irrigate massive amounts of land in the West.[69] Today, that appears to be something of an understatement.

The Bureau, like the Corps, needs to undergo a serious external evaluation of its projects to see if they still meet the needs of the American people.[70] The agency's budget for 2011 was slightly over $1 billion and included funding for a laundry list of projects. Some of them would clearly meet the test, but others, conceived in an era when men stood behind animals to plow the ground and the federal government had plenty of money to throw around, are no longer in the public interest.

Keep in mind that the two major water agencies in the United States were created for missions that are completely irrelevant today. The Corps was organized to clear snags so troops could move up and down rivers. The Bureau was created to settle the West. Since then both agencies have gone through convulsive changes, always provoked by external critics. It is worth considering whether these two agencies are still the best way to manage America's rivers. Are they Oldsmobile agencies in a Prius age? Surely some of their great works are worthy of our highest praise— testaments to spirit and resolve. But some are the legs of Ozymandias.

There have been innumerable attempts to eliminate or fundamentally reorganize each agency, usually spearheaded by the White House.[71] All of these have failed because Congress did not want to lose its pork delivery system. But this nation can no longer afford the status quo. A huge national debt, climate change, population growth, and changing public values have made old concepts and aging agencies not just wasteful but dangerous.

In the following chapters we will encounter the harbingers of change, as well as those who continue to cling to the past. We will meet people with starkly different visions for America. They all mix together in the grand milieu of rivers.

I end this chapter with a story from a water conference in 2005 when I was on a panel about the future of the Colorado River with Commissioner John Keys. I was there to talk about river restoration and the damage done to America's rivers by traditional water development.[72] Commissioner Keys was there to speak on behalf of the Bureau. After I had finished my presentation, Mr. Keys expressed his frustration with such talk. "Look, we have got to figure out a way to get the resource out of the river."[73]

I expressed a different perspective. "I think you are missing the point. The river *is* the resource."

Rivers cannot be all things to all people. This book is about the earnest struggle to democratize rivers and turn them toward public use—what I call the River Republic. In a larger sense, it is the struggle between past and future, between extractive use and sustainable use, and between the Floyd Dominys and the Dan Beards who populate the arcane habitat of river politics.

II

DISMEMBERMENT

Water hubris reached its apex in the hundred years that followed the American Civil War. Rivers became beasts of burden, harnessed to a yoke that pulled along the nation's growth and development. The general sense that land, resources, and water were available in infinite supply gave rise to the notion that rivers should be disassembled and parceled out to whatever special interest put in an order with the U.S. Congress. Federal water policy was dominated by a concept of water development that conceived of rivers as servants to particular interests. In this part of the book we will look at five uses of rivers that diminished the value of rivers as a whole and instead allocated them to narrow, extractive uses. The chapters will examine agriculture, hydropower, barging, flood control, and pollution.

4

HANDOUT HORTICULTURE

Farming and the Feds

As if the bright fringe of herbs on its brink
Had given their stain to the waves they drink
—FROM WILLIAM CULLEN BRYANT, "GREEN RIVER"

Farmers are perhaps the only people in modern America who still understand and appreciate the value of demanding physical labor. They value work not just for its productive capacity but also for its ability to build character. The virtues that we typically ascribe to the quintessential American—honest, hardworking, thrifty—are derived from the archetype of the family farmer. It is not surprising that Thomas Jefferson and many of his contemporaries felt that farmers would be the essential backbone of democratic government—that a government without the benign imprimatur of the yeoman farmer was doomed. Of course, at the time of this nation's birth, farmers comprised more than 90 percent of the population, so promulgating such beliefs was politically popular. A half century later, small farmers went to war against the South's slavocracy in the name of free labor and the Union. When the North triumphed, the *London Times* effused that a "nation of farmers and traders" had kept "the jewel of liberty in the family of freedom."[1] Such was the stuff of farmers.

It is still politically popular to trumpet the values of the farmer. The only trouble is that the classic family farmer that so enthralled Jefferson and the *London Times* is largely a relic of yesteryear. There are only 2.2 million farms in a nation with a total population of 311 million (in 2011). And most farm production comes from giant agricultural corporations and very large family farms.[2] Most important, the legendary thrift and self-reliance of farmers have been replaced by a system of subsidies and handouts so byzantine, enormous, and illogical that it defies the imagination. What does all this have to do with rivers? A great deal, since agriculture consumes 34 percent of the nation's water; that figure climbs to over 80 percent for the arid western states.[3] In addition, agriculture is now one of the largest sources of water pollution in the nation (see chapter 8). And agriculture is one of the largest users of the nation's barge channels, which have had devastating impacts on riparian habitat, wetlands, and riverine species (see chapter 6). The extractive use of the nation's rivers for a single industry competes directly with other river uses, such as fishing, hydropower, urban amenity use and potable water, recreation and tourism, and endangered-species protection. As a result, agriculture plays a role in most river restoration efforts. The impact of agriculture on our waterways is so great that a true rebirth of America's rivers will require fundamental changes in federal agricultural policy. A little history and some political context are in order to understand the challenge.

THE AGRO-INDUSTRIAL COMPLEX

My father grew up in the farming community of Dana, Indiana, where his parents ran a small general store on Main Street.[4] When I was a boy, my Dad would often tell stories about the Great Depression and the crushing impact it had on Dana. Once a fairly prosperous town composed of Jeffersonian yeomen, Dana became destitute almost overnight.[5] Farmers would show up at the McCool store, hat in hand, and ask for credit. They were loath to go into debt, but they could not feed their own families, much less sell enough crops to make a decent living. My grandparents dutifully extended credit to them—their neighbors and friends—even though it nearly drove them out of business. Many years later, after the worst had passed and President Franklin D. Roosevelt had helped rescue rural America, farmers would come into the store and ask how much they owed from those dark years of the Depression. My grandfather would dutifully pull a card out of a shoebox full of credit vouchers and tell them. And they would pay. Now that, my father would say, is real integrity.

One of the reasons that many Dana farmers got back on their feet was the various assistance programs devised by the Roosevelt administration and its attendant alphabet-soup agencies. Foremost among these was the Agricultural Adjustment Act, administered by the Agriculture

Adjustment Administration. It dealt directly with a problem that had plagued American farmers for generations—an oversupply of basic food crops. In effect, farmers were too productive and too numerous. In most years, they flooded the market with their products, which drove down prices. In a free-market economy, this would have forced the least efficient farmers, and those producing crops for which there were no buyers, to find some other way to make a living. But in the Depression, there was no other way to make a living in an economy with 25 percent unemployment. So Roosevelt paid them to reduce their crop acreage; he rewarded them for doing less work. In the short run, that was a great idea. In the long run, it's a runaway freight train full of pork-barrel subsidies.

The Depression ended in 1941, but by then the price supports, loan guarantees, crop insurance, and numerous other assistance programs had become an article of faith, the holiest of Beltway sacred cows.[6] No one wanted to take on the icon of *American Gothic*, even though most families had been bought out by corporate operators; by 2007, 75 percent of all farm production came from just 125,000 farms.[7] These big corporate farmers hire legions of lobbyists and make enormous campaign contributions. Corporations got into farming because it was an easy way to "cultivate" federal handouts and make a very handsome profit. The crops that bring in the most federal money are corn, wheat, cotton, and rice, but there are also big payouts for barley, sorghum, soybeans, dairy products, sugar, and a smattering of other grains and oilseeds. In the 1980s federal payments to farmers increased by 234 percent; in the 1990s they grew by 120 percent.[8] Between 1984 and 1994, agricultural subsidies totaled $135 billion.[9] From 1995 to 2010, taxpayers shelled out an additional $261.9 billion, according to the Environmental Working Group (EWG).[10] Yet farm-subsidy advocates in Congress complain that they aren't getting enough of our money and should not be cut in the budget-cutting year of 2011: "Funding for farm policy, including crop insurance, over the last five years averaged $12.9 billion per year, a 28 percent reduction from the 2002–2006 average of $17.9 billion per year, and a 31 percent reduction from the 1997–2001 average of $18.8 billion."[11]

The irrationality of this giveaway is heightened by the fact that much of the subsidized cotton and rice is grown in the desert. Growing overproduced tropical crops that consume scarce water in arid regions does not make economic sense—it's like grazing cattle in the Arctic. No rational business person would do that—unless you can get taxpayers to cover much of your business overhead and guarantee a profit.

The subsidy programs are so large and complex that the Department of Agriculture is not even sure who is being paid. A 2004 GAO report found that the Department had paid hundreds of millions to people who were not even eligible farmers.[12] A *Washington Post* investigation in 2006 found that $1.3 billion was given to people who don't even farm; suburb developers of former farm acreage advertise that home owners can collect government checks on the acreage in their yards.[13] A 2007 GAO report discovered that between 1999 and 2005 the USDA had paid $1.1 billion to dead farmers.[14] Chicago voting starts to look clean compared with this operation.

No significant changes to the farm-subsidy program took place until the passage of the 1996 Federal Agriculture and Improvement Act, which was meant to wean farmers from the federal dole. It also initiated a new land-conservation program that has resulted in some acreage being returned to a natural state. The conservation reserve program has been a success, but the goal of eliminating the subsidies by providing "emergency" and transition payments to farmers only resulted in more big payoffs. By 2002 these temporary emergency payments had cost $71 billion, and farmers were still growing the same overproduced crops. In 2002 Congress gave up trying to bring capitalism to farm country, and the 2002 farm bill brought back the New Deal–style subsidies, bigger than ever, and committed to $81 billion in new subsidies for the next ten years.[15] President Bush's 2004 budget continued the largess, earmarking $174 billion over ten years for the euphemistic "farm safety net," a program that would make Stalin smile.[16]

By 2000 the farm industry had become one of the biggest recipients of federal handouts, making it what the Heritage Foundation called "America's largest corporate welfare program."[17] A 2011 GAO analysis included direct farm programs in a report on wasteful government programs, noting that the payments went primarily to the largest, richest "farmers," even in years of record farm income.[18] The industry is so well entrenched that it is not an exaggeration to refer to an "agro-industrial Complex," much akin to the military-industrial complex that Dwight Eisenhower warned about in his farewell address. Taking a page from the defense industry, big agriculture has finely honed its skill at "defeat-proofing" massive farm subsidy bills by adding programs for nearly every state and congressional district in the nation. There are food stamps for the poor, school lunch programs for public schools, and support for organic farmers. Even Washington, D.C., gets price-support funds for growing crops on some postage stamp–sized lot.

Everyone gets a piece of the pie, but the lion's share goes to a handful of megacorporations. Twelve Fortune 500 companies received large subsidies in 2002; one of them, John Hancock Insurance Company, collected $2.3 million that year from American taxpayers. Kenneth Lay, the deposed chairman of the defunct Enron, pocketed $18,000.[19] Even a Tea Party candidate raked in farm handouts.[20] The Environmental Working Group study cited above found that two-thirds of America's farmers received nothing or a paltry amount, but 10 percent of the beneficiaries consumed 72 percent of the payments. A study by the Department of Agriculture in 2004 found that just 8 percent of the recipients collected 78 percent of the money, while the other 92 percent of farmers were given an average of less than $1,000.[21] Another USDA study found that in 2001 nearly 60 percent of all government farm subsidies went to farms with a net worth of more than $600,000, noting that "payments tend to be concentrated among the larger farms."[22] A few large farms in California's Central Valley collect millions from multiple types of subsidies.[23]

There are both acreage and income limits on subsidies, but corporate farmers have found clever ways to avoid them by creating false front farms that mask their true ownership and income. In California's Central Valley a number of big farms set up dummy partnerships to get around the 960-acre limitation.[24] One of them, the 23,000-acre Boswell Farms—a big user of subsidized water—collected nearly $11 million between 1995 and 2006, according to EWG figures. A study by the Heritage Foundation found that one farm in Arkansas, Tyler Farms, collected $32 million in subsidies in just five years by dividing itself into sixty-six separate corporations.[25]

A study by the *Washington Post* in 2004 found that farms exceeding the income limit still got $312 million.[26] A 2008 GAO report found that the USDA had paid 2,702 farmers who exceeded income limits.[27] As the *New Republic* put it, "Agricultural corporations have proven themselves adept at circumventing existing limits on farm subsidy payouts."[28]

Subsidies are also concentrated in the states and districts of the legislators who serve on the agriculture committees in Congress. An EWG study discovered that farmers in just 19 congressional districts (out of 435) collected half of all subsidies between 2003 and 2005.[29]

It is necessary to understand this larger context of farm history and subsidies to gain an appreciation of how difficult it is to challenge the status quo on rivers that are diminished or compromised by subsidized agribusiness. With so much federal largesse at stake, there are increased conflicts between river restoration and agriculture that uses rivers for barge transportation or subsidized hydropower or diverts water that would otherwise maintain the health of river ecosystems. It means there are conflicts between endangered species and farmers, fishermen and farmers, cities and farmers, recreationists and farmers, and water-quality efforts and farmers—"farmers" meaning corporate agriculture in most cases.

Many of these conflicts exploit an image of the farmer as a lonely outpost of Americana, fighting against alien forces—the banks, big-city high rollers, foreign owners, or faraway government agencies. It is an iconic image: intrepid locals confronting faceless authority in a saga reminiscent of a John Steinbeck novel. There are certainly instances of that in American history, but many water conflicts are dominated by large agribusiness or energy corporations working in partnership with those "faceless authorities" in the government. They rule the roost. In the American West, agriculture consumes about 80 percent of the water that is diverted, making irrigation districts powerful and reactionary. This is in spite of the fact that in most western states agriculture is a small fraction of the economy, and the most produced crop is water-sucking hay. For example, in Oregon, agricultural production is only 2.26 percent of the state's economy, and much of this is due to subsidies; in 2002 the state's farmers collected over $80 million from the feds, according to Environmental Working Group data.[30] In my home state of Utah, agriculture consumes about 85 percent of the water (90 percent of which is used to grow hay), but produces only 0.3 percent of personal income.[31] In the arid state of Arizona, agriculture—mostly cotton production—is only 1 percent of the economy but absorbs 70 percent of the water.[32] In short, irrigated agriculture is a powerhouse in the West and uses most of the water but produces only a small part of the economic output.

This is not for lack of effort. There are 55 million acres of irrigated farmland in the United States, mostly in the West. These farms consume

90 million acre-feet of water annually, according to the National Agricultural Statistics Service (NASS).[33] The biggest crop, in terms of water consumption, is hay, which is a low-value crop that requires a great deal of water. Next on the list is corn, which is an overproduced commodity crop that generates federal subsidies. Other heavily subsidized—and irrigated—crops are cotton, wheat, rice, barley, sugar, and soybeans, according to NASS survey data. Thus, in most western states, agriculture has a wildly disproportionate impact on water supplies and the local economy. There is one state, however, where big farms are big business, and that is California.

California irrigates more land than any other state, nearly 8.2 million acres according to the NASS Farm and Ranch Survey. These farms produce a veritable cornucopia of crops, including many high-value products that are unique to the California desert environment and have tremendous market value. These include fruits and vegetables, notably specialty crops such as artichokes, almonds, raisins, wine grapes, and winter lettuce. But, amazingly, the California desert is also cultivated to produce vast amounts of cotton and rice, both of which are surplus crops that are price-supported by the federal government. And much of this rice and cotton is grown using heavily subsidized water from the Bureau of Reclamation's Central Valley Project (CVP).

The CVP, like other Bureau projects, was supposed to provide water to small farmers, who would then repay the costs of the project. It didn't turn out that way. The project was one of the Bureau's biggest, with twenty dams and 1,600 miles of canals and drains, built at a cost of $3.6 billion. After sixty years of operation, the project's farmers had repaid only 11 percent of their allocated cost. However, they do consume one-fifth of all the water used in California, and they receive irrigation water that is heavily subsidized, according to an EWG analysis.[34]

The CVP delivers about 7 to 8 million acre-feet of water per year to irrigate 3 million acres. Such a large project has an enormous impact on the environment, from downstream cities to in-stream species. With burgeoning suburbs clamoring for more water, direct environmental degradation to the Kesterson and the Kern National Wildlife Refuges, and the endangered delta smelt, the CVP became a target of reforms and thus a central player in the effort to restore the San Francisco Bay Delta and its source waters from the Sacramento and San Joaquin rivers.[35] In

1992 reformers in Congress, led by George Miller in the House and Bill Bradley in the Senate, forced through some significant reform measures with the Central Valley Project Improvement Act. These changes were incremental and did not eliminate the huge subsidies, but the 1992 Act did begin the long process of fundamentally reshaping the CVP and the Sacramento River.

FISH TRAPS

What comes to mind when you think about Central California? Wine? Artichokes? Almonds? How about rice? How about salmon? When I first began reading about the Sacramento River and efforts to restore anadromous fish runs, I was very skeptical. California is famous for glitz and grapes, not natural riverine environments. But in fact there is an enormous effort to restore some species to the Sacramento and its tributaries. This effort has included the removal of Saeltzer Dam on Clear Creek and the restoration of parts of Butte Creek, both in Central California, but also involves the Trinity River in the north, the Sacramento and its tributaries, the San Joaquin River, and the delta on the east side of San Francisco Bay. The politics of these projects is even more complex than the hydrology.

The Self-Removing Dam

Serge Birk was the Ecosystem Coordinator and Environmental Director for the Central Valley Project Water Association (CVPWA), and he walked a diplomatic tightrope. By training, Serge was a fisheries biologist, an expert on the living habits of the salmon and steelhead that inhabit the Sacramento River—which just happens to run through his backyard. But he did not work for an environmental organization; he did not represent any fishing groups; he was not employed by the government's fish and wildlife agencies. Rather, he served at the pleasure of the board of directors of the CVPWA, a nonprofit umbrella organization that represents the interests of over seventy-five Central Valley irrigation districts that have contracts for water from the massive CVP.

A few moments after I arrived at Serge's house, he told me he had allocated exactly one hour for our interview.[36] For the first hour, he talked spontaneously; I could hardly squeeze in a question. Then he looked at his watch, and I knew this was my only chance to put some hard questions on the table. He obviously felt like he had been challenged, and he forgot about the time. Instead of just answering his own questions, he began to think out loud about how agriculture and fish can coexist in the Central Valley. His goal was not to save fish per se, but to save fish to the extent necessary to save agriculture—the people for whom he worked. After nearly three hours we were still talking.

Serge was proud of the work that had been done on behalf of fish, and he suggested we get on the river and take a look. So we jumped into his aluminum skiff and headed up the Sacramento River toward Red Bluff Diversion Dam, part of the CVP system. Red Bluff Dam is unique in that it can essentially "self-remove," because it consists of eleven huge water gates that can be raised. On the day we visited the dam, all the gates were down, causing the water upriver to rise 13 feet higher than where we sat in the skiff. But that was a rare sight; in order to provide sufficient water for fish passage, the gates were kept open all but four months a year. This was good news for spring-run and winter-run chinooks and green sturgeon, but it increasingly impinged upon the ability of the dam to deliver water to 150,000 acres of agricultural land. A solution had to be found that eliminated the dam but still provided water to farmers.

The Bureau of Reclamation engineered just such a solution, called the Fish Passage Improvement Project, which provided traps and screens to protect the fish, and massive pumps to pull water out of the river and onto farmers' fields. As our skiff neared the dam, Serge pointed out four large metal grates on the west side of the dam, which were the intakes for four pumps. Each of these sucks in massive amounts of water—and fish. But unlike traditional pumps, these can screen out the fish and spit them out downriver. The water then goes to irrigated fields, and the fish continue their journey to the sea. With the pumps in place there is no need for the dam.

Serge used to work for the Bureau of Reclamation as a project manager at Red Bluff. As he put it, "I bought the revolving fish traps you see just below the dam. They are my traps." Serge clearly felt a sense of ownership, although it was the taxpayers' money that actually bought the traps. As our skiff neared the pump intakes, he decided to get a closer look at the dam and take some pictures.

As we idled near the intakes, Serge noticed two U.S. Fish and Wildlife employees cleaning one of the revolving fish traps (the ones Serge purchased when he was with the Bureau here). Serge decided to tie up at the trap they were cleaning and show me the inner workings of the dam. As we approached the men cleaning the trap, I got the bowline ready, but one of the young men warned us off, telling us it is against regulations to allow boats upstream of the fish outlet (a concrete device in the middle of the river where the fish that are sucked into the pumps are spit back into the river). The young man spoke in a forceful, don't-mess-with-me voice. Serge was taken aback; after all, these were "his" fish traps. Serge worked at this dam for years, knew everyone, and helped build this project. He is a fixture, and he clearly expected recognition and the privilege of bending rules when he pleased. Yet these young yahoos didn't even recognize him, wouldn't even let him access the traps that he knew so well. As we pushed off, he yelled out, "Say hello to Jim Smith for me," just to show these young know-nothings that he knew their boss, that it was somebody important that they had just brushed off. He did not identify himself, making his request to "say hello" impossible to fulfill. As we drifted back downstream, Serge made repeated comments about being rebuffed; it clearly ruffled his feathers. "I would have done the same thing—they were just doing their job," he said. But he was bothered by the lack of recognition, the inability of these young men to show proper respect for someone who's been around a long time and knows everybody in the business.

But Serge did not need to take the rebuff personally. What had really happened was a result of the plodding evolution of river policy and the estrangement between the old water establishment and the dawning River Republic. Serge was part of the old Bureau, but even a stodgy old agency like Reclamation, stuck in the steam-shovel age for so long, has evolved in recent years in an effort to avoid becoming an archaic, obsolete token of a West that no longer exists. Serge knew all the old water buffaloes, the old dam builders—the fellas who reworked rivers with a sense of confidence and certainty that is impossible in today's complex politics of water. But the Old Guard of the Bureau is retiring, or dying, or simply quitting in frustration. Or, like Serge, they quit the "new" Bureau, the Bureau of Dan Beard and "environmental mitigation," and went to work for their traditional constituents.

In the meantime, the new Bureau goes on without them. The fish passage project at Red Bluff Dam got a boost when President Obama awarded the project $110 million in the American Recovery and Reinvestment Act of 2009.[37] The project and the accompanying federal funding are now hailed as a great step forward by a wide coalition of interests. It is one element in the complex effort to save rivers and endangered species in Central California while also protecting irrigation interests in the Central Valley.[38] However, the innovations at Red Bluff Dam do not change the basic water calculus; there is insufficient water for all demands, and still more water must be allocated to fish if the endangered species of the Sacramento are to survive.[39] In 2009 the California legislature passed the Water Conservation Act in an effort to resolve the pending crisis over Sacramento water and other water issues; only time will tell if this latest effort will be successful.[40]

DYING 4 WATER

In the Central Valley, farmers use not only water from the Sacramento Basin but also water from the Trinity River, which is in the Klamath Basin to the north. In 1955 Congress authorized, as part of the CVP, the construction of two dams on the upper Trinity with a transbasin diversion tunnel into the Central Valley. The purpose of the project was to divert "surplus" flows into the Central Valley for irrigation. Section 2 of the Trinity River Division Act (1955) directed the Secretary of the Interior to "preserve and propagate" the fish and wildlife in the Trinity, in part because of treaty obligations to the downstream Hoopa Valley and Yurok Indian reservations.[41] The two dams, Lewiston and Trinity, were completed in 1963, and the giant Westlands Water District immediately signed contracts for the water. The CVP began moving Trinity water— lots of it—to the Central Valley and apparently forgot that the authorization act stipulated that only surplus water was to be taken. In some years it diverted 90 percent of the river. Predictably, the fish populations fell by 60 to 80 percent.[42]

Commercial fishing organizations, environmental groups, and both Indian tribes began pushing for a reallocation of Trinity water and succeeded in getting Congress to pass the Trinity River Basin Fish and Wildlife Management Act in 1984, with the objective of restoring fish runs to the river. The river received another boost when Congress passed the CVP Improvement Act in 1992, which guaranteed a minimum flow. In 2000 the Bureau of Reclamation completed an Environmental Impact Statement (EIS) and issued a "Record of Decision" (ROD), which is a planning document that lays out what the agency is going to do to restore the fish. The ROD proposed a variety of stream-rehabilitation techniques, but the hot potato was the decision to reduce water flows to the Westlands Water District and other districts that receive CVP water. The districts immediately sued, which led to a long series of appeals and countersuits. Since then there has been a consistent effort to increase flows in the Trinity and bring back its once-magnificent salmon and steelhead runs. High flows in 2011 aided this effort, resulting in a somewhat amusing announcement from the Bureau of Reclamation that the "official water year designation for the Trinity River is 'wet.'"[43] Thank goodness water is wet.

The Trinity River cuts through the rumpled green corduroy of northern California. One mountain range follows another, with the river taking a tortuous track between them, first westward, then picking up the south fork and heading north to meet the Klamath. The terrain is rough and vertical, and although homes are scattered throughout the countryside, the region is not heavily populated. The small number of inhabitants made the river an easy target when the CVP went looking for more water. Local people resent that and feel that there are serious social justice issues as well as habitat issues with the Trinity diversions.

Arnold Whitridge is a former Trinity County supervisor and a long-term advocate for re-watering the Trinity River. He explained the social dynamics of the Trinity diversions from the perspective of people who live in that valley. "We are like a colony here. When the dam was built, it was just Indian tribes and a few residents. Admittedly, the project had gigantic benefits for the state in hydropower and agriculture. But they took our water. Our water goes to Westlands Water District, which is dominated by zillionaires; they are absentee landowners of great wealth."[44] In contrast to this image, Westlands presents itself as a collection of family farms with an average size of 900 acres.[45] That does not necessarily negate Whitridge's point. For Mr. Whitridge, recent efforts by West-lands to delay the restoration project are simply a stall tactic so it can use as much water as possible for as long as possible. "My belief is that the writing is on the wall; the legal obligations are clear; all we have now are people delaying the implementation, primarily by objecting to proce-dural details."

The delays are indeed impressive. The ROD was issued in 2000; the project did not really get under way until 2003 and is still only partially completed, in part because of limited funding. But there is an ironic twist in all this. If Trinity River diversions into the Sacramento River are re-duced to save fish in the Trinity, the reduction could harm efforts to save fish in the Sacramento. The irrigation districts have jumped on this as a way to maintain their Trinity water supplies. As a result, the restoration of flows has been a very slow process, and that frustrates some people.

Tom Stokely made a career out of his passion—restoring the Trinity River. He was a planner for Trinity County, and restoring the river was his job.[46] When I interviewed him in 2003, he reminded me of a professor who has given a lecture so many times that he can deliver it in his sleep—without notes or a pillow. He was a veritable compendium of arcane river data, ranging from flow statistics to budget figures and the Public Law numbers of every relevant act passed by Congress. Stokely derived energy from the battle over the Trinity the way some people are energized by danger or drugs. During our interview we had to pick up his daughter at the small high school in Hayfork, California. We hopped in his van, retrieved his daughter, and returned to his office without so much as a 30-second hiatus in Tom's Trinity River stream of consciousness.[47]

I asked him how the restoration effort got started; who were the instiga-tors? The answer came to him immediately. He gestured toward some old photographs on the wall. "By the early seventies the fish populations in the river had been decimated. A lot of local people used to fish that river. So the students in the high school held a mock funeral for the fish. It was only a class project, but it got national attention." That small act of defiance set in motion a movement to challenge the mighty CVP and regain some lost water for the fish, and the people, in the Trinity River Valley.

During our long conversation, Tom described how a small county with few residents could win against a powerful irrigation district and the Bureau. "It's David versus Goliath when a public interest group fights the government and the status quo. If you want to win in such a situation, you have to have the federal government on your side." One way to do this is to engage the Indians. In the case of the Trinity River, it was the Hupas, and later the Yuroks, who brought in federal power and money on behalf of restoring the river. This is the reverse use of the "Indian blanket," an old strategy of cloaking highly questionable irrigation projects in the guise of an Indian project to get it through Congress. Such projects rarely had many Indian benefits, but they made bad projects look like they had a social justice function. In this case, however, the Indian tribes were working against a project and for a river that was the very heart of their homelands.

As I drove into the tiny burg of Hoopa, California, on the lower Trinity River, I noticed a hand-painted sign near the road. It said "Dying 4 water" with ethereal images of disappearing salmon. This was my first indication that this isolated Indian community is in a struggle for a resource that is more than a protein source, more than a sport, more than a source of revenue. To the members of the Hoopa Valley Tribe, fishing for salmon is an elemental ingredient in their cosmos, as much a part of their psyche as it is a part of their natural environment. Unfortunately, the federal government either did not understand the importance of the salmon to the Hupa or did not care.

"I can tell you horror stories about our fishing rights." That is how Mike Orcutt, Fisheries Director of the Hoopa Valley Tribe, began our interview.[48] He then proceeded to do just that. Mike is a college-educated expert on fish biology, but his job requires him also to be an expert in politics and the arcane world of water law and Indian law. His division of the tribal government does everything from counting fish to lobbying public officials, and it is obvious to all of the people working in Mike's division of tribal government that their success in the latter will determine the former.

The Hoopa Valley and Yurok tribes have been instrumental to the success of the Trinity River Restoration Project. They have kept up a constant harangue to the Bureau and the U.S. Congress, reminding them they have a trust obligation to the tribes, and that means restoring the salmon. To them, it is worth "dying 4."

The Trinity restoration is an example of the popular "adaptive management" approach to natural resource problems. Basically, that means

Sign on the Hoopa Valley Indian Reservation, California.

that all parties must work within the constraints of physical and political reality. In this case, there is no talk of dam removal; this is a below-dam project. With creative management, the fish runs can be restored without removing Lewiston and Trinity dams. The restoration project received $4.5 million in federal stimulus funding in 2008; with adequate funding, the project could be complete by 2012.[49] However, recent budget cuts make that unlikely.[50] The increased water flows brought about by a 30-year political struggle have the potential to dramatically increase salmon in the river while still maintaining a substantial flow of water to irrigation districts.[51] That could mean less rice grown in the Central Valley of California, but that is the trade-off when fish collide with agriculture.

CROPS, CAPITALISM, AND CONTEXT

No one who eats is opposed to agriculture. But it is crucially important to understand that agriculture involves trade-offs with other uses of water and land. The restoration of America's rivers will not succeed on a large scale until those trade-offs are carefully analyzed and choices are made about which uses are best for the country as a whole.

The first step in bringing rationality to farm policy is to introduce capitalism. If farmers, especially large agribusinesses, paid market price for their water and had to make a profit without government handouts, a fundamental shakeout would occur. The most efficient farmers, growing the crops most in demand and utilizing natural resources carefully and prudently, would do fine in an open market. This would have several immediate impacts on rivers. First, it would reduce water usage. As water prices rise to market levels, users always find ways to conserve; it becomes profitable to be efficient. Second, it would save taxpayers billions of dollars that could then be used for more productive programs or, better yet, simply returned to the people who earned the money in the first place. Third, water quality in the United States would dramatically improve (more about that in chapter 8).

Under the current system, the American people are paying via taxation for the destruction of their own rivers. If we would simply quit subsidizing activities that degrade or diminish rivers, that alone would go a long way toward restoration. For example, in California's Central Valley, one of the biggest water users is the Farmers Rice Cooperative. Its farmers grow rice by flooding it to an average depth of five inches. It also happens to be one of the biggest recipients of federal farm subsidies, pulling in a whopping $146 million between 1995 and 2006, according to EWG data.[52] Does California really have a water shortage, or is it simply an ecologically and economically foolish place to grow rice? Another big farm in the area, Dublin Farms, received $17 million during the same time period. It grows cotton and wheat. In 2007 the biggest handout recipient in the nation was another Central Valley operation, Sandridge Partners, which took in a little over $1 million that year, primarily for cotton.[53] Right behind Sandridge were two other Central Valley megafarms. If the government did not subsidize these crops and provide a cheap supply of scarce water resources, these companies would either move to a location where rice and cotton grow without massive water projects, or they would grow another product. The fact is, growing rice and cotton in the desert is inane economics. It's like trying to sell copies of *The Pregnancy Workout* at an AARP convention—wrong product in the wrong place. At some point, there has to be a reckoning of economic rationality with farm subsidies.

The Farm Bureau argues that Americans want cheap food, and that justifies the huge subsidies. But the government is not funded by generous extraterrestrials; it is the American people who foot the bill for those subsidies. The food is not cheaper; the costs are simply hidden by the labyrinthine process of government. And we compromise the health and public utility of our rivers in the process.

It may seem politically impossible to make real progress against farm subsidies, but there has been a growing movement in recent years to challenge the status quo. President George W. Bush tried valiantly to reduce subsidies, despite the fact that Texas collects more federal farm money than any other state. Bush tried to impose a $250,000 cap on each recipient in the 2008 farm bill, which had a jaw-dropping price tag of $600 billion over nine years.[54] The president vetoed the bill, arguing it "continues subsidies for the wealthy and increases farm bill spending by more than $20 billion, while using budget gimmicks to hide much of the increase."[55] Congress overrode the veto, but not without protest from some members of both parties. President Obama, in his first budget, took direct aim at farm subsidies. Like his predecessor, this took courage; Obama's home state of Illinois is ranked third in federal subsidies. Obama proposed the same income cap as Bush—$250,000 per individual—and proposed to eliminate direct payments to farms with more than $500,000 in annual sales.[56] Congress did not accept these reforms, so Obama tried again with his second budget.[57] With a $14.5 trillion national debt, there is increased pressure to finally reform the farm subsidy program.[58] There was a movement in the 112th Congress to cut farm subsidies, but legislators from both parties resisted.[59]

The second step toward a more progressive farm policy is to look at it in holistic and systemic fashion, taking into account long-term environmental, social, and economic consequences and what economists call "negative externalities," which refer to costs and damages done by an individual or company to other people, the environment, and society. These include such factors as the pesticides, herbicides, antibiotics, and chemical fertilizers that taint our rivers, the unemployed fishermen along the Oregon and California coast who have gone broke because of a lack of salmon, the rivers that have dried up because of diversions or excessive groundwater pumping, and the recreational benefits that have been lost because of inadequate water. We should also consider the health impact of eating high-fat, high-sugar foods, which tend to be the ones that are heavily subsidized; would the nation be better off if we ate more salmon and trout and less beef, dairy, and grain?[60] One-third of the population is medically obese and another third overweight—a condition that is even worse among children and costs America an estimated $147 billion a year.[61] So, the nexus between the outdoors and our health-care costs is important. This is cost-benefit analysis writ large, but the nation can no longer afford to take a myopic view of either rivers or agriculture, lest we all end up "dying 4 water."

Just before my father passed away, we took him back to Dana, Indiana, for one last look at his hometown. By then his memory had faded to a fractured tableau, but he still remembered details about Main Street and the people who had lived there in a bygone era. His halting hand pointed to buildings, now boarded up, and he told us of their former glory. He pointed out the modest storefront that used to be the McCool store. Just before we left, he wandered into the middle of Main Street, which was devoid of traffic. He looked forlornly at the bleak, crumbling buildings and the empty street, trying to recall happier days when this same street was filled with farmers and friends. Then, as though he were asking these buildings a question, he said, "What happened to this town? It used to be such a nice place."

There is a powerful lesson in his words. There are many places in the United States that "used to be such a nice place." Some of them are neighborhoods, some are landscapes, and some are rivers. We no longer have the luxury, if we ever truly did, of flippantly discarding places and resources, even though time drives us in new directions. If we take care of what we have and restore at least some of what we used to have, we can enjoy this planet for a long time to come.

5

FALLING WATERS

Hydropower and Renewable Energy

With a sudden flash on the eye is thrown,
Like the ray that streams from the diamond-stone
—FROM WILLIAM CULLEN BRYANT, "GREEN RIVER"

A t the visitor center at Glen Canyon Dam there is a real-time meter that clicks out a dizzying succession of numbers—too fast to actually read anything but the numbers on the far left. That is, of course, by design. The flashing numbers indicate how much money is being made by selling electricity generated by the dam's turbines. They make a stark visual impression and remind the visitor that this dam is what the Bureau of Reclamation refers to as a "cash register dam." One gets the sense that this is "free money," courtesy of gravity and the Bureau. This accounting of benefits is impressive, but it is not accompanied by a similar accounting of costs or losses. Hydropower is always touted as "clean energy," which it is—if one's concept of "clean" is limited to air.

Hydropower plays an important but limited role in the nation's energy production, creating about 7 percent of our power. Only a small portion of the nation's dams—about 3 percent—actually produce electricity, but they manage to churn out about 100 gigawatts (GWs).[1] That is roughly equivalent to the power needs of 28 million homes. Many hydro dams are concentrated along the West Coast states of Washington, Oregon, and California, but there are also numerous hydro dams in Idaho, New York, and nearly every other state. In the Northwest, hydro produces 70 percent of the region's electricity.[2]

There is an ongoing debate about whether hydropower should be considered "renewable" energy and lumped in with other green sources, such as wind and solar. On the one hand, the "fuel" for hydro is water, which renews itself as precipitation. On the other hand, dams have a finite life, reservoirs fill with silt, and hydro dams can have devastating effects on riverine environments.

Many of the attributes of hydro that make it attractive as an energy source also have harmful environmental impacts. One of the great advantages of hydro is that it can be ramped up or down rapidly, producing power that matches the demand curve throughout the day or season. But these surges of water are quite different from the natural pulses of the river—the hydrograph—so downstream ecosystems suffer. Also, most hydro dams must store water to produce "head," the pressure that forces water through penstocks. That requires holding back water in the spring and releasing it slowly during the rest of the year. That flow regime is exactly the opposite of natural river flows, which produce a large spring runoff that nourishes downstream flora and fauna.

Dams also trap sediment, which eventually fills reservoir basins.[3] They also deprive the downstream corridor of this natural building material. For example, the beaches and riparian corridor in the Grand Canyon are disappearing because the natural sediment load is trapped behind Glen Canyon Dam; according to the Glen Canyon Institute, the equivalent of 30,000 dump-truck loads of sediment are trapped behind the dam every day.[4] Sediment accumulation also presents problems when dams fail or are removed. All dams will some day come down—the average age of dams in the United States is fifty-one years—but none of these dams were built with any thought about how to remove them or what the consequences would be when they outlived their usefulness. As a result, huge sediment loads trapped behind dams represent a significant future threat to rivers and greatly complicate the removal process.

Additional ecological problems are caused by changing water temperatures. Water trapped behind dams is sometimes colder and sometimes warmer than natural temperatures. In the Colorado River, the cold water released from Glen Canyon Dam has created a hostile living environment for four native fish species that thrive in warm desert rivers. The opposite problem occurred in the lower Klamath River; salmon need cold water, but the four dams on the lower Klamath heat the water, making it a hospitable environment for diseases that afflict the fish. This is one of the reasons that those four dams are slated for removal.[5] Dams also trap pollutants, which can become embedded in the sediment, as at Milltown Dam in Montana, or cause unnatural plant growth, as in Upper

Klamath Lake.[6] And of course, dams interfere with fish migration, perhaps their most inimical impact. Thus hydropower offers an alluring set of charms but presents a mixed character of good and evil.

Nonfederal hydropower in the United States is regulated by the Federal Energy Regulatory Commission (FERC), which was created by Congress in 1920 to assist in the development of hydro and other energy sources. Operators of private, municipal, and state hydro dams must get a license from the FERC, which is renewed for either thirty or fifty years. The relicensing process takes about five years and provides an opportunity for all stakeholders to participate and provide input. In its early years, the FERC functioned primarily as a promoter of hydro, but the Federal Power Act was amended in 1986 to require power companies to give "equal consideration to . . . energy conservation, the protection, mitigation of damage to, and enhancement of fish and wildlife . . . the protection of recreational opportunities, and the preservation of other aspects of environmental quality."[7] That is a dramatic departure from past FERC policies, but "equal' is a relative term. Asking the FERC, with its long institutional history of promoting hydropower, to balance that with other uses is like asking Rush Limbaugh to balance nuance and diatribe. Like the Corps and the Bureau, the FERC has been slow to move to a more balanced perspective, but there has been progress.

The power of the FERC over the fate of rivers is enormous. Because so many dams were built in the heyday of the big-dam era in the mid-twentieth century, literally hundreds are coming up for renewal. The FERC regulates over 2,000 nonfederal dams.[8] From 1935 to 2010 the agency issued 1,012 licenses.[9] Each relicensing is an opportunity for members of the public to participate in the future of their local river and possibly alter that future.

The FERC is just one big player in the hydro drama. Numerous other federal agencies play a direct role, including the U.S. Fish and Wildlife Service, National Oceanic and Atmospheric Administration (NOAA) Fisheries, the Corps and the Bureau, and even federal land-management agencies such as the Forest Service and the Bureau of Land Management. In addition, there are four power-marketing administrations in the Department of Energy that sell power from federal dams. The two largest are the Western Area Power Administration, which markets water from fifty-six hydropower plants in fifteen western states, and the Bonneville

Power Administration (BPA) in the Pacific Northwest, which sells energy from twenty-nine dams.[10] These power marketers, like the FERC, have built-in incentives to give priority to energy production, which often places them in conflict with other river users.

In short, like everything else about water, hydropower creates mixed costs and benefits and complicated politics. There are numerous disputes over hydro-generating dams in the United States, but some of the most vitriolic conflicts are in the state of Washington, which derives 74 percent of its power from dams. The dominant river in that state is the Columbia, which is big, beautiful, and under heavy leash. The Columbia Basin extends from beyond the Selkirk Mountains in British Columbia, South to the arid valleys of northern Nevada, and east into Montana's Flathead River Valley. The river itself crosses the Canadian border, wanders erratically south across eastern Washington, then flows west to form the border between that state and Oregon, and finally collects itself for a massive surge to the sea. Its greatest tributary is the Snake River, which begins in the Yellowstone Country before meandering across southern Idaho and joining the Columbia in east-central Washington. In historic times the salmon and steelhead runs were estimated to be between 10 and 16 million, and perhaps as high as 50 million, but by the 1970s they had plummeted to 2.5 million and were still falling.[11] The upper Columbia's spawning habitat was blocked by Grand Coulee Dam, so the Snake River, with its abundance of pristine streams and clear mountain water, makes up much of the remaining habitat for anadromous fish in the basin.

The Columbia River of today would scarcely be recognizable to Lewis and Clark. There are 250 dams in the Basin, including eleven main-stem dams in the United States (there are additional dams in Canada), but most of the attention in recent years has focused on four dams on the lower Snake River. The fight over these four dams has become the focal point of the conflict between fish and hydro dams. The politics and legal history of this issue are so complex as to defy the imagination.

SMOLTS AND VOLTS

The Snake River and its tributaries were once a thriving salmon fishery. The lower river cut through a broad canyon with dozens of rapids and clean, swift water. It was a ribbon of wild fishing heaven amidst a plain of rolling hills. That changed when the Corps of Engineers decided to extend the Columbia River barge channel up the Snake River to Lewiston, Idaho. That required four dams, each backing up a reservoir that reached to the foot of the next upstream reservoir. Ice Harbor was completed in 1961, just ten miles from the confluence of the Snake and the Columbia. Three more dams followed in quick succession: Lower Monumental, Little Goose, and the farthest upstream, Lower Granite, which was completed in 1975. With the completion of those dams, salmon and steelhead had to run a gauntlet of eight dams to get to their spawning grounds—some of them 1,000 miles upstream. And the juvenile smolts had to negotiate the warm, sluggish reservoir water and turbine blades to return to the sea. The fish runs, already abysmally low because of decades of abuse and mismanagement, declined even further. By the mid-1990s thirteen species in the Columbia Basin had been declared threatened or endangered, four of which spawn in the Snake. But wheat farmers had a cheap way to ship their product. Lewiston, Idaho, 730 miles from the sea, became a seaport. And more cheap electricity was added to the Northwestern grid.

I drove through miles and miles of undulating yellow wheat fields called the Palouse Country. The name comes from an Indian tribe that used to fish at Kettle Falls—now beneath the waters backed up by Grand Coulee Dam. The farm homes in this area look like something out of an old *Life* magazine. Many of them sported a "Save Our Dams" sticker on the mailbox or window. Mule deer and pheasant were plentiful along the roadside. A lot of the older houses stood empty, and I passed abandoned schoolhouses and a boarded-up 1902 Grange hall—testimony to a society that does not stand still for very long.

It was harvest season when I visited, and the grain trucks were out in legion. The trucks on this road were on their way to the loading docks at Almota, just downstream (or more accurately, down-reservoir) from Lower Granite Dam. As I drove past Almota, I saw a mountain of wheat (a variety known as soft white), waiting for the next barge. The grain will travel to Portland and then, most likely, to the Orient.[12] In other words, the harvest from these quintessentially American family farms may soon be feeding people living in the shadow of the Great Wall of China. The Palouse Country may look like it did when federal agencies were building big dams around here, but in fact it has changed dramatically in response to changing world markets.

I knew from the map that I was close to the river, but all I could see were dry hills in all directions. Then suddenly the Snake River appeared, impounded behind the dams that transformed the river into a 14-foot-deep shipping channel. Lower Granite Dam, just upstream from where I broke through the hills, is farthest upstream, backing up water into Idaho. The dam sits in a wide valley, surrounded by hills that alternate between the vivid ivory of ripened wheat and outcroppings of chocolate-brown volcanic rock. A smattering of small trees adds a flourish of green. On the day I visited Lower Granite, five of the eight spillway gates were open, allowing a great deluge of water to churn through the dam and provide water for fish migration. I arrived at the dam just in time to drive across before the gates closed at 5:30 p.m. This is a remote area, but every corner of America has been touched by 9/11, and security was tight. At night the dam is off-limits to the public, and I had to show a picture ID just to drive across. Although Lower Granite Dam (there is no Upper Granite Dam) is not large compared with the dams on the lower Columbia, it is still an imposing sight. The Snake River is a big watercourse, and it takes a fairly impressive structure to hold it in place, even briefly. The dam includes a modest powerhouse, a lock that is small by Mississippi River standards, and a long earthen section that is now the focus of a bitter battle over possible breaching of the dam.

Dam proponents often characterize this as a fish-versus-people conflict, but a more accurate characterization pits those who make a living off the fish against those who make a living off endangering the fish. When the four dams on the lower Snake River were first proposed, they provoked a bitter debate. Sportsmen and commercial fishermen from Idaho to Alaska opposed the dams, arguing that they would imperil the great salmon runs that course through 900 miles of river to spawn in remote lakes and streams in Washington, Oregon, and Idaho. But at that time the United States was still enthralled with dams, and opposing them was equated with opposing progress. The dams had state-of-the-art fish ladders, but no one gave any thought about getting the smolts—the juvenile salmonids—downstream through the dams, so they were chewed up by turbines or slowly sickened in the warm, sluggish water.

The dams created jobs in the barge and farming industries, but for those who made their living off the salmon, the dams meant a loss of livelihood, the end of a way of life, and the passing of an era. For Indian people in the Basin, they meant even more—the breach of treaties, a dire threat to their culture, and an end to economic well-being. Today, people in this region talk as though the dams have always been here, but they are in fact fairly recent additions to the Basin, among the last of the great dams built during the heyday of dam building (the last was built in 1975). By the 1960s and 1970s, even as the dams were being constructed, many people living in this area began to question whether the four Snake River dams were worth the sacrifice. They helped push for new laws such as the 1980 Pacific Northwest Electric Power Planning and Conservation Act, which required the protection of fish and wildlife. Combined with the Endangered Species Act, these laws began to turn the tide against big dams. In 1991 the Snake River sockeye salmon was listed as endangered; twelve more species in the Columbia Basin followed shortly. Local and national interests began exploring ways to reverse the decline of the salmon and steelhead of the Columbia River Basin.

Save Our Wild Salmon was formed in 1991 to help coordinate the work of dozens of environmental groups, sportfishing interests, and commercial fishermen. The Northwest Sportfishing Industry Association was formed two years later. Trout Unlimited, Idaho Rivers United, and American Rivers became actively involved in the Columbia Basin. These various groups do not see eye-to-eye on everything, but when it came to restoring the Columbia Basin salmonids, they were all on the same page. In 1994 the Idaho Department of Fish and Game sued NOAA Fisheries for an inadequate biological opinion, called a "BiOp." The BiOp is required by the Endangered Species Act and is the official multiagency plan to restore endangered species. The court ruled against the NOAA, and this initiated a long, convoluted path through numerous legal actions and agency responses.[13] The battle was on, and right in the middle of it were the Corps of Engineers and its four dams.

Eye of the Storm

Greg Graham is chief of the Corps' planning branch that deals with the four dams on the lower Snake River. In 2006, when I interviewed him, he had worked for the Corps for twenty-seven years.[14] He seemed somewhat apprehensive when I met him at the new Corps headquarters building in Walla Walla, Washington. When I introduced myself, he shook my hand and said, "A pleasure to meet you, at least so far." He had the air of a man who had been attacked and pilloried on more than one occasion, and he was not sure whether I was there to deliver one more thrashing. I assured him that I wanted to get the story straight, and that was why I was there for an interview.

"Where do I begin?" he asked, his eyes wandering all over the room.

"Tell me where the idea of breaching dams originated," I asked. That was all he needed.

"Let's start by talking about the concept of drawdown."

The decline of the salmon had been a long time coming, but by the late 1980s it was apparent that something had to be done, or the salmon and steelhead of the Columbia River Basin were going to disappear. In 1990 a "salmon summit" was held in Portland, Oregon, in hopes of bringing diverse agencies and experts together to formulate a strategy.[15] One of the ideas that came out of that meeting was "drawdown," meaning that the water stored behind a dam should be drained relatively quickly to allow smolts to proceed rapidly downriver rather than loiter in sluggish water. As Graham explained, it is a valid concept, but it simply would not work on the Snake River because all the dams are run-of-the-river dams, meaning that they only briefly store water to produce hydraulic head for the turbines and to maintain the 14-foot-deep channel for barges. Despite these problems, the Corps experimented with drawdowns at Little Goose and Lower Granite. These experiments helped convince the Corps that drawdown was not the answer.

"These dams have about a 3–5-foot operating range; they were designed to operate only in that range. If you draw the reservoir down below that, you effectively render the dam inoperable, and that includes everything from the fish ladders, to the irrigation intakes, to the turbines. Drawdown just won't work," Graham explained.

In 1996 the Corps wrote an interim report on the operation of the lower Snake River dams and how they affected juvenile fish passage.[16] The report, authored by Graham and his colleagues, noted that, given the problems with drawdown, it would actually be better to simply breach the dams since they would no longer be operationally effective anyway if the reservoirs were lowered. In effect, "drawdown" had morphed into "breaching" in the debate.[17] The word *breaching* is used instead of *removal* because the simplest solution is to remove only the earthen portion of each of the four dams; the concrete structures would remain in place in the river, perhaps to be used in some future scenario if salmon restoration failed and the Corps wanted to return to the current system. "The idea of breaching came from us, not the environmental groups," Graham emphasized.

The Corps' interim report came out about the same time at which numerous other agencies and stakeholders were studying the Snake River. What resulted was something of a war of competing scientific reports from government agencies, as well as stakeholder groups—what one local characterized as "biolitics," or the combination of biology and politics.[18] Taken as a whole, this body of research tended to favor dam breaching. But change of that magnitude never comes easy. Once the idea of breaching reached the stage where it was being taken seriously, the dams' defenders went into action.

The Corps' Environmental Impact Statement (EIS) process began soon after the publication of the 1996 interim report. The Corps' goal, according to Graham, was to be as inclusive as possible and to generate data, debate, and discussion. "We wanted to be honest brokers of information. We had a lot of public meetings and roundtable workshops. We set up a website, we prepared newsletters and developed an extensive mailing list. We created a traveling display that we took to agencies, conferences and schools. We produced educational videos and set up a speakers bureau."

The one clear message that came out of all this activity was that the public cared deeply about this issue, but there were bitter divisions within the region. During the comment period on the draft EIS, the Corps received 230,000 written responses. Hundreds of people showed up for public hearings, so many, in fact, that the Corps had to schedule double hearings, one in the morning and another in the evening.

"We started measuring the inputs [from the public] by the pound. Usually, we might get a couple thousand comments on our proposed EISs, but nothing else has come close to generating this type of interest," said Graham.

The Corps' draft EIS did not include the selection of a preferred alternative, but the final EIS, which was published in 2002, did, and it did not select breaching. Graham explained: "Everyone can agree that dams are not good for fish, but the question we had to answer was, do these dams have to be breached to save the fish? Relying on the NOAA Fisheries data, we decided there were other tools in our toolbox that had less significant economic impact than breaching."

The preferred alternative was labeled the "adaptive migration approach." Instead of breaching, the Corps proposed to make "major system improvements that would improve effectiveness and increase flexibility for optimizing migration routes within seasons and years. Surface bypass collectors, behavioral guidance structures, and removable spillway weirs could be installed at one to four dams, if testing warrants, to maximize adaptive migration capabilities."[19] To the Corps, this was a way to save the dams and the salmon. To its critics, it was a giant step backward from the 1996 interim report and a potential death knell to the salmon.

The Corps' EIS had a marked bias against breaching. This should come as a surprise to no one. As Edward R. Murrow put it, "We are all prisoners of our own experience," and the Corps has a 200-year tradition of building dams. Of course, all parties involved in such conflicts tend to interpret data in their favor—a tendency even more pronounced in studies that involve complex ecosystems. Steve Pettit, a career fish biologist, explained. "Whenever you put a living organism into the equation, there's room for variation, and that can be manipulated. There is always a gray area. Consultants have become experts at exploiting this gray area."[20] Another scientist working in the area put it this way: "Everyone is looking for the scientific fig leaf to cover their decisions."[21]

Perhaps the greatest potential for interpretation of data is in the economic analysis. Research by the Institute for Fisheries Resources calculated that the loss of salmon and steelhead runs cost $500 million a year.[22] A study by the Oregon Natural Resources Council found that restoring the salmon runs would generate $87 million per year, which it claimed was greater than the benefits produced by the four dams.[23]

Another study, called "Revenue Stream," also by pro-fish groups, concluded that removing the four dams would save taxpayers and ratepayers between $2 billion and $5 billion over the next twenty years and generate $9 billion in new revenue.[24]

Hydro interests quickly countered with a critique charging that the environmentalists and fishing interests had used incorrect methodology, and thus their conclusions were inaccurate.[25] Dam proponents argued that "many tens of thousands of jobs along the rivers would be affected. Communities and businesses that depend on navigation and irrigated agriculture could be decimated."[26] The Corps claims that the four dams generate $500 million in revenue each year.[27] However, these revenues must be balanced against the costs of the fisheries recovery program; the BPA plans to spend $45 million a year on restoration through 2017.[28]

Samantha (Sam) Mace of Save Our Wild Salmon claims that "the Corps was doing backflips in the economic analysis to avoid concluding that breaching is the best option."[29] There are certainly assumptions in the report that tend to make breaching look less attractive. For example, the Corps assumed that breaching would not necessarily solve the problem and would result in only "moderately reduced extinction risks" for fall chinook and steelhead and merely "slightly reduced extinction risks" for spring/summer chinook.[30]

Other such assumptions by the Corps can be found in its economic analysis regarding power, farming, barging, and recreation. Hydropower is the largest of these. Power from all the federal dams in the area is sold wholesale by the BPA—the "800-pound gorilla on the river," as one sportfishing advocate put it. The Corps' analysis assumed that all power lost to breaching would be replaced by new power plants powered by natural gas. But a much more attractive and less expensive alternative would be to invest in improved efficiency and renewable forms of energy. The four dams on the Snake River produce only 5 percent of the region's electricity, which could easily be replaced through conservation, which would in the long run save consumers money. According to a 2002 report by RAND, switching to alternative energy sources would not harm the local economy and would actually create 15,000 new jobs.[31]

The impact of breaching on barging is also the focus of much debate. The Columbia River system ships nearly 17.5 million short tons a year, mostly grains and petroleum products.[32] To get a better understanding of how the system operates, I met with Arvid Lyons in the "seaport" of Lewiston, Idaho.[33] Lyons has been working with grain his entire professional life. After he earned a degree in economics from Oregon State, he became a grain inspector for the U.S. Department of Agriculture and then worked at a grain elevator. When I met with him in 2006, he had been working for twenty years as the manager of the Lewis-Clark Terminal, one of the private facilities at the port that handles grain shipments on the river. His livelihood depends on the continued operation of the four dams on the lower Snake. His facility ships grain, nearly all of it wheat. Although a railroad track passes right by its grain elevators, it has no way of loading train cars; it can ship only by river. After some discussion, Lyons noted that the terminal had lost a lot of business to the railroads when the railroads began running "unit trains," which are 100-plus-car trains that are exclusively devoted to hauling commodities.

"At our high point, we were shipping 22 million bushels/year; the whole port [of Lewiston] was doing about 45 million bushels/year. Now, due to losses to the railroads, we do about 16.5 million bushels/year, and the whole port does about 30–35 million bushels/year."

To be sure, the port of Lewiston has a major impact on the local economy, doing $120 million in sales annually and supporting 1,900 jobs directly or indirectly. However, the fact that many growers have switched voluntarily to shipping with the railroads makes it clear that trains provide a commercially viable alternative to barging. Even Potlach, the local wood products company, has shifted much of its shipping from barging to rail. Yet the Corps' EIS assumes that nonbarge shipping would cost an additional $38 million/year.

In regard to recreation, the Corps did include moving-water recreation in its analysis—perhaps the first time the Corps acknowledged such values in an economic analysis. The Corps' EIS assumed that "river-oriented" recreation would generate $71 million/year along the lower river.[34] An even greater economic stimulus would occur upstream in Idaho. A report commissioned by pro-breaching groups concluded that a restored salmon and steelhead fishery would result in an additional $544 million in economic activity in Idaho alone.[35]

In regard to farming operations, the EIS assumes that irrigation pipes would have to be extended to reach the river. However, the EIS did not include the cost to American taxpayers of farm subsidies for wheat growers and other overproduced commodities. Nor is there a discussion of the farmers using heavily subsidized barging. I asked a local environmentalist why environmental groups did not point out that reducing agricultural production, especially wheat and corn, would save the taxpayers millions of dollars. His reply was revealing. "We want to work closely with the farmers to help them find an alternative way to get their crops to market. No one here talks about subsidies; in western Washington, just bringing that up is a death knell to the discussion."

Implicit in the Corps' approach is the assumption that salmon decline is due to causes other than dams. The Corps' Witt Anderson, chief of the district support team in the Northwest Division, succinctly presented that perspective:

There is no question that those fish are in trouble; otherwise they would not be listed. Clearly, in the Columbia Basin hydropower has had a material effect on that population, and our hydropower dams are a limiting factor to recovery. Having said that, it gets complex from there. There is the legacy effect of commercial harvest, habitat changes, hatchery programs, and four million plus people living in the basin; these are all limiting factors to recovery.[36]

One possibility is that clear-cutting and other forms of riparian degradation have robbed the salmon of sufficient spawning beds. I asked Rod Condo, former director of the Idaho Game and Fish Department, about that. "There is more than a thousand miles of pristine salmon habitat in Idaho. It's not lack of habitat that is killing the Snake River fishery."[37] Much of this habitat is in federally designated wilderness areas. In total, Idaho has approximately 3,700 miles of spawning habitat; that amount has not appreciably declined since the 1960s, but the fish numbers certainly have.[38]

Another claim is that overharvesting is the real culprit. But the NOAA Fisheries' 2000 Biological Opinion concluded that the endangered fish species "are not affected, or are affected only minimally, by today's much reduced fisheries."[39]

Another alleged culprit is sportfishing. I asked Liz Hamilton, director of the Northwest Sportfishing Industry Association, about that claim.

"To be sure, we are a big business; we generate $3.5 billion a year, and employ 36,000 people. But our take from the resource is actually quite small. Sports anglers average only one fish for every four days of fishing. In other words, we generate a lot of economic activity per fish."[40]

I was quite surprised that, given the declining numbers of fish, there was still a viable sportfishing industry. Hamilton explained, "It's important to remember that a lot of people lost their livelihoods when those four dams were built on the Snake River. As fish numbers decline, more and more people will lose their jobs, and I don't just mean in this area. Sports and commercial fishermen from California to Alaska depend on Columbia River salmon. We're talking about a major economic catastrophe."

Perhaps the best conclusion that can be derived from these competing economic claims is that dams reallocate both wealth and water; breaching dams does the same thing, creating winners and losers. All of this was on my mind when I had an opportunity to tour the Ice Harbor Dam. My host was Gina Baltrusch, a public affairs specialist with the Corps. In previous phone conversations she had made it clear that she was an ardent supporter of the Corps' past work and current "adaptive migration approach," the Corps lingo for anything-but-dam-breaching. It is not always easy to get Corps people to talk on a personal level, but at a lunch break on the way to the dam, I asked her what she thought of the proposal to breach the four dams on the lower Snake River. Her answer was honest and forthright. "I'm for balance, and that means solving the salmon problems with the dams in place. Look, I'm an army gal; I was active service before I took this job with the Corps. We should keep the Corps dams."[41]

At Ice Harbor Dam, Gina and I met with Mark Plummer, the project fisheries biologist at the dam. He showed me many aspects of the dam's operations, but by far the most interesting was a device called a removable spillway weir. In government everything is known by its abbreviation, so this is an RSW. It is one of the numerous strategies described by Greg Graham to improve salmon transit (and thus survivability) through dams. Corps biologists had noted that smolts often gathered upstream of a dam, in an area they call the forebay, and loitered, often for hours, before passing through the dam's turbines, or fish bypass system, or spillways. Perhaps they were sensing that conditions simply were not natural. Indeed, none of these passage routes are conducive to the normal habit of the fish, which is to remain near the surface and ride the flow downriver. The RSW was designed to mimic these natural conditions so that smolts would quickly and safely pass through the dam. It is, in effect, a giant, movable cam-shaped waterslide that directs surface water through a spillway in such a way that it appeals to the natural tendencies of the salmon. The objective is to pass the fish through quickly and safely while using as little water as possible. This may sound simple, but it is not. The volume of water involved, and the finicky nature of smolts, required that the RSW be a massive structure, weighing 1.7 million pounds, and be capable of moving thousands of small fish without harming them. RSWs are also expensive, at about $14 million per unit.

As we toured the dam, we passed by the six generators. They were small compared with those at Grand Coulee, and only one of them was in motion. We then stepped into the control room, where two Corps employees watched a wall of dials and meters. I asked why only one generator was online. "That's all we can do under the court order and the requirement for spill," one of the men replied with disdain. "That court order rules how we operate this dam." "Spill" is water sent through a dam without passing through the penstocks in order to increase flow for fish.

The "court order" he referred to is the latest in a long-running legal battle. In most areas of the country virtually no one knows the name of the local federal judge, but in the Northwest, Judge James Redden is a household name. If the BPA is the 800-pound gorilla on the river, Redden is the 900-pound sumo wrestler. He is either reviled or beloved because he has been a forceful advocate for strict enforcement of the Endangered Species Act. After the NOAA was successfully sued over its 1994 BiOp, it rewrote its plan in 2000. That revised BiOp came before Judge Redden in 2003, and he cut it to pieces in a strongly worded decision.[42] By then the Columbia fish issue was national news. President George W. Bush stood on Ice Harbor Dam and announced that there would be no dam removals while he was president.[43] His appointees in the NOAA wrote another BiOp in 2004, arguing that the dams had actually become part of the natural ecology of the river. Judge Redden savaged that BiOp as hopelessly ineffective and essentially began to micromanage the NOAA's recovery efforts.[44] He warned ominously that if the NOAA didn't submit an effective recovery plan, "the courts would be required to 'run the river.'"[45]

Proponents of the dams have argued that increased spill through the dams would not have a significant impact on smolt survival rates because dams are not really the problem and the fish are doing just fine. The Pacific Northwest Waterways Association, the lobby group for the barge industry, claims that "there are more fish in the river than at any time since the first dam was built at Bonneville in 1938."[46] Thus spill is just a waste of water.

That argument was dealt a serious blow when the Fish Passage Center, the official "counter" of fish in the river, studied the impact of the increased spill and concluded that it had a clear positive impact on fish survival rates.[47] This was good news for salmon advocates, but for those who dislike Judge Redden's "interference" in dam management, it was seen as justifying his position. Hydro interests went to Senator Larry Craig of Idaho, one of their staunch supporters, and complained. In response, he inserted language in a House conference report that forbade the BPA from providing funding for the Fish Passage Center. It looked as though the messenger had been duly killed off. But a lawsuit was filed to save the Fish Passage Center, and the court granted a stay while the parties debated the legality of Craig's conference language. In 2007 the Ninth Circuit rebuked the BPA for accepting Craig's conference report prohibition against funding the Center.[48]

To get a better understanding of the mountains of data about the Columbia River fishery, I visited the much-beleaguered Fish Passage Center in 2006 when the Center was in Larry Craig's bull's-eye. I interviewed the manager, Michele DeHart, and immediately noticed the bare walls, stripped of everything but the nails that formerly held frames.

"We packed up everything on the day we were supposed to be defunded, and then the judge issued a stay," she explained in a remarkably chipper mood, given the precariousness of her position. It was clear at that time that she had no idea how long the Center would remain in existence, and thus there was no need to rehang pictures.[49]

I asked about the claim that overfishing is the real reason that the salmon are in trouble and the somewhat contradictory claim, made by the same parties, that fish numbers are actually up and the problem has been solved. She showed me reams of data and more graphs than I could count. There are many complicating factors, but the basic outline looks something like this: Fish numbers declined precipitously before the dams were built on the lower Snake River. Some of that decline was due to habitat destruction, and some to overfishing, but those days ended long ago. Other factors are now clearly at work because of the dams. Since the four dams were completed, the numbers have remained far below natural levels, although there have been increases in the past twenty years for most species. But without an effective recovery program, it is not clear that fish populations are sufficient to sustain the species in the long run. Recent data confirm these trends.[50] Indeed, salmon numbers in 2010 set records, but they are still a tiny percentage of historical runs.[51]

The Fish Passage Center is still in operation, which surely must make former Senator Craig toe-tapping mad. This was not the only case where government officials engaged in a process of killing the messenger. In the late 1980s the Idaho Department of Fish and Game took a strong stance in favor of salmon restoration. It played a large role in the salmon summit and supported the idea of drawdown. The Department's perspective was presented in a report titled "Idaho's Anadromous Fish Stocks: Their Status and Recovery Options," which came out in May 1998. The report concluded that the "natural river option," meaning the removal of the earthen portions of the four dams, was the best option.[52] The state's legislature, which was dominated by farming and timber interests, took notice, according to Steve Pettit, the Department's fish-passage specialist for thirty-one years. "When our 1998 report came out, every legislator in the state was livid and they promised retribution. They passed a bill removing responsibility for endangered species from the Department and put it in the governor's office. They hired an ex–Forest Service supervisor to run it. The Department can no longer speak regarding the state's position on endangered species; only the governor's office can do that now."[53] Such is the nature of the "fish wars" in the Columbia Basin.

During the debate over the four dams, salmon advocates have fought against the impression that saving salmon meant doing without electricity. Hydro interests have been quite successful in characterizing the choice as "smolts versus volts." In its cost calculations, the BPA considers all water used for salmon—for spills, bypasses, and other measures—to be water lost to hydropower and therefore an "opportunity cost." It even puts this figure on some of its customers' bills to remind them that they are paying higher power costs because of salmon. Its customers have some of the cheapest rates in the country, but the message still carries resonance. However, water "lost" to irrigation and barging is not figured as an opportunity cost, even though it has the same effect on power generation. And critics like to point out that the BPA does not own the river; as a navigable watercourse, it belongs to all the people.

To most of the parties involved in this long-running debate, it is a question of whose economic ox is going to be gored. But for the Indian tribes in the Basin, there is much more at stake. Jeremy Five Crows works for the Columbia River Inter-tribal Fisheries Commission, a consortium of four traditional fishing tribes on the Columbia and lower Snake rivers (Yakama, Warm Springs, Umatilla, and Nez Perce). I met with Jeremy, a Nez Perce, at the Commission's headquarters in Portland.[54] He is mild-mannered, yet he can talk at a staccato pace when he describes what the tribes have lost, and how important it is to bring back the salmon.

"We have our own salmon restoration plan called 'The Spirit of the Salmon.'[55] It is our take on gravel to gravel management, meaning we look at this as a whole life-cycle issue. We need to look at the river ecosystem and the salmon's habitat holistically." Mr. Five Crows pointed out that no agency on the river has that kind of comprehensive management authority. Yet this is perhaps the most important reform that needs to be made if the salmon recovery program is to be successful. According to Jeremy, not everyone welcomes the tribes' proposals for change.

"Some people seem really angry that 'Indians' are trying to take away their dams. At the hearings in Clarkston a lot of Nez Perce people testified. According to the hearing rules, elected officials get to speak first at the hearings. That includes tribal elected officials. This angered some local whites, who said the tribal leaders should not be allowed to speak along with the white elected officials."

In 2008, four salmon-fishing tribes (Yakama, Warm Springs, Umatilla, and Colville) signed an agreement with the government that provides nearly $1 billion in funding to improve fish habitat; in exchange, the four tribes agreed to drop out of litigation designed to force breaching the four dams, at least for ten years. The Nez Perce Tribe did not sign the deal.[56] The withdrawal of the four tribes from the antidam crusade weakened the coalition, but there are still dozens of environmental and fishing groups that are continuing the fight.

As the emotional content of the debate ratchets up, both sides reach for hyperbole and more creative name-calling. At a pro-dam rally in 1999, Washington state senator Dan McDonald exclaimed, "We are not going to let a few Seattle ultraliberal environmental zealots destroy what took generations to build."[57] At the public hearings held by the Corps in 2000, a pro-dam speaker argued that dam breaching was a plot by "the moguls, the socialists, the Marxists in our universities," who were working with "Bruce Babbitt and his Bolshevik bandits" to "destroy the infrastructure of America. Reduce it to make it more easily integrated in an international global government run by socialists."Another speaker, a member of the Nez Perce Tribe, had a decidedly different take on dam breaching: "These dams have become a symbol of death, the death and the ultimate extinction of the salmon."[58]

The fish wars in the Columbia Basin are about more than fish; they are about economic dislocations, changing social values, demographic upheavals, and sometimes race and class. At this point, the parties cannot even agree on goals, much less how to achieve them. NOAA Fisheries issued yet another BiOp in 2008— actually three of them, which the agency claimed would "provide comprehensive, far-reaching plans for the protected salmon species."[59] The new BiOps covered fifteen endangered species but did not consider dam breaching. In an "issue summaries" paper, the NOAA explained that instead of breaching, "most of the sovereign parties focused on better defining the key elements of aggressive non-breach . . . strategy to improve the survival and recovery prospects for ESA-listed salmon and steelhead."[60] Several parties argued that this BiOp was a big improvement over previous plans, but the state of Oregon and a coalition of fishing and environmental organizations criticized it as yet another Bush administration whitewash.[61] In 2008, the same year the BiOp was released, two other events occurred that changed the political scene. Senators Larry Craig and Gordon Smith, both longtime dam supporters, left office involuntarily. And of course, America elected a new president with a substantially different energy agenda than that of the Bush administration. It was in that context that Judge Redden once again turned his gaze to the BiOp. In May 2009 he sent a letter to the lead lawyers, signaling his "serious reservations" about the NOAA's claim that the endangered fish were "trending toward recovery." Judge Redden indicated that there was sufficient flexibility in the "adaptive management" approach to permit the federal agencies to amend their inadequate efforts and still make the 2008 BiOp work.[62] This led to considerable discussion of settlements and compromises, even on the question of dam removal.[63] Despite Redden's misgivings, the Obama administration supported the 2008 BiOps but indicated that dam removal should be studied as a "contingency of last resort."[64] It then produced yet another BiOp in 2010, in response to Judge Redden's instructions, that included adaptive management—but not dam breaching.[65] That same year the Republicans took over the House of Representatives, and Congressman Doc Hastings, an ardent opponent of dam breaching, became chairman of the Natural Resources Committee. Dam breaching was off the table, at least until the political winds changed. The federal cost of implementing the BiOp is pegged at nearly $18 million for FY 2012.[66] Judge Redden, at 82 years old, issued his last opinion on the case in August 2011, concluding that the 2008/2010 BiOp was still arbitrary and capricious because it failed to identify mitigation plans beyond 2013. The judge then surprised everyone in December of that year by requesting that he be removed from the case.[67]

The Obama administration's position on breaching surprised and angered some groups. Several parties, including the National Wildlife Federation, the state of Oregon, the Nez Perce Tribe, and others, filed yet another court challenge to the BiOps, arguing that "overstating the costs and ignoring the benefits of measures to protect salmon violates the Northwest Power Act."[68] In a reply redolent with frustration, the federal government lashed out at these opponents and provided a succinct summary of just how complex the effort to restore the fishery has become:

> The collective body of work compiled over the last five years, including the Biological Assessment, Comprehensive Assessment, Supplemental Comprehensive Assessment, three Biological Opinions for FCRPS, Upper Snake, and *United States v. Oregon*, the Fish Accords and Estuary

MOA, Adaptive Management Implementation Plan, 2010 Biological Opinion, and extensive public comment and collaboration throughout comply with this Court's remand.[69]

That is a testament to the vitriolic drama of salmon in the Columbia River Basin.

Despite all the politics and court cases, the discussion always gets back to the salmon—the icon, the "signature wild creature" of the Northwest.[70] As Timothy Egan put it, "In the Northwest, a river without salmon is a body without a soul."[71] To Sam Mace of Save Our Wild Salmon, the goal is clear: "We don't want a museum fishery—doing just the least amount to save just a few fish from extinction. That's the lowest bar possible. The issue should be about restoration, and that's what dam removal would do, bring the fishery back to 1960s levels." Arvid Lyons of the Lewis-Clark Terminal also sees the situation in fairly stark terms, but from a vastly different perspective: "If the dams are breached, it will kill a lot of industry and destroy the value of the land. People would have to close their doors." Dam breaching would indeed close some doors, but it would also open many others. It is not uncommon for people to view the conflict in stark, polarized terms. It is also typically American for people to think that they can have it all—if the government just spends enough money. Jeremy Five Crows succinctly described this tendency. "People want to hear that these techno-fixes will work and we don't have to make any sacrifices. The Corps has made a policy of gold-plating the dams. The fish ladders didn't solve the problem, so they did juvenile bypass, but that didn't solve the problem, so they did this brushing system for the surface bypass. Now they're doing the removable weirs."

Liz Hamilton of the Northwest Sportfishing Industry Association thinks that the numerous studies conducted by various government agencies are just a stall. "Every time the data say the fish are in trouble, we get a new process. The federal agencies are like addicted gamblers; they keep rolling the dice until they come up with the numbers they want." To Hamilton and the people she represents, the Columbia fishery is teetering on the brink of a great disaster. "If this fishery goes, it will be the next New Orleans."

To many people of the Northwest, the salmon is an integral and essential part of their environment, like air or water. To lose it is to sacrifice who they are. To the farmers and loggers of the region, the goal is to preserve their accustomed livelihood without incurring additional costs. To the Indian people, the goal is survival as a separate political and cultural entity. The future management of the Columbia River is the basis of both their hopes and their fears.

THE SCENIC AND THE SACRED

When salmon and steelhead smolts encounter the Snake River dams, they are still hundreds of miles from the ocean. If they survive those dams, there are four more on the main-stem Columbia to impede their progress. But not all anadromous runs must face such formidable obstacles. Some, such as the Elwha River on the Olympic Peninsula, flow directly into the sea. It is a direct route to freedom for fish that spawn in the Elwha. The only problem is that there are no spawners in the Elwha because of two dams on that river. The Elwha Dam was built in 1912. Glines Canyon Dam was built farther upstream in 1927 and was later included within the boundaries of Olympic National Park. Both dams produce a small amount of hydropower. There is no fish passage and hence no anadromous fish—yet.

Doug Morrill is a big man, well over 6 feet tall and stout as an oak tree. He towered over me as we walked beside the Elwha River. He is the fisheries-management specialist for the Lower Elwha Klallam Tribe, a traditional fishing tribe that has lived at the mouth of the Elwha River since time immemorial. Morrill is non-Indian—originally from Los Angeles—but has worked for the tribe since 1992. He described his job as rewarding and committing and has become a forceful advocate for the tribe's fisheries.

We began our interview at tribal headquarters.[72] I tried to engage Doug in a conversation about the politics of removing Elwha and Glines Canyon dams, but with little success. I asked him about the politics that led to the passage of the 1992 Elwha River Ecosystem and Fisheries Restoration Act, which provided for the federal purchase and eventual removal of the two dams.[73] He mentioned something about fish habitat. I finally shut down my laptop and asked if he could show me some areas that would be affected by the removal of the two dams. At that moment he lit up. We jumped into a tribal SUV and worked our way along a rutted dirt road that wound through a dense forest of red cedar, western hemlock, and alder and popped out on a gravel embankment beside the Elwha River, near what Doug described as an ELJ—engineered log jam.

All the while, Doug talked of the new science of removing dams and restoring fish runs. "The dams changed everything, from the flow rate, to the river bottom, to the river's course, to the vegetative make-up of the banks, and most importantly, to the fish."

He explained how the lower Elwha had become a victim of the alterations made by humans. The two big dams on the river had cut off both deadfall—trees and vegetative debris that used to fall into the river—and sediment and gravel. As a result, the river no longer formed braided, meandering channels that slowly transported water to the sea. Instead, it flowed swiftly and without restraint through an ever-deepening channel that self-scoured because of the high flow velocity. The lower Elwha had become an open pipeline, not a natural river coursing through its floodplain.

"Look at this," Doug said as he picked up a stone the size of a bowling ball. "These big rocks cover the bottom of the river bed, but these are not good habitat for anadromous fish; they need small rocks and a gravelly bottom." Doug spoke in measured tones, without ebullience or emotion, but I sensed his displeasure at the damage done. "Fish can't spawn in this. The tribe will have to make a tremendous effort to restore the fish runs to this river."

In the midst of a forest, next to a clear, swift-running stream, everything looked fine to the untrained eye. But the river's health is directly related to the overall heath of this ecosystem, and it is essentially in a state of gradual decay. The first step in reversing that decay is the removal of Elwha Dam and Glines Canyon Dam.

As we drove upriver toward Elwha Dam, we talked about the proactive effort it will take to restore a semblance of the natural fishery—seven species—in this watershed. River restoration is seldom a straightforward back-to-nature process, in part because nearly every watershed in the United States is developed to at least some degree. Allowances must be made for human occupation. To restore the Elwha, it will be necessary to build a higher levee around the small Indian village that forms the heart of the Lower Elwha Klallam Reservation.

"This road we're driving along is actually a levee, built by the Corps of Engineers," explained Morrill. "The Corps will raise the levee 3 to 5 feet to make sure that the newly freed river will not flood the village."

We continued along the forest road past a state fish hatchery and broke out onto a local highway. We passed by the Elwha Dam RV Park, so I knew we were approaching the dam. A narrow, winding blacktop road led to the south side of the dam. We stopped at a point where the dam could be viewed in its entirety, and Doug explained how it was built:

When they started constructing this thing, they did not dig all the way to bedrock. This caused problems; the whole base of the dam started to wash away. So they dynamited the sides of the canyon and pushed the resulting debris into a pile along the bottom of the dam to stop the hemorrhage of the bottom material. So, unlike most dams, this dam is not very deep on the upstream side because there is so much fill material.

We drove a little farther and parked next to the dam. The gate that allowed access to the top of the dam was unlocked, so we strolled across the top of the dam on a broad wooden walkway and got a close look at its features. Some dams, such as Hoover or Glen Canyon, have clean, symmetrical lines—a simple statement of grace and strength. But Elwha Dam, 105 feet tall, is perhaps the homeliest dam I have ever seen. It is a sloppy maze of convoluted concrete, wood, and metal. Penstocks, pipes, and wires seem to go off in all directions, and an ancient-looking crane, encrusted in rust, hovers overhead. There are two different types of flow gates in the dam; one set is operated by a chain winch that moves along a track on top of the dam. The chain winch and the gates are made of riveted steel and look very much like they were cobbled together with parts from a nineteenth-century steam locomotive. It is hard to imagine that contemporary electrical generation could come from such an archaic device.

It was clear that Doug Morrill had a certain amount of respect for the men who built this dam nearly 100 years ago. But as a fisheries biologist he was acutely aware of the damage done. He pointed to a pool of turbulent water beneath the dam's face. "Fish come up this stream—

Elwha Dam, on the river of the same name, in Washington, to be removed in 2012.

salmon and steelhead—and this is as far as they get. I've seen them butt repeatedly against the concrete at the base of the dam, trying to get upstream."

I asked why any fish would come up this stream, since they were spawned elsewhere. I had always heard that fish return to the stream in which they were spawned.

"A certain percentage of anadromous fish engage in what we call straying; they explore new streams during their spawn, looking for a better place to lay their roe. So, with each run, some fish come up the Elwha. There are no dams below this point, so they can access this river directly from the Strait of Juan de Fuca."

In other words, the Elwha is a very attractive breeding site for anadromous fish, especially for "strays" that seek out new territory, much the same way settlers used to spread out across America in search of new opportunities. The water in the Elwha runs clear and swift, and the bottom, under natural conditions, is gravelly and variegated. And, unlike the upper Columbia and the Snake River, it offers direct access to the sea without passing through numerous dams, making the Elwha an obvious river to be targeted for restoration. But Elwha Dam is only half the problem.

If Elwha Dam is a statement in convoluted architecture, then Glines Canyon Dam, a few miles upstream on the same river, is simplicity in the raw. It is a featureless concrete arch, 210 feet tall, that plugs a deep, narrow canyon. The dam's dark, mossy wall is 270 feet across the top, but so narrow at the bottom that it just sort of disappears into the depth of the canyon. On the day I visited, a spillway gate had been opened, and a great plume of water jetted beneath the spillway gate and plummeted 200 feet into the narrow slot below, a great man-made waterfall. Glines Canyon has another interesting attribute: it lies within the boundaries of Olympic National Park. Although the park was created after the dam was built, the National Park Service has a strong interest in restoring the upper Elwha River to a relatively pristine condition; the river's basin makes up 19 percent of the park's area. The Elwha Valley would then be a complement, rather than a glaring exception, to the natural beauty of the Olympic Range.

Restoring the Elwha involves two great challenges. The first is engineering. There is little precedent for removing a massive concrete-arch dam and restoring the site to a relatively pristine state. Questions regarding downstream silt deposition, the disassembly and removal of tons of concrete, and site restoration will challenge the best minds in the field.

The second challenge is biological; how do you "re-nature" a badly damaged stream and bring back the anadromous fish runs? When a river is restored, there are two choices available regarding fish runs: wait until strays repopulate the restored section of stream, or assist that process by actively transporting fish to the newly restored stretch—a technique called outplanting. The plan for the Elwha is to use both methods to restore salmon and steelhead runs as rapidly as possible. There are seven distinct runs of anadromous fish that are native to the Elwha; planners hope to reestablish all seven.

To accomplish a successful outplanting program, the Lower Elwha Klallam Tribe will use its existing fish hatchery and build a new one to increase the number of smolts it releases into the Strait of Juan de Fuca. The idea is to flood the area with fish to help these native species reestablish themselves in the Elwha River after nearly a hundred years of

Glines Canyon Dam, Elwha River, Washington, to be removed in 2012.

blockage. That ambitious program is being led by tribal member Robert Elofson, the Elwha River Restoration Program Coordinator. Robert has college degrees in physics and biology and worked for many years as a commercial fisherman. I asked him how the movement to restore the river got started; who were the instigators?

> It's an embarrassment to have two dams on the Elwha; there has never been a time when the tribe did not want the river in its natural state. The actual idea phased in with the relicensing; it was kind of a joke at first—that we could take out the dams. But when James River Paper Company [the primary user of the power] realized they had to provide fish passage, things changed and removal became an attractive alternative. And we've had very strong support from the environmental organizations.[74]

So the instigators in this case were tribal members, including Robert.

I asked him about early opposition to dam removal. "At first we had 'the establishment' against it, but as time went on and information went out, people changed their minds. The majority of non-Indian people in the beginning were against removal. But now most people are in favor; it is now the way things are. Because of the downturn in logging and fishing, Port Angeles is now focusing on tourism."

Robert's reference to the new "way things are" required a fundamental and dramatic change in local attitudes, which is never easy to accomplish. The town of Port Angeles, adjacent to the Indian reservation and the location of the timber mill that used Elwha power, was the focal point of resistance to dam removal, but it has come full circle. I met with Russ Veenema, director of the Port Angeles Chamber of Commerce. The Chamber has gone from outright opposition to full-on support. "We've had very little negative feedback about dam removal. What we are finding is that dam removal will be a bit of a business windfall. We expect it to be a big tourist draw, and we're having a discussion with the Park Service to have a nice pull-out area so people can see the dam removal site and learn about how we are doing that. People are coming to this area just to see how dam removal affects the ecology."[75]

The tribe and Port Angeles are just two of the many players that are involved in restoring the Elwha. Heather Weiner, of the National Parks and Conservation Association, explained:

There are many stakeholders. In addition to the tribe and the town, there's the surfers, via the Surfriders Foundation, due to the coastal features that have been altered. And a lot of people are interested in the estuary and coastal zone. For example, the Orca Conservancy. The 300,000 to 400,000 salmon that used to run the Elwha were the main food source for the orca whales. The town of Sequim is involved. The National Park Service and the U.S. Geological Survey are doing a lot of the research. The Bureau of Reclamation is doing the planning and subcontracting for the removal. NOAA is doing commercial fisheries and the EIS. The U.S. Fish and Wildlife Service is involved. Clallam County has an aggressive salmon-restoration program funded by the state. These are all stakeholders in the Elwha restoration project.[76]

The power plants at the dams were shut off in June 2011, and actual dam removal of both dams began in September 2011 and will require about three years.[77] Costs are now estimated at $351 million, including the cost of two water-treatment plants as well as work on the new fish hatchery.[78] The restoration of the Elwha River is taking place in a fishbowl; the whole world is watching. How do you remove two large dams, restore an entire habitat, and lure several hundred thousand stray salmon to a new environment and, at the same time, bring smiles to the faces of surfers, park rangers, local businesspeople, and the first people to call this place their home? The political hurdles have been daunting, but the Elwha will undoubtedly become a case study of what can be accomplished when people of widely divergent opinions and values come together with a new vision of their collective future—a vision of the future that contains powerful elements of a long-lost past.

BUCKET OF ASHES

The take-home lesson from these case studies is that hydropower makes a lot of sense in most places. But there are rivers where other values trump hydro, and we should recognize that fact and react accordingly. This requires making some significant changes to hydropower policy.

The most important reform that should be introduced to hydropower policy concerns externalities, a term that is a fancy way of describing what happens when a business or government agency finds a way to transfer some of its costs and problems to other people or other businesses. Hydropower looks cheap only if the externalities are excluded from the production costs—the loss of fish, recreation, habitat, and other uses that serve both the public and private business. The numbers look even less attractive when the costs of subsidies for farmers, barge companies, and power customers are figured into the mix. The true cost of hydro will be apparent only when dam operators assume responsibility for their externalized costs. When all costs are calculated for all sources of energy, alternative energy looks quite attractive compared with hydro.

Unfortunately, the popular media often portray alternative energy as a sort of wishful but implausible dreamscape. Detractors in the fossil-fuels and hydro business liken it to a cottage industry that could never be a major player, as though all alternative energy is produced by people pedaling bicycles connected to discarded Chevy alternators. But alternative energy is becoming so mainstream, and so big, that we need to find another label for it. The term *clean energy* makes more sense, especially if we include conservation and efficiency as "sources" of new energy.

The economic and engineering viability of clean energy will play a big role in future decisions about hydropower. In the debate over the Snake River dams, the antidam groups argued that improved efficiency would more than make up for the loss of hydropower and in the long run save customers money. The BPA and other dam supporters calculated the cost of removal on the basis of the assumption that the lost power would be replaced with high-cost natural gas, as though clean energy does not exist. A vice president at the BPA recently made this statement: "If it's not dams, it's coal, natural gas or nuclear."[79] This is despite the fact that the BPA itself is investing heavily in wind power.[80] Is it really possible that efficiency and clean energy could take the place of some big dams?

Let's make some comparisons. The four dams on the lower Snake River produce 1,022 megawatts of power, which is less than the wind power produced in just one state, Wyoming.[81] Windmills in Iowa produce three times that amount of power—3,053 megawatts.[82] Montana has the potential to produce 4,700 megawatts from wind.[83] The BPA in total produces a little over 7,000 megawatts from hydro; that is less than the wind power produced in Texas—nearly 9,000 megawatts.[84] A recent study by the U.S. Department of Energy predicts that 20 percent of our energy—300,000 megawatts—will come from wind by 2030.[85] All hydro in the United States produces one-third that amount. A 2006 study found that just three states—Kansas, North Dakota, and Texas—have enough wind potential to equal the electrical consumption of the entire nation.[86] A similar potential is found offshore, according to the Department of Energy.[87] The total estimated onshore potential for wind energy on federal lands alone is 350,000 megawatts.[88] Total installed wind-energy capacity for the United States as of 2011 was 35,000 megawatts.[89] For comparison, Glen Canyon Dam produces 1,320 megawatts. A 2009 study found that wind energy in the contiguous United States could "produce excess electricity during large parts of the year."[90] And the cost of a wind-power project is roughly equivalent to that of coal.[91]

Solar power also has enormous potential. Partisans of limitless economic growth seem to assume that resources are infinite and can support growth without end. But in fact, only one resource is virtually limitless: the sun. A study by the Energy Foundation found that rooftop solar panels, using today's technology on existing rooftops, could generate 710,000 megawatts; desert cities should be net producers, not consumers, of power.[92] A 2006 report by the Western Governors' Association predicted that solar energy in the western states would total 8,000 megawatts by 2015.[93] That figure now appears to be too small; just in Arizona there are proposals for solar power that, if built, would create enough electricity for 25 million homes.[94] In just one year—2010—the Bureau of Land Management approved leases for solar energy with a capacity of 3,682 megawatts, with an additional 100 permits with a capacity of 61,000 megawatts awaiting approval.[95]

Even geothermal power now has significant productive capacity. The United States produces more geothermal power than any other nation, at 3,152 megawatts.[96] For comparison, that is more than the annual production of power at Grand Coulee Dam. The potential for total geothermal development on federal lands is 12,210 megawatts.[97] A recent review of all sources of alternative energy, including geothermal, wind, and solar, found that most states could meet all their electrical demand with alternative energy.[98]

Increased efficiency is another alternative to hydropower. In the Pacific Northwest, for example, efficiency measures have resulted in saving 3,700 megawatts annually—almost four times the amount of energy produced by the four Snake River dams.[99] Nationally, hydropower produces about 7 percent of the nation's electricity. Efficiency measures proposed in the Save American Energy Act would reduce electrical consumption by 15 percent—twice the amount created by dams. Research by a private company suggests that increases in efficiency could reduce our nontransportation energy use by 23 percent—more than three times the amount generated by hydro dams.[100] Studies suggest that America could reduce its electrical use by 4.5 percent just by switching from incandescent bulbs to high-efficiency fluorescent lighting.[101] Another 2.5 percent could be saved by using more efficient appliances.[102] Those two numbers together total almost exactly the electricity produced by hydro dams.

The real "problem" with clean, efficient energy is that it lowers the profits of power producers. If all citizens switched to fluorescent or LED bulbs, insulated their houses, and put solar panels and vertical windmills on their roofs, profits for power companies would plummet. An agency such as the BPA is perversely incentivized to discourage efficiency measures that dramatically reduce customer demand for its product. Every dollar saved by using less energy is a dollar less in the coffers of energy producers.[103]

All predictions are that the clean-energy sector will grow dramatically, especially given the support it has received from the Obama administration.[104] I am not suggesting that clean energy will make hydropower unnecessary; there will be a role for hydro in the future. But these data make it clear that we can live comfortably without hydro dams that create significant problems.

There is a theory of history that great social and economic change coincides with the advent of a new source of fuel. According to this interpretation, the switch from wood to coal ushered in the Industrial Revolution, and the switch from coal to oil brought humankind into the modern technical age. This explanation may be a bit simplistic, but it is worth keeping in mind as the United States and other developed nations move into a new era of clean fuels and superefficiency. As with all such revolutions, the transition will involve pain and resistance, unknown consequences, and the promise of a much brighter future.

I believe that the United States will eventually accept this transition with the same gusto as the transition into the digital age. When I began my academic career, every university department employed a small army of typists to turn professors' scribbling into manuscripts. To assist the typists in my department, I purchased the latest typewriter so I could type my own drafts; it even had an erase function. But in a remarkably short time I junked the typewriter in favor of a newfangled computer that had a screen that seemed to be specially designed to make me squint. It was a terrific improvement, and I welcomed the change. As soon as everyone in my department received a computer, the department fired all the typists. Some of them had worked there for years, and typing was their only skill. They were joined in the unemployment lines by thousands who had worked in typewriter factories. It's a good thing they did not have a powerful lobby, or we would all still be pecking away on Selectrics. Such adjustments are painful, but that is the nature of change.

America is overdue for fundamental change in our energy policy. Much of our current energy use, including some hydropower, is unsustainable.

Some energy producers are still typing on those old Selectrics. We must, as Edmund Burke counseled, "obey the great law of change."[105] It will not be easy, especially when we begin tearing down dams, coal-fired power plants, and other forms of obsolete energy production. The difficulty and the necessity of accepting such a jolt were expressed powerfully in Carl Sandberg's poem "The Past Is a Bucket of Ashes":

> The woman named To-morrow
> sits with a hairpin in her teeth
> and takes her time
> and does her hair the way she wants it
> and fashions at last the last braid in coil
> and puts the hairpin where it belongs
> and turns and drawls: Well, what of it?
> My grandmother, Yesterday, is gone
> What of it? Let the Dead be dead.[106]

6

RIVERS INTO WATERWAYS

Barging, Locks, and Dams

From dawn to the blush of another day,
Like traveller singing along his way
—FROM WILLIAM CULLEN BRYANT, "GREEN RIVER"

W hen the first steamboats began chugging up the Mississippi
River in the 1820s, they marked a crucial development in
the growth of the Union. Historians Charles and Mary
Beard wrote that as a result of the opening of the great river to transport,
"the far country was brought near. The timid no longer hesitated at the
thought of a perilous journey."[1] None of this would have surprised Thomas
Jefferson, who predicted in his *Notes on the State of Virginia* (1781–87) that
the Mississippi "will be one of the principal channels of future commerce
for the country."[2] Rivers were seen as a way to tie the country together,
expand the empire, and grow the economy. Everyone was certain that
goods and people would always travel best by water, so the government
gave the Corps of Engineers its marching orders.

The only problem was that rivers did not always go directly to markets
or destinations. To remedy that, both public and private parties went on
a canal-building binge in the early nineteenth century, digging nearly
4,500 miles of canals. The canal craze quickly faded as the costs and disadvantages became apparent. As a 1908 report put it, there was a "popular impression that such systems of public works had done more harm
than good." By 1908 over half the canal mileage had been abandoned.[3]
America's experience with canals clearly demonstrated that transportation choices made in one era did not necessarily make good sense in
subsequent eras.

Despite the failure of most canal schemes, America pursued its dream
of a nationwide waterway system. Rivers were so important to commerce

that in the midst of the Civil War, President Lincoln made it a priority to open the Mississippi River to northern ships. When he learned that General Grant had taken Vicksburg and opened the river, he exuberantly exclaimed, "The Father of Waters again goes unvexed to the sea."[4] That was perhaps the last "unvexed" moment for the river, for it has been the focus of intense controversy, tragedy, and hope ever since.

Our collective experience with the Mississippi, either directly or through story and song, has shaped our national identity and our destiny, from Huck Finn, the newly freed slaves singing the blues while working the delta bottoms between the Mississippi and the Yazoo, and the horrendous floods of 1927 and 2011 to the nearly endless calls for government spending to change the lower river into a canal and convert the upper river into a series of slack-water reservoirs. The Mississippi, with its great tributaries, the Ohio and the Missouri, is as much a part of the American fabric as the flag itself. Jack Kerouac styled it "the great brown father of waters rolling down from mid-America like the torrent of broken souls."[5] Mark Twain's friends called it "The Great Sewer."[6]

Shipping on the lower Mississippi and the Ohio was extremely important in the early years of the American Republic. The lower Mississippi has the combined flows of the Ohio and the Missouri, as well as the upper Mississippi. The upper river is considerably smaller in volume and breadth, and water levels would often fall to the point where it was no longer navigable by commercial boats and barges. This meant that while the lower Mississippi needed only channelization, the upper Mississippi would have to be dammed, locked, and dredged its entire length to assure water access as far north as Minneapolis/St. Paul.

River traffic in the Mississippi Basin faded as railroads and other forms of transportation expanded in the late nineteenth and early twentieth centuries. The great era of commercial passenger service on riverboats came and went. On the upper Mississippi, shipping traffic plummeted.[7] It looked as though the river highways were destined for the same fate as the canals, which today exist only as quaint reminders of a long-lost era. But the federal government, prodded by congressmen and senators and fearful of railroad monopolies, decided that river shipping was deserving of great public investment. And the money came.

Beginning in the early twentieth century, the Corps went on an inland waterway bender, converting over 11,000 miles of river into barge channels and building 230 lock sites with 276 lock chambers.[8] These waterways carry about 10 percent of the nation's cargo.[9] Some of the locks were built as recently as the early 1990s, although most were completed during the first half of the nineteenth century, and the oldest dates from 1839. The average age of Corps locks is 58 years.[10] The largest cargos by weight are crude petroleum and coal, followed by chemicals and "crude materials" such as rock and sand; much of this material is actually used to maintain the barging channel.[11] Farm products are also heavily dependent on barge shipping, especially commodities such as grain—the crops that receive the largest agricultural subsidies. The largest single type of cargo on the Mississippi, Illinois, and Snake rivers is "food and farm" products.[12] Much of the upstream cargo consists of fertilizers for these crops. Thus subsidized crops are transported via subsidized barges to subsidized ports, provoking the antitax advocacy group Taxpayers for Common Sense to label the Mississippi waterway system a "river of subsidy."[13] Like all successful pork-barrel policies, the benefits of the waterway system are spread far and wide to garner many votes in Congress. Forty-one states, including every state east of the Mississippi, have federal waterways.[14]

Until 1978 barge companies paid nothing for the construction or maintenance of these waterways. The effort to get barge companies to pay just a small share of the expenses began in the 1930s, but it took over forty years to overcome the well-organized resistance of the industry and its allies in Congress.[15] In 1978 a modest fuel tax was imposed, but by 1990 it covered only about 8 percent of the cost of maintaining the waterways.[16] The tax rate was gradually raised over the ensuing years, but the Inland Waterways Trust Fund, where such funds are collected, is used only to pay a 50 percent share of new construction and rehabilitation; none of it is used to cover annual operation and maintenance costs, which total $1 billion a year just for dredging.[17] In 2003 the Bush administration attempted to allocate some trust-fund monies to cover a small portion of operation and maintenance but met a wall of resistance from the barge industry, which preferred to let someone else pay for the upkeep of its waterways.[18]

With such a large direct subsidy, there was little thought about economic rationality; the Corps simply built waterways wherever there were rivers. Some of them proved to be quite useful, but others were grossly underutilized. In order to justify the construction of some waterways, the Corps inflated the projected traffic flow; this happened most dramatically on the Tennessee-Tombigbee and on the Columbia-Snake system—the one that is creating so much controversy over salmon and dam breaching. About 90 percent of all barge traffic is found on just four waterways; other waterways, such as the Ouachita and the Red River, see virtually no traffic.[19] The Pearl River waterway, with three locks and dams, often didn't see a single barge in a year; the Corps was finally forced to place it in "caretaker" status.[20] And some rivers, such as the lower Missouri River, look like they have some traffic until it is revealed that most of the barges are carrying material just to maintain the channel.[21]

In 2000 the Corps published a selective sample of waterway traffic projections and claimed that only a few were far short of projected traffic. However, it excluded many of the worst projects from the sample, and—surprise—used questionable methodology.[22] A coalition of public interest groups critiqued the Corps report and found that numerous projects failed to live up to the promised traffic projections.[23] Another critical study found that eighteen of the inland waterway segments carried less than 3 percent of the traffic (in ton-miles).[24] Amidst these big industrial barges, a handful of riverboats still transport passengers.

TALLSTACKS

Imagine a long line of gleaming white riverboats slowly cruising by on a crisp autumn evening, lit up like Las Vegas on water. Their paddlewheels flash with moonlight as they churn brown water into froth. Steam whistles hoot at the crowd onshore, where men in stovepipe hats and women in hoop dresses parade smartly on the wharf. St. Louis in the 1870s? Good guess, but it was Cincinnati in 2003. The fifth Tallstacks Festival was under way, and 700,000 people showed up to see a piece of history come alive again. Seventeen riverboats, gathered from the far reaches of America's midwestern rivers, were in town for five days to celebrate the human passion for moving over water. The oldest boat, the *Belle of Louisville*, built in 1914 and still powered by a steam engine, seemed to have

Tallstacks Festival, Cincinnati, Ohio.

emerged out of a faded black-and-white photograph. But most of the riverboats gathered here were built in recent years and are part of a river renaissance that is sweeping the country.

Cincinnati is the perfect place for the Tallstacks Festival.[25] It is a leader in the movement to turn back to the rivers that once bound the nation together and to enliven riverfronts once forsaken but then rediscovered. Like many cities, Cincinnati's river corridor was for many years neglected and dilapidated. But now it has a whole new look, dominated by the "wave wall," an undulating ribbon of contoured concrete that also functions as a levee.

The Tallstacks Festival harkens back to an era when hundreds of steamboats plied the Mississippi and Ohio rivers, each one spewing twin columns of smoke from sculpted smokestacks. Back then the rivers were free flowing and ever changing, and it took a man of consummate skill and an intimate knowledge of the river to pilot a boat through the shoals and snags. Today, a riverboat pilot must be equally skilled, but the challenges are different. The Ohio and the upper Mississippi rivers are no longer rivers; they are linked reservoirs, damned, dredged, and channeled to support the barge industry. Nineteen locks and dams on the Ohio and twenty-nine on the upper Mississippi have converted these rivers into a series of pools, maintained at a depth that allows huge conglomerations of barges to be pushed up or down the river by towboats with 5,000 horsepower.

The Ohio River has a long and colorful history replete with tales of conquest, daring, and romance, but it also witnessed the more prosaic tasks associated with the rudimentary need to get large amounts of goods and travelers from one place to another. The Ohio has long been an artery of commerce that sustained a growing nation. On that river, the government followed the early settlers and tradesmen. The story of how the upper Mississippi River developed is quite different. In most cases of government action the citizens demand a solution to a problem, and government responds with a policy. In the case of the upper Mississippi, the government created a solution to a nonexistent problem, assuming a build-it-and-they-will-come posture.

Riverboat traffic on the upper Mississippi peaked in the 1860s and then declined precipitously. Newly completed rail lines siphoned off some of the shipping, but even more devastating was the utter depletion of the massive white pine forests of Minnesota, which had been the riverboat's principal downstream cargo and source of fuel. Despite the sudden drop-off in river traffic, Congress ordered the Corps of Engineers to construct a 6-foot-deep channel the entire length of the upper Mississippi from St. Paul to St. Louis in 1907. A former river pilot, writing at that time, expressed skepticism that such public works were necessary: "Millions of dollars have been spent in the work, and its preservation costs hundreds of thousands annually. All this outlay is today for the benefit of a scant score of steamboats between St. Louis and St. Paul."[26]

This policy of overbuilding the river far beyond economic rationality has been standard practice ever since. In 1922 the Interstate Commerce Commission stated that "water competition on the Mississippi River north of St. Louis is no longer recognized as a controlling force but is little more than potential."[27] Undeterred by lack of demand, promoters of navigation on the upper Mississippi pushed Congress to fund a complex system of locks and dams. The proposal to dam, lock, and dredge the entire length of the upper river to St. Paul had to compete with a starkly different vision for the river. Some people in the area recognized the unique beauty and tourist value of the river and sought to create a linear national park, to be called the Mississippi Valley National Park.[28] But the park idea had one great political drawback: it would not bring in millions in federal construction dollars. In contrast, the waterway proposal would spread tax dollars like butter on bread. Five states and numerous congressional districts would be recipients. The call for big federal spending drowned out the call for a national park.

The Corps of Engineers dutifully responded by doing an economic analysis in 1929. Supporters of the locks and dams assumed that the project was in the bag; they knew the Corps' propensity for construction. They were stunned when the agency concluded that the proposed project was not economically feasible. Furious politicians then lobbied President Hoover, who fired the author of the study and replaced him with someone more willing to work the numbers creatively.[29] Suddenly, the economics of navigation on the upper Mississippi looked very promising—if only the taxpayers would bear all the costs of developing the river into a highway for barges.

Using the "revised" Corps study, Congress passed the 1930 Rivers and Harbors Act, authorizing a 9-foot-deep channel and twenty-six locks and dams that reduced the river to a series of slow-moving reservoirs. The project was completed in 1940. Later, three more locks and dams were added. This massive, taxpayer-funded outlay on behalf of the barge industry finally made it "profitable" (if the public investment was not counted) for barge companies to operate on the upper Mississippi River.

But just as the barge industry was hitting its high point, an awareness of the immense environmental costs of navigation began to affect policy. In 1986 Congress passed a water resources act that was quite different from the innumerable make-work pork-barrel water authorizations of the past. As explained in chapter 2, this water authorization act was used to develop not just new projects but a newfound awareness of the cost and environmental consequences of the nation's concrete-oriented water policy. However, Congress was not yet ready to abandon its past traditions, and as a result, the 1986 act is a schizophrenic combination of projects that waste money and destroy the environment and admonitions to the Corps and other agencies to quit wasting money and stop destroying the environment.

One of the best illustrations of this is the clause in the 1986 Act characterizing the upper Mississippi as "a nationally significant ecosystem and a nationally significant commercial navigation system."[30] The act stipulated that the Corps must set up a long-term environmental monitoring system for the river, but it also buttered the bread of the barge industry by tossing in authorization for a second, extra-capacity lock at Lock and Dam 26 on the upper Mississippi. This in turn led to a demand to enlarge other locks and dams on the river.

In the 1980s, when barges started backing up at some of the locks on the upper Mississippi and Illinois rivers, the Corps responded in classic fashion; the problem could easily be solved by building longer locks that would hold more barges. End of story. All the Corps needed was another billion dollars from the taxpayers, and it could start pouring concrete again.

The Corps produced a proposal to extend seven locks, essentially doubling their length. Before construction could begin, a feasibility study had to be prepared, including a finding that the benefits exceeded the costs of the proposed project, known formally as the Upper Mississippi, Illinois Waterway Navigation System Feasibility Study.[31] Project supporters assumed that the study would be a formality with predictably positive results. But a voice of protest welled up from the warren of offices deep within the agency. A career man with the Corps, with a doctorate in economics, drew a line in the sand.

COOKING THE DATA

The Corps colonel stared grimly. "You have a family to support, don't you?" It was about as subtle as a porno flick. Dr. Donald Sweeney, who had been with the Corps for twenty-two years, was being threatened by his boss: either change the outcome of your cost-benefit analysis, or we'll charge you with insubordination and then demote you. Sweeney did indeed have a family to support, but he also had scruples. He refused to fudge the numbers. Sweeney knew what he was doing—he had spent his entire professional life doing analyses for water projects. But he was not prepared for the high-pressure campaign of deception that was thrust in his face.

When Sweeney and his team of economists were ordered to do a cost-benefit study for enlarging seven locks on the upper Mississippi, they dutifully ran the numbers. Sweeney was the manager of the economics work group for the study, with a bevy of other well-trained economists working for him. The $57-million Upper Mississippi Navigation Study marked the first time that the Corps was studying an entire river system involving multiple districts, not just an individual project, and early on it became obvious to Sweeney and his crew that the existing methodology was biased toward approval. His team devised more realistic criteria for assessing the economics of spending billions of dollars so that the barge industry could get their barges through the locks faster.

He came to the conclusion that it wasn't worth it. In fact, it wasn't even close. What happened next stunned him. He was told to go back to his office and cook the data until he came up with a positive cost-benefit ratio. Sweeney's bosses in the Corps had not always been so mendacious. Indeed, the study had gone well for about the first two years. But halfway through the study, the Corps reorganized, and a new division, the Mississippi River Division, took control. That was when things started to go south. Some of the leadership in the Division put ambition and politics above sound economics. Sweeney explained: "During that reorganization, some people saw this as an opportunity to expand the program, and they were less concerned with the science, and much more interested in expanding the locks and dams and pleasing the navigation companies."[32]

Sweeney's superiors at Division headquarters pushed hard. They demanded a positive cost-benefit ratio, or else. "They threatened me with insubordination. Then I got suspended, then they removed me from the economics team and replaced me with an engineer. Then they demoted me, and made me an assistant for the engineer that replaced me."

The Division thought that it had its yes man in place and awaited the results. But the engineer could not get the economics team to go along. Frustrated, a colonel from the Rock Island District—with no training in economics—put himself in charge of the economics team. He sent an e-mail to the team, specifying exactly the parameters he wanted in the study—parameters that would guarantee the desired conclusion.[33]

At that point Sweeney decided that something had to be done to stop the charade. In early 2000 he sought whistle-blower status and produced a forty-four-page affidavit that is a tour de force of economic reasoning in the face of belligerent irrationality. He described in detail how "the benefit and cost analysis has been intentionally and deliberately altered to produce a seemingly favorable recommendation for immediate large-scale expensive improvements (essentially doubling the length of seven system locks)."[34] The affidavit was accompanied with hundreds of pages of supporting documentation, including copies of incriminating e-mails. Sweeney submitted his accusations to the U.S. Office of Special Counsel (OSC). It was a stunning inside look at how the Corps really operates. It revealed that Major General Russell Fuhrman, the Director of Civil Works, had overtly stated that the Corps was to become "the Federal Government's advocates for inland waterways."[35] The Corps was now working for the barge industry, not the American people. Top Corps officials held numerous closed meetings with barging interests and farming organizations (no public interest groups were invited) and assured them that the agency would find a way to justify the lock extensions.[36] So many e-mails were flying around with incriminating information that one Corps official cautioned his colleagues: "As a further suggestion, avoid sending E-mails around which suggest our recommended alternative will be determined based on assumptions that appear most likely to justify large scale alternatives in the near term."[37] That's engineer talk for: don't let people know that we skewed the study to arrive at a predetermined conclusion. But it was too late; the Corps' cat was out of the bag. It was like the political version of the *Sultana* explosion.[38]

Sweeney's revelations produced a flurry of media coverage. Then the OSC made public its review of his allegations and concluded that "the information that Dr. Sweeney provided established a substantial likelihood that Corps officials and employees had violated applicable laws, rules and regulations governing the conduct of such cost-benefit analyses, and that a gross waste of funds would occur as a result of these violations." The OSC then referred the charges to the Department of Defense for further investigation.[39] The Army's Inspector General investigated Sweeney's charges and confirmed his most serious allegations that his superiors had manipulated study methodology to ensure a positive cost-benefit analysis. But the Inspector General went even further and found serious problems with the Corps itself, writing that the agency's "Grow the Corps" program had created "an institutional preference for construction." The report also strongly criticized the Corps for becoming the handmaiden of the barge industry: "Treating the barge industry as a customer created a conflict with the Corps' role as an honest broker in the study. It also led to granting the barge industry preferential treatment in terms of exclusive access and involvement in development of the economic analysis."[40]

These investigations convinced the U.S. Congress's House Appropriations Committee to look carefully at the proposed lock expansion on the upper Mississippi. The Corps realized that it was in big trouble; it had been caught red-handed in the act of ignoring reality in order to pad its budget and please its barging clientele. In an effort to stem the controversy, the Secretary of the Army requested that the National Research Council, the research arm of the National Academies of Science, assess the Corps' analysis of navigation on the upper Mississippi River. The Council's ninety-page report is a blistering critique of the Corps.

Most government reports are filled with mundane, guarded phrases, but the Council's report is replete with powerful, direct language. It begins by critiquing the models used by the Corps, finding that "many of the assumptions and data used as input to these models are flawed."[41] The Council was particularly harsh in its criticism of the Corps' analysis of future barging demand, writing that it had "serious reservations about methods used to forecast traffic volumes" and concluding that "indeed, the shortcomings [of the Corps' predictions of waterway traffic] are so serious that the current results . . . should not be used in the feasibility study."[42] The Council noted that the Corps' forecasting model, used to predict future demand, failed to accurately predict demand for grain exports between 1995 and 2000.

To drive the point home, the Council reminded the Corps that its projections for barge traffic on the Tennessee-Tombigbee Project (another massive, heavily subsidized Corps project, often called the Tenn-Tom) were way off; the Corps had predicted that by 1998 the canal would handle over 40,000 tons of freight. In fact, less than 10,000 tons were shipped via the Tenn-Tom.[43] The Corps also "greatly overestimated" barging demand when it replaced Lock 26 on the Mississippi. In 1975, and again in 1982, the Corps predicted that nearly 124 million tons would be shipped through the lock. In fact, only 78 millions tons were shipped.[44]

Regarding the Corps' projections of grain-transportation demand, the Council had "no confidence" in the Corps' forecasts, noting that the Corps' analysis "ignores a downward trend in U.S. grain exports since 1980" and offers "little empirical basis" for the Corps' estimates.[45]

In other words, the Corps significantly overestimated barge demand. To make matters worse, the Corps' analysis ignored market factors: "The Corps has stated that no study resources were directed toward identifying the distribution of willingness to pay for commodity movements. This is a crucial omission." The Corps' economic assumptions "should be based on data, not just unsupported theory." The Corps' "failure to include alternative markets" was therefore another major flaw.[46] The Council's critique made it clear that the Corps' methodology and data had been manipulated to make the proposed lock extensions look like a sensible investment.

Another harsh critique involved the Corps' treatment of environmental factors. In essence, the Corps either ignored environmental values or failed to incorporate them into the larger analysis of the proposed project. The Council noted that the Corps' study "does not describe the relations between the river's environmental resources and the substantial economies (tourism and recreation) that depend on those resources."[47]

As a result, the Corps "failed to complete a systematic study aimed toward integrating engineering, economic, and environmental issues."[48] The Council made it clear that a lot was at stake:

> The vast river-floodplain ecosystem of the Upper Mississippi River basin also provides a range of ecosystem services, including drinking water, food (fishes and waterfowl), groundwater recharge, purification of polluted waters, and flood retention. The Upper Mississippi River ecosystem is a storehouse of biodiversity, which produces social benefits today (e.g. food and fiber), and may produce additional benefits in the future (e.g. medicines).[49]

The U.S. Fish and Wildlife Service also weighed in on the Corps' treatment of environmental factors, citing five deficiencies in the Corps analysis and reminding the Corps of the ecological importance of the rivers under study:

> The Upper Mississippi River System is a complex mosaic of bottomland forests, wetlands, and aquatic habitats which are home to over 150 species of fish and 44 species of Unionid mussels. Over 40 percent of North America's migratory birds use the Mississippi Flyway. There are 265,000 acres of the National Wildlife Refuge system along the Upper Mississippi River mainstem, and another 10,000 acres along the Illinois River. The five state agencies manage over 140,000 acres on the Upper Mississippi River and the Illinois DNR [Department of Natural Resources] manages approximately 60,000 acres on the Illinois River.[50]

In short, the Corps egregiously overestimated the value of, and demand for, its traditional construction-based services while shortchanging environmental values.

In stark contrast to the exaggeration of benefits and the trivializing of environmental damage, the agency seriously underestimated the cost of its proposed project.[51] For a point of comparison, the Council noted that the Corps' estimate of $350 million to rebuild Lock and Dam 26 turned out to be but a small portion of the final tally of $1.05 billion. The Council also believed that the costs of future rehabilitation would be "far greater" than the Corps' estimate.[52] These expenditures would be in addition to the $905 million the Corps had already spent in the past twenty years rehabilitating its navigation works on the upper Mississippi.[53]

The National Research Council's report rather blandly concluded that "the Corps did not always spend its money wisely." Yet even its critique, as harsh as it was, was incomplete, because it did not address the problems created by massive subsidies. The Council merely noted that Congress had decided to subsidize navigation and left it at that.[54] However, any economic analysis of shipping on the upper Mississippi must include the costs of subsidies if it is to make any sense whatsoever. Thus an economic analysis that does not include barging and farming subsidies excludes much of the costs. If those costs were incorporated into the analysis, the resulting ratio of costs to benefits would be an embarrassment. I believe that it would show that barging on the upper Mississippi makes about as much economic sense as trying to market pink pickup trucks and BYU beer mugs; it is simply irrational. At a time when the nation is facing a $14.5 trillion national debt, such expenditures are hardly prudent. Perhaps the issue is best summed up by Taxpayers for Common Sense, which charges that Corps subsidies pay people to "violate their own common sense."[55]

Because of the serious questions raised about the Corps' feasibility report, the Assistant Secretary of the Army in 2001 ordered the Corps to start over. This time its objective would be to create a plan "designed to achieve the sustainability of both the navigation system and the ecosystem of the Upper Mississippi River System."[56] The Corps could not bring itself to the point of actually considering the abandonment of the expanded lock project, but at least it was willing to consider the environmental damage done by the locks, dams, and barges.

The result of this new process was the "restructured" study, which used the age-old strategy of handing out goodies to everyone to buy off the opposition.[57] This new approach offers billions of dollars for new locks and billions more for ecological restoration. In other words, the Corps would both damage and repair the river simultaneously and in theory make everyone happy. The cost, of course, skyrocketed. Sweeney described the price inflation: "The original project to enlarge seven locks and dams had an estimated cost of $1 billion. Now the cost is estimated at $2.4 billion just for the navigation, and $5.6 billion for the environmental restoration project. It went from being a $1 billion project to nearly $10 billion. The Corps bought off the environmental groups with the massive restoration program; that way they can continue to build the big navigation project."[58]

This greatly expanded project, like the original project, attracted criticism. Nothing had really changed to make the lock expansions economically feasible. Indeed, the passage of a couple of years had made it obvious that the original Corps projections for increased traffic were, as Sweeney pointed out, unrealistic. The National Research Council was again asked to review the Corps' study.[59] In a series of three reports, the Council lauded the Corps for taking a more comprehensive adaptive management approach that included ecological factors, but reiterated that it agreed with all the major recommendations that had been made in the original 2001 study, and it criticized the Corps for continuing to downplay inexpensive nonstructural alternatives, for using unrealistic predictions of future barge traffic, and for an inadequate methodology for valuing ecosystem benefits.[60] A recent analysis by a consortium of environmental groups estimated that "based on two decades of flat or decreasing barge traffic, the proposed new locks will likely result in a loss of 80 cents for every dollar provided by taxpayers."[61]

After reading the relentless criticism of the Corps and its apparent compulsion toward bigger locks and dams, I wondered what drove it to such extremes. I posed that question to Don Sweeney, and he did not hesitate to answer: "Colonels don't get promoted to general by saying no to projects. They get promoted by saying yes to what politicians want." It should come as no surprise that Sweeney is no longer with the Corps. He and the Corps reached a confidential settlement, and he now has a faculty position at the University of Missouri. He is quite happy to put the whole Corps thing behind him.

After reading about Sweeney's allegations, I resolved to visit the upper Mississippi and ask some questions of my own. As great fortune would have it, a friend introduced me to her brother-in-law, who just happened to be a riverboat captain on the upper Mississippi.[62] I asked him if I could come along on one of his trips. It turned out that he had a big one planned.

"How would you like to ride with us from Cincinnati to St. Paul?"

DINNER BUCKETS AND PARTY BARGES

The *Harriet Bishop* was one of the seventeen riverboats at the 2003 Cincinnati Tallstacks Festival. The captain, John Halter, invited me to ride along as it deadheaded back to its home port in St. Paul by going down the Ohio River and then up the upper Mississippi. John is a big man with a powerful build, thick salt-and-pepper hair, and a love of the river that seems to run through his veins. He spent his younger years in the Coast Guard, working the Great Lakes. He realized then that his fate was intimately linked to boats and rivers. He knows his trade well and piloted the *Harriet Bishop* with a confidence that can only be attained through a lifetime spent in a wheelhouse. His many years of experience have provided him with a wealth of river stories, most of them humorous, some of them grim, all of them fascinating. Halter loves not only the river itself but also the assortment of rivermen and roustabouts who engage in commerce and transport on the river. He has a special respect for the barge pilots and views the navigation channel as an essential part of the river environment. The *Waterways Journal*—the voice of the barge industry—is, by his account, "the bible of every true river man."

The *Harriet Bishop* (named after Minnesota's first schoolteacher) is 100 tons of steel, but it is designed to look like an old wooden steam-powered riverboat. The tall black smokestacks and the big red paddle-wheel at the stern are useless simulations. Gingerbread filigree—also in the classic style—decorates the uprights and the railings. These cosmetic accoutrements, explained second pilot Robert Wilson, give the boat "dock appeal." It is typical of the new class of commercial pleasure boats on the Mississippi River system. The *Inland River Record*, a compilation of all the licensed professional boats on the Mississippi, lists seventy-six excursion boats, which are passenger boats without berths and are lovingly referred to by their crews as "dinner buckets" or "party barges." The *Record* also lists thirteen passenger and packet boats, which offer overnight accommodations.[63] These boats are commercial recreational craft and offer passengers an opportunity to relax, view the slowly passing scenery, and, as pilot John Halter said with a gruff laugh, "enjoy the ambiance and pageantry of America's historic waterways."

Our journey from Cincinnati to St. Paul was 1,351 river miles and consumed ten days and nights of constant travel. Throughout that time the twin 450-horsepower engines never stopped and consumed about 4,000 gallons of fuel in the process. The entire length of the river route is marked by channel buoys that indicate a safe depth, and shore signs that indicate location. During the day the channel seemed to lazily unroll directly in front of the boat, mile after mile. At night, following the channel is more like stumbling into a dark room with a flashlight in your mouth while carrying a load of dishes. Huge searchlights on the roof of the boat sought out buoys, which glowed like dim stoplights when the beam found them. The color of the buoy indicates whether it is to the left or right of the 9-foot-deep channel. Tom O'Rourke, working as a deckhand for the first time, explained how to remember the color code: "Red, right, return," meaning that when the boat is going upriver (returning home), the red buoys are on the right of the channel.

Soon after we left Cincinnati, the Ohio River coursed between forested hills on both sides of the river. When we steamed (yes, they still use that term, even though we were actually dieseling) through this section of Ohio, it was late October, and the trees had that colorful postcard look. Occasionally the forest screen is broken by small towns consisting chiefly of quaint, weathered brick buildings, small churches, and modest but well-kept houses. A few hilltop mansions, some with their own private cemeteries bounded by picket fences, peek through the trees. The riverside attractions also include derelict buildings built too close to the river— stark reminders of the lessons learned from invading floodwaters. The bucolic scenery, the river's hypnotic movements, and the slow pace provided a perspective on the landscape that few Americans experience. It invites contemplation, a chance to gaze at the river corridor in detail and to think about both its history and its future. This section of the Ohio could easily qualify as a national scenic waterway, similar in stature to the Columbia River Gorge. With sufficient care and preservation, the Ohio River could become a nationally significant tourist resource, drawing millions to its soft, genteel beauty.

But not all of the riverbank is scenic; the river also has a working side. In a few places fields of corn and soybeans come down to the water's edge. An occasional dredge, rusting and listing, sits in midstream, sucking sand from the bottom. There are innumerable power plants along the way, cooled with river water and fueled with coal shipped in giant barges. Gravel operations are common, and a few huge, garish casinos have popped up on the Indiana side of the river because of a bizarre law that allows gambling in the state, but only if it floats. These "boat" casinos could not hoist anchor if the jackpot depended on it. The whole presentation is an illusion: fake riverboats beside false-front buildings built to look like Roman architecture but made of pressboard and plastic. For the casinos, the river is there only as an excuse, as though the river's presence somehow sanctifies a game of chance.

On the river itself a whole industry moves by. Massive sets of barges lashed together and known as tows (even though they are pushed) dominate the river. Most of the tows consist of twenty to twenty-five barges (but the record is eighty-four barges, according to Captain Halter), pushed along by a stubby, flat-nosed tug called a pushboat. Even the medium-sized tows cover an area of water that would rival a small aircraft carrier.

The tows consume the river channel; other boats, including ours, give them a wide berth, using a signaling system that was created in 1857 by Isaiah Sellers, a famous steamboat captain. Back then a boat blew its whistle to signal the side on which it wished to pass: one whistle indicated a turn to starboard, two whistles indicated a turn to port. Today pilots discuss their plans on the radio, but they still speak of passing on the one-whistle or the two-whistle.

Modern barging on the Ohio is possible because the U.S. government spent the last 100 years converting the river into a series of pools with a minimum depth of 9 feet. In 1910 Congress authorized a system of locks and dams for the entire length of the river. By 1929 the Corps had built a chain of forty-six wicket dams that created the longest controlled waterway in the world at that time.[64] But this was judged to be inadequate, so in the 1960s the Corps replaced all but two of the existing dams with nineteen much larger dams that created longer pools.[65] Incredibly, some of these dams are not visible; they lie just under the surface and allow boats to pass over them with a motion that resembles a mellow waterslide.

As we made the turn up the Mississippi, barge traffic increased and the channel narrowed as we passed Cairo (pronounced KAY-ro by the locals), a town that was once predicted to grow into a huge megalopolis with "domes, towers and monuments."[66] As barges crowded our boat, John Halter's expertise really paid off. The Mississippi was unusually low that year, and long strands of sandy embankments had emerged, revealing hundreds of "wing dams," which are rock jetties built into the river to force current into the channel. The navigation channel is quite narrow in places, and it was John's duty to avoid the barges that show up around nearly every bend.

This section of the Mississippi lacks the elegance of the Ohio. In the old days it was known as the "graveyard stretch" because of many rocky shoals. Perhaps the Corps overreacted to the fears of boatmen and engineered it to the point of sterility. It looked more like a canal scooped out with earthmovers than a living river. The Corps lined the banks with riprap—crushed rock spread on embankments to reduce erosion. At low water, much of this riprap is exposed, along with the wing dams that are constructed of the same materials. As one of our crew put it, the Corps "uglified" this section of river.

Just below St. Louis the limestone hills on the west bank are dotted with either gravel operations or stately mansions overlooking the riprap. As we approached St. Louis, the industrial activity increased. The Gateway Arch was visible 6 miles out; at 5 miles per hour we had plenty of time to watch the setting sun wreath the arch in iridescence.

Above St. Louis and the Chain of Rocks Canal, the river takes on a much healthier aspect. Much of the surrounding land is wildlife refuge. Even though the river here is simply a reservoir between dams, forested islands and rolling hills give this stretch a natural look, and it is not hard to imagine how it appeared when the *Virginia* puffed its way up the river to Fort Snelling (present-day Minneapolis) in 1823, becoming the first steamboat to reach that far north. One of the passengers on that boat, an Italian adventurer named Giacomo Beltrami, described the river and its corridor: "Never had I seen nature more beautiful, more majestic, than in this vast domain of silence and solitude."[67] Mark Twain also experienced this beauty: "We move up the river—always through enchanting scenery, there being no other kind on the Upper Mississippi."[68]

The silence and solitude still exist to some degree. Above the confluence with the Illinois River (which was also turned into a dam-and-lock navigation channel) we saw very few barges, even though it was the harvest season for the crops that make up most of the downriver freight. As we passed through the Ted Shanks State Wildlife Area and some of the units of the Mark Twain National Wildlife Refuge Complex, late October duck hunters sat shivering in blinds scattered throughout the many channels and sloughs.[69] However, we (and presumably the hunters) saw more decoys than actual ducks, despite the fact that the river is a fly-way for 40 percent of North America's waterfowl. Indeed, throughout our trip we saw waterfowl, but no other wildlife.

The human history of the river is also part of the experience. There are graceful new bridges of spiderweb cables that fan out across the river. There are ancient-looking iron railroad bridges; some have sections that can be raised, while others pivot around a piling to permit boat passage. Along the banks are many small fishing cabins, built on stilts inside the levee at the water's edge. And no one can pass the town of Hannibal without recalling the rich legacy of Mark Twain and the dozens of other river writers who followed. "River people" are never short of stories. Some of them are even true.

The Des Moines River joins the Mississippi at Keokuk, which is the location of one of the two former rapids on this stretch of river (the other one was at Rock Island). In 1837 the U.S. Army sent an obscure army engineer to this area with orders to blow up the rocks in the rapid and clear a channel. Lieutenant Robert E. Lee dutifully set the charges and created a channel.[70] Today there is no sign of rapids, but the lock at this point has the largest lift, at 38 feet, of any lock between St. Louis and St. Paul.

Most of the locks, despite constant repair and rebuilding, looked worn and archaic. Giant geared wheels move the gates on the locks. Crumbling concrete forms the walls. A few of the control stations have been replaced with faceless concrete structures, but some are original Depression-era architecture with an appealing decorative look that was possible only when labor was cheap and people took pride in craftsmanship. These buildings are of such an age that they have value as historic structures as well as architectural statements.

The northern half of the upper Mississippi is a long, sinuous passage through what locals call "the bluff country," composed of limestone folds, perhaps 200 feet in height, that parallel the river channel. Much of the riparian area here is still wooded because it is too steep to plow. Eagles are a common sight, and seagulls play and feed in the wake of boats. A railroad runs alongside the river, reminding barge pilots of their competition. After a long series of locks, the river opens up into a natural lake called Lake Pepin, which is rimmed with summer cottages and stately homes.

Just upriver of the lake is the burg of Red Wing, which is something of a microcosm of the multiple uses of the river. The town center consists of stolid redbrick buildings, including the St. James Hotel, refurbished to its historic splendor. It is flanked by enormous grain elevators and industrial structures. Along the waterfront are hundreds of pleasure boats, ranging from Jet Skis to yachts. The northern end of town abuts the well-manicured Bay Point Park, which juts into the river. It is a modest park, but its history reveals one of the essential truths about American politics. In the 1980s a woman named Lilly Merkl worked as a city administrator for the town of Red Wing. The city dump was right on the river and was nearing capacity. Merkl, a classic instigator, floated the idea that the dump could be capped with soil and converted into a city park. City officials scoffed at such a preposterous idea. Some elected officials thought that the idea was so ludicrous that they pressured Merkl to resign from her job with the city. Feeling the hostility, she eventually quit her job and left town, but her idea remained and was championed by other individuals willing to buck the status quo. Eventually Lilly Merkl's crazy idea became a widely accepted concept. Thanks to her courage and insight, Red Wing has Bay Point Park, rather than a dump, on its waterfront.[71]

As the river channel entered the Minneapolis/St. Paul area, we passed through Ford Dam (also known as L&D No. 1). Unbeknownst to most people, this stretch of the river was characterized by boiling white-water rapids before the dam was completed in 1917. These rapids, if exposed today, would create a complex of rapids in the heart of an urban area—a truly unique resource that would enhance the $1.2 billion already spent each year on recreation on the upper Mississippi.[72] A small group of instigators has proposed restoring this dramatic gorge, creating a unique water park in the heart of the twin cities. Mike Davis, one of the leaders of the restoration effort, summed up the situation well: "It is time for an alternative vision with a view towards the future."[73]

After ten days on the Ohio and Mississippi rivers I felt a sense of hope for these rivers. I grew up in the Midwest, not far from the Ohio, but big rivers were not part of my consciousness, other than when I glanced out a car window from a bridge. Like much of modern society, rivers seemed disconnected from daily life. I always thought that Indiana consisted solely of hog farms, cornfields, and suburbs. Indeed, nearly all of it does, but the stretch of land along the north bank of the Ohio is another world—a world of limestone cliffs, dense forests, and picturesque towns. The world looks different from a boat that is slowly meandering through America's heartland along the last thread of a landscape that was once considered an "endless" wilderness. A trip down a big river is a chance to reconnect to the earth and get a feel for both natural and human history.

However, the tourist and recreational potential of these rivers is hardly recognized. Until we reached the approaches to St. Paul, there was a conspicuous absence of small pleasure craft on both rivers. We saw perhaps two dozen motorboats in ten days, and only two sailboats. Perhaps the river has been so industrialized along much of its length that it lacks appeal to recreationists. Also, it is dangerous to mix it up with a line of barges that are, at best, difficult to maneuver. Squash any romantic notions you might have of a Tom Sawyer–like trip down the Mississippi. The Corps requires motors on all craft. However, a few hardy souls still attempt to float the river in homemade contraptions.[74] On our trip we passed a young man on a raft that looked to be constructed of bailing wire, scrap lumber, and good intentions. He haled us in a confident voice, saying he was going "all the way." In a sense he had already reached his destination.

Another great feature of the river is the towns along the way. Most of them have that quaint, redbrick look of an America that has long since passed into memory in most areas. Any plan to restore the upper Mississippi must include a coordinated effort to assist these towns in preserving their unique river culture and architecture. They are as much a part of the environment of the Mississippi as the wildlife and the riparian forests.

Also much in evidence along the river is the innate human desire to reside close to water. Many wealthy people have built homes overlooking the river. Most of them are tastefully done in traditional architecture; a few have the garish look of nouveau riche starter castles that blend into the scenery like dandelions at a rose festival. But the banks of the Father of Waters also afford a place of respite for many modest homes and cabins; every socioeconomic stratum can find a place along the river. These many and diverse dwellings clearly indicate a universal attraction to bodies of water and their environs. A comprehensive restoration plan

Upper Mississippi River at dusk, taken from the deck of the *Harriet Bishop*.

must include an effort to accommodate these admirers of the river; their decision to buy property along the river is a clear statement about the river's importance to them. A well-planned restoration program would enhance property values along the river. A smart restoration program is an investment from which everyone can benefit.

Although much of the upper river corridor is surprisingly attractive and relatively pristine, the section below St. Louis is clearly a working-man's river (I did not see a single woman working on the river). That section of the river is a corridor of commerce, an engine of the economy that is far more valuable as a transit route for goods than as a natural or scenic area. But the Mississippi above its confluence with the Illinois still has a great deal to offer in terms of scenic beauty and its potential for recreation and tourism. Using it as a barge channel is like mopping the floor with a finely embroidered silk scarf.

All over the American heartland rivers that have been abused as a convenient sewage dump or industrial corridor are now being recognized as thin threads of pastoral beauty. Some city riverfronts have been born again, but others languish in neglect. Some sections of these great rivers have been sacrificed to the industrial age, a necessary expedient in a consumer society. But the potential for preservation and restoration of some stretches of these rivers is enormous. Sometimes—not always, but sometimes—nature should trump navigation.

In the Midwest there is virtually no public land or large biologically sustainable natural area. It is still possible to find open space, riparian forests, and fish and wildlife, but only on or by the rivers. These linear landscapes/waterscapes are biologically rich and geologically unique; there is much to please the eye, interest the mind, and entice the soul. But the upper Mississippi is not wilderness; rather, it is a slice of both natural history and American history. Thus any effort to protect it must encompass the historic structures as well as the special culture and heritage of the "river people" who have made these rivers the stuff of legends and lore. This river and the culture that grew up around it are worthy of our best efforts to preserve what is left and to consider restoring what can be salvaged.

A NATIONAL DUTY

In 1908 President Theodore Roosevelt gave a memorable speech to the Conference of the Governors of the United States. Roosevelt organized the conference to discuss the problem of resource depletion and the need for conservation. In his speech, titled "Conservation as a National Duty," he noted that President George Washington organized a conference between Virginia and Maryland to discuss the use of waterways because he realized that commerce, shipped via rivers and canals, was the best means of cementing the various states into a solid union. Roosevelt warned his audience not to judge the founding fathers too harshly because "they could not divine that the iron road would become the interstate and international highway, instead of the old route by water."[75] In other words, even 100 years ago there was a realization that moving goods by water had serious deficiencies.

Teddy Roosevelt was no lackey of the railroads; the year before the Conference of the Governors he appointed an Inland Waterways Commission to study the feasibility of developing more inland barging channels as a way of competing with the railroads. But he also saw the limitations of waterway transport and looked to a future where railroads carried the bulk of the nation's domestic cargo. What he could not foresee was the vast expenditure of public wealth to convert nearly every major river in the United States into a barging channel, regardless of cost, economic feasibility, or environmental impact.

Today those disadvantages are much more evident. Some of the economic trade-offs are simply illogical. What is the value of the immensely popular recreation on the upper Missouri reservoirs compared with marginal barging on the lower Missouri? What is the value of the salmon runs in the Columbia/Snake system compared with a modest reduction in shipping rates for farmers and timber companies? And does it really make sense for the American people to subsidize the eating habits of people in Asia? Nearly 80 percent of the food barged down the Columbia/Snake system is destined for foreign markets.[76] Nearly half the cargo barged down the Mississippi River consists of food for foreign markets, primarily in Asia.[77] Sixty percent of the grain exported by U.S. farmers is transported on barges.[78] A realistic cost-benefit analysis of waterways, including subsidies, the loss of ecological services, and opportunity costs, would assist the nation in deciding which waterways are worth the annual maintenance, and which ones should be returned to a natural river.

It is time to consider returning some channelized waterways to a more natural state. There should be an open discussion about whether the idea of a Mississippi Valley National Park should be revived. The river channel could be kept clear for recreational boats and pleasure cruises without the expense and environmental degradation of twenty-nine locks and dams and a 9-foot channel. The towns along the river, which have experienced economic decline for years, would have a new prominence in the amenity economy. The same question should be asked regarding the channelized portion of the Missouri (from its confluence with the Mississippi River to Sioux City, Iowa). Barging on the Missouri sacrifices an $85-million recreation industry to provide water for a barge channel that generates $9 million.[79] Should it be restored in the same fashion as the Kissimmee? An honest appraisal would, I believe, result in the restoration of many hundreds of miles of rivers, while some waterways, such as the Ohio and the lower Mississippi, would continue to serve as valuable shipping channels.

Today, the inland waterways ship about 622 million tons of cargo.[80] If some of them revert back to natural rivers, how will the nation make up for the lost shipping capacity? Proponents of barging point out that waterways transport the equivalent of 58 million truck trips.[81] Of course, no one proposes that river cargo should be packed wholesale onto trucks. Instead, two things would happen.

First, the railroads would assume much of the burden. Railroads have been the principal competitor of river shipping for most of our nation's history. Railroad shipping is more expensive, but the differential is largely explained by two factors. Railroads have to pay for the construction and maintenance of their roadbeds, but barge companies pay only a fraction of the costs to build and maintain the nation's waterways.[82] If barge channels were not so generously subsidized, the price differential would be minimized or eliminated. Also, our railroad system is overdue for a radical modernization. There was a time when our rail system was the envy of the world; today it is the envy of no one. We have lost about 100,000 miles of track since mileage peaked in 1930. That is unfortunate because railroads offer many advantages over both barging and trucking.[83] Unlike rivers, railroads make a beeline to markets and ports. For example, a farmer in the Midwest who ships his crop to China via waterways would have to barge his grain down the Mississippi, across the Gulf of Mexico, down the coast of Central America to the Panama Canal, and then across a great expanse of the Pacific. By rail, the farmer could send it directly to the port of Seattle for shipment directly across the Pacific. Which is more efficient?

A second reason that lost barging opportunities would not result in millions of new trucks on the road is economic rationality. If shippers had to pay the full cost of transportation, they would make more efficient choices. Few would buy large-volume products, such as coal or gravel, from distant sellers if they had to pay full transportation costs and instead had a local supplier; all buyers would have an incentive to minimize their transportation costs, whether by barging, rail, or trucking. Subsidies incentivize people to be economically irrational; eliminate the subsidy, and it is amazing how efficient businesses can become. The old adage "Buy and sell local" would become an important part of the market strategy of everyone who, at present, ships on barges.

A modern, clean, efficient rail system is not only important to restoring the nation's rivers but also critical to our efforts to revitalize the economy, reduce our carbon footprint, and enhance the sustainability of our transportation system. America's waterways will always have a role in that system, but the health and vitality of our rivers must be a prime consideration in making transportation choices. It is also crucial to eliminate barging subsidies, which will lead to greater market rationality. The Bush administration, in its proposed budgets for FY 2008 and FY 2009, tried to replace the barge fuel tax with a "lockage-based user fee," but barging interests managed to stop it.[84] President Obama is continuing the fight to replace the fuel tax with a lockage user fee.[85] Opposition to the higher fee is being led by Iowa senators Tom Harkin, a Democrat, and Charles Grassley, a Republican, continuing the long tradition of bipartisan support for river pork barrel.[86] Their opposition is understandable, given how successful they have been at funneling tax money to the barge companies. According to Taxpayers for Common Sense, "Currently, the inland waterways system is publicly subsidized at approximately 90 percent, including 50 percent of the costs for new construction projects and rehabilitations projects, and 100 percent of all operation and maintenance expenses. This is by far the most subsidized freight shipping mode in the country."[87]

Nevertheless, the Corps and the barge industry continue to argue for even more waterways and dams, despite the fact that inland waterway usage was static for twenty years and has declined significantly in recent years.[88] On the Missouri, barge traffic, which was never great, has declined precipitously.[89] The upper Mississippi saw a 33 percent decline between 2007 and 2008. One of the locks on the upper Mississippi, number 5, was down 59 percent from its high point in 1996.[90] This did not stop the Corps from pursuing—and getting—authorization for expanded locks on the upper river, but half the money for that construction must come from the Inland Waterways Trust Fund, which had a balance in 2010 of $38.2 million for new obligations—not even enough money to start planning.[91]

The inland waterway system needs to be carefully evaluated with the BRAC-style review that was discussed in chapter 2. Some waterways would make the cut; others would become rivers again.

At the end of our voyage on the *Harriet Bishop*, Captain John Halter invited me to a celebratory dinner in St. Paul with his extended family. My fondness and admiration for John had grown enormously during the trip, and I felt privileged to be part of such a special occasion. It was a big group of Irish Catholics—close-knit, fun-loving, exuberant. There was a lot of food, even more drink, and even more laughter. During our river trip I had made vague references to some of my ideas about river restoration. John listened respectfully. Most important, there was no diminution in his warmth toward me, even though it was obvious that we had some differences of opinion regarding the future of the upper Mississippi. Toward the end of our dinner I made an oblique reference to the changing economics of barging on the river. John tilted his beer and looked wistfully in the direction of the river. "Yeah, I know, things have to change. But I hate to see it go. I just hate to see it go."

Such is the nature of change at the end of an era. It is difficult and convulsive, and it is human nature to yearn for what is being lost. We all understand that at some level. Tom O'Rourke, Captain Halter's shipmate, made that point well: "The river is a community. I don't want to lose that." But the river community will not be lost if the lock-and-dam system is replaced by a world-class one-of-a-kind national park and refuge system. Instead, it will expand to include new and diverse groups of people; the river will still be the focus. And there will always be a place on the river for true rivermen like John Halter.

7

BLACK WATER RISING

The Myth of Flood Control

And thy own wild music gushing out
With mellow murmur of fairy shout
—FROM WILLIAM CULLEN BRYANT, "GREEN RIVER"

Floods are, quite literally, biblical in their iniquity and reach. The specter of drowning in one's own home or watching it wrench from its foundation and disappear into muddy swirling water is sufficiently malignant to conjure up the starkest religious imagery. But we lack Moses' ability to part the waters or Noah's option to ride out the storm in an ark, so we do the next-best thing: we seek higher ground. Or, if we want to tempt nature's wrath and indulge in optimistic water hubris, we stay low and build a wall—a levee—between us and the river. Of those two options, the United States has historically chosen the latter rather than the former. We walled in entire rivers and built our civilization on the "safe" side of the wall. This allowed us to inhabit low-lying areas but created an Achilles heel; large swaths of population are vulnerable to the fickle, rambunctious vagaries of rainfall as it collects and batters against the rampart.

A levee is simply a riparian embankment, nearly always made of earth, of sufficient height to stop a rising river from overtopping it. A levee directs high water elsewhere and thus essentially makes your flood problems someone else's flood problems. A high-water river that is constrained by levees on both sides does two things. First, water piles up at the head of the levee and spills out over the land to an extent that is greater than would be the case in the absence of the levee. Second, it increases its velocity as it roars through the constrained channel, turning a fast current into a boiling rush that can eat away levees, scour the bottom of the river, and when finally released at the end of the levee, burst out with unnatural ferocity. The river's immense power and relentless surge are concentrated in an artificial channel, like directing the Running of the Bulls down a narrow hallway. Every levee is an act of faith—in God, in nature, and the Corps. Flood "control" is essentially an act of bravado, raising a middle finger toward the upstream and gambling on man rather than nature. If the levee breaks, with no natural floodplain to temper the river's volume and velocity, the area will look like a fleet of roof-shaped ships floating in formation on an inland sea.

The construction of a levee is relatively simple, as is the politics of levee building. If one area along a river builds a levee, then everyone else along the river needs to build a levee. This means that every congressional district along a river will demand a levee—and get it. Then, if disaster strikes, every congressional district along the river can demand emergency funding, followed by loud cries for bigger levees to catch the next big one.

In a sense, there is no such thing as a flood in the natural world; all rivers vary dramatically in volume and breadth, depending on season and precipitation. As Toni Morrison put it, "All water has a perfect memory and is forever trying to get back to where it was."[1] It is only a "flood" if someone has been foolish enough to build a house or a city within the spatial variance of a river. The federal government warns us that floods are "America's number one natural disaster," but the only thing natural about them is the rise in water level; the resulting destruction is by our own hand.[2] But that is small consolation to those who, perhaps unwittingly, make their homes in the path of a moving wall of water.

So, we build levees.

THE STORM OF GOOD FRIDAY

The year was 1927, and America was in the midst of an ebullient thrust of extravagant styles, frivolous ventures, and social experimentation. Americans were a nation of Charles Lindberghs, ready to break the next record. But their confidence was shaken a bit when spring rains throughout the Midwest pounded the nation's midsection for weeks. Maybe they had been a bit too cocky about their rivers, their homes, and the relationship between them.

The Mississippi River levee finally broke on April 16, the day after the Storm of Good Friday, which had dumped 15 inches of rain in less than twenty-four hours. Up to that point, despite the unprecedented rainstorms, few really thought that the levee would break. There were 1,000 miles of levees up and down the river, and the Mississippi River Commission and the Corps had assured the public that the levees would hold. As long ago as 1861, a self-assured army engineer, Captain (later General) Andrew Humphreys, had written an influential report on the river claiming that levees were the best protection against floods. His report was so influential that he was appointed chief of Army Engineers. If the Army engineers said that the levees were safe, then surely they were.[3]

They were not. The city of New Orleans was in imminent danger of inundation. In a controversial move, the city fathers in New Orleans blew holes in the levee downstream to reduce the amount of water pressuring the New Orleans levees. This saved the city by sacrificing the lives and property of others. Not a good way to run a river.

The great flood of 1927 was of monstrous proportions. The entire Mississippi Delta was submerged by a wall of water that may have been 100 feet high in places.[4] Along the length of the river and its tributaries, 27,000 square miles were under water, and 330,000 people had to be plucked from rooftops. Half a billion dollars in property damage occurred, including the destruction of 41,487 buildings and the flooding of 162,000 homes. At one point the river was 70 miles wide.[5]

Despite the obvious dangers and limitations of levees, the U.S. Congress did not learn a lesson from the great flood of 1927. Instead, it decided that even more levees should be the principal weapon in the war against floods. This attitude was embodied in the Flood Control Act of 1936, which gave the Corps of Engineers primary responsibility for national flood control. Flood control turned out to be an excellent vehicle for pork-barrel spending, and every congressman worth his salt (and later, her salt), especially in the South and Midwest, routinely put out the call for more federal spending.

Southern legislators were especially adept at bringing home the water bacon because of a historical artifact. After the Civil War and Reconstruction, no one in the South could be elected from the party of Lincoln; all elected officials were white Democrats, hence the phrase "the Solid South." So the incumbent Democrats just kept getting elected, year after year, decade after decade. Southern legislators thus gained tremendous seniority in Congress, which is the criterion for chairing important legislative committees. This placed a great phalanx of southerners in the most prominent positions of power in both the House and the Senate. They used that power to accomplish two things: stop civil rights legislation for a century and funnel federal money into the South for water projects, including navigation channels, reservoirs, and flood control. Much of the Corps of Engineers developed a southern accent. The departure of the Dixiecrats from the Democratic Party began in 1948 with the dispute over civil rights and was virtually complete with the election of Ronald Reagan. Reconstruction ended in 1876, and Reagan took office in 1981; that is 105 years—enough time to build a lot of levees and barge channels. Dutifully, the Corps built 25,000 miles of levees and floodwalls and walled off 30,000 square miles of floodplain, much of it in the South.[6] Today, 43 percent of the U.S. population lives in counties with levees.[7]

There were two lessons that should have been learned from the 1927 floods. The first one is that levees make floods worse in other places. When rivers are constrained, they ultimately lash out violently. There must be numerous outlets to the floodplain to allow the water to dissipate its volume. Second, there is no such thing as flood "control." High water can be directed, but it cannot be stopped. Communities can be walled off like an ancient city, but no set of defenses is foolproof. Sooner or later, it's Jericho all over again. The best protection against floods is to minimize construction in the floodplain and instead use the floodplain for farmland, forests, parks, wetlands, habitat, a natural water filter, and floodwater dissipation. But that strategy has one great political liability: it is regulatory policy rather than pork-barrel policy. Few voters want to hear a candidate promise that if elected, he or she will restrict development. The Corps often gave lip service to nonstructural approaches, but Congress paid it to build, not limit.[8]

So, we build levees.

GALLOWAY

Even though the Corps did not learn those two key lessons from the 1927 flood, it did learn how to make better levees. Over the decades, this gave people the confidence to continue building in low-lying places. Their confidence and their wallets were further boosted by the creation of a national flood-insurance program, which essentially rewards people for building their homes in perilous locations.[9] Also, people tend to have short memories and a belief that disaster always befalls someone else. But if there is one immutable law about rivers, it is that they will flood again; it's only a matter of time.

Sixty-six years after the great flood of 1927, the Upper Mississippi River System went on a rampage. The flood of 1993 was a surprise to people who had heard the term *flood control* all their lives; what they saw was not flood control but a flood out of control as both the Mississippi and the Missouri violently reclaimed much of their floodplain. The Mississippi was above flood stage for 152 days, the Missouri for 116. The flood did between $12 and $16 billion in damage and inundated seventy-five towns and cities. Nearly 100,000 homes were damaged. The government spent $6 billion on flood response and recovery, including $94 million from the federal flood-insurance program.[10] Over 1,000 levees failed.[11]

But this time, the response was different. Unlike past flood events, which had simply resulted in more calls for more endless levees, the Clinton administration put together a research team to assess what went wrong in the 1993 flood, and what direction the nation should take to avoid future flood catastrophes. Brigadier General Gerald Galloway was chosen to lead the effort. It was an auspicious choice. Galloway had spent thirty-eight years within the confines of the Army Engineers. If Jimmy the Greek were laying odds, he would have predicted that Galloway would regurgitate the company line about building more levees and pleasing more riverside congressmen. But in his own way Galloway was an instigator. He was not powerless or unrecognized; quite the contrary, he held a prominent position with all the trappings of power. But first and foremost he was an independent thinker, and he could see that a major reorientation of the nation's flood-control policy was desperately needed.

The resulting report, referred to as the Galloway Report, was a radical departure from past Corps policies and set the stage for a new era. The focus now was on "floodplain management," not "flood control." General Galloway got right to the point in his report. He began by noting that although the nation had known about the effectiveness of floodplain management as an alternative to structural solutions (i.e., levees, walls, and dams), it had "not taken full advantage of this knowledge." He then proposed a whole new approach to floods: "The Review Committee supports a floodplain management strategy of, sequentially, avoiding inappropriate use of the floodplain, minimizing vulnerability to damage through both structural and nonstructural means, and mitigating flood damages when they do occur."[12] Galloway's report went even further and acknowledged ecological values of rivers and their floodplains. In confronting the dangers posed by floods, the nation should also establish the goal of "concurrent and integrated preservation and enhancement of the natural resources and functions of floodplains."[13]

This was a radical departure from the existing mind-set in the Corps of Engineers—a mind-set that had been established in the 1850s. General Galloway was finally updating General Humphreys (the army engineer who had argued for levees at the beginning of the Civil War). In a sense, General Galloway's report was a classic example of how to turn a crisis into an opportunity. The ideas embodied in his report had been around for decades, championed by Professor Gilbert White and others.[14] But the 1993 floods were an opportunity to pull White's ideas together with the latest environmental thinking and produce a new approach.

Galloway's report pushed the Corps to think in new directions. The evolution of the agency's thinking can be traced through what might be called "terminology evolution." For over a century the Corps did "flood control." After Galloway, this term was supplanted by the "national flood damage reduction program," which later morphed into "flood damage reduction," which in turn was replaced by "flood risk management." This latest term conveys the idea that there is always risk; no structural device can totally eliminate that.[15] But it is more than just a nonstructural approach. It includes evacuation planning, public education, and an awareness among everyone who lives in a floodplain that the risk is never totally eliminated, so each individual must accept some responsibility for his or her own safety.

But change comes slowly to flood-control policy. According to General Galloway, the problem today is not the Corps but the Congress: "The Congress manages water by earmarks."[16] Such a myopic, district-centered vision will always be an impediment to long-term floodplain management. A flood-control policy that has been in place for 100 years is much like a great ocean liner; it cannot be turned with ease. In the meantime, the environmental damages and other disadvantages of traditional flood control accumulate. The results can be catastrophic.

HOUSE-A-FIRE

There are flood-management structures all over the nation, but nowhere is the titanic struggle between low-lying people and high-flying water more dramatic than in the Lower Mississippi River Valley. The story of flood "control" along the Mississippi is a long saga of incessant pleading for more money—a clamor that is never quelled—interspersed with fierce pronouncements of Jeffersonian independence and minuteman defiance and, on occasion, desperate calls for help.

When I first began research for this book in 2002, I visited the Corps of Engineers' headquarters in Washington, D.C., to get a sense of the agency's attitude toward river restoration. Jim Johnson, who had been with the Corps since 1969, knew the agency as well as anyone, and he had experienced many big changes at the Corps. He made it clear that the Corps considered river restoration an important component of its mission, but that the agency had no plans to abandon its two principal missions of flood control and navigation. "We call it river ecosystem restoration and incorporate it into our other primary missions. . . . But restoration is probably never going to be more than a third of the budget." I asked Johnson which Corps ecosystem projects were the most important. He quickly mentioned the Kissimmee Project, but then he said something that really intrigued me: "The next big one is the coastal Louisiana study. That area has the nation's richest wetlands resource, and we're losing them fast."[17]

Obviously I could not write a book about restoration without including the massive effort to restore the receding coastline of southern Louisiana. In December 2003 I spent a week there, interviewing people, walking the levee, and visiting the remnants of wetlands. I asked questions about endangered species, commercial fisheries, and habitat, but most people seemed to have something else on their mind. I kept hearing about grave threats to New Orleans. The predictions were so extreme that I didn't believe them; they surely must be exaggerating the threat of flooding in New Orleans. One of the people I interviewed, from an environmental organization, warned darkly of dire threats and punched his finger theatrically at the back cover of a magazine called *Water Marks*, exclaiming, "That's what we're in for!" He was pointing to a canned question-and-answer exchange between the magazine's editor and Professor Bob Thomas of Loyola University in New Orleans:

QUESTION: There's been a lot written about coastal wetlands forming a buffer zone that protects the urban population of New Orleans. Everyone agrees that part of the protection has been lost, but have the losses been large enough to really make the city vulnerable?

THOMAS: Vulnerable is an understatement. This city, its people, its economy and its culture are all exposed to a potential—if not inevitable—catastrophe. . . . The consequences are difficult to imagine.

QUESTION: Give us a little glimpse.

THOMAS: . . . The storm surge and after-effects from a major hurricane would put 20 feet of water in downtown New Orleans.[18]

Yeah, sure, and the sky will fall in with it. Crazy professors. As I walked away from these interviews, I truly felt that there must be something in the drinking water that was making people paranoid. Besides, the Corps said that New Orleans was safe. I chuckled to myself and muttered a phrase that my grandmother always said when she thought someone was overstating a danger: "Oh, house-a-fire, house-a-fire!" That was in 2003, two years before Katrina.

Southern Louisiana is essentially one giant delta. The Mississippi River has meandered all over this land for millions of years, bringing with it a sediment load of continental dimensions. Much of the soil that used to blacken the fields of the American Midwest, as much as 2 million tons a day at its height, turned into river muck and built up successive deltas and natural levees. Each of these river deltas would reach out into the sea, building height until it formed a natural dam between the river and its destination. The river would then seek a new course to the sea, repeating this process through the eons. This accumulation of soil and water created an estuarine paradise swarming with waterfowl, crustaceans, fish, and a whole assortment of swamp creatures that scared the hell out of early settlers and convinced them that the only good wetland was a drained wetland.

The first 200 years of Louisiana history consisted of a man-against-water struggle; locals built levees to constrain rivers, drained swamps to create farmland, and declared victory, at least until the next big flood or hurricane made it abundantly evident that there is a swampland version of the dust-to-dust apothegm. Today there are perfectly rectangular lakes in southern Louisiana, startling in their geometric precision. These tracts used to be plantation fields, cleared of cypress, surrounded by earthen levees, and farmed until nature reclaimed her own. Lessons are learned hard in this country.

But Americans are nothing if not persistent. Some of the cotton fields were eventually displaced by towns and subdivisions built entirely below sea level and surrounded by levees. Some neighborhoods in New Orleans are 15 feet below the Mississippi River. The only things that stand between this lowland house of cards and a watery end are the levees, which are not invulnerable—a lesson learned the hard way from a vicious taskmaster called Katrina.

The effort to control the Mississippi River is inextricable from another great American struggle, the Civil War. In the nineteenth century manliness was equated with boldness and tenacity, and the same men who fought one another in the Civil War led the fight to control the river. It was Robert E. Lee who attempted to blow the rapids at the Des Moines River. That same year, 1837, he successfully redirected the river channel in front of St. Louis. An early technical report on how to control the river was written by Charles Ellet Jr., who was later killed in the war while serving aboard a Mississippi River gunboat at Memphis. Andrew Humphreys, mentioned earlier, was a corps commander in George Meade's Army of the Potomac who penned a history of that army's last campaign. His assistant in writing the Mississippi report was Gouverneur K. Warren, the savior of Little Round Top at Gettysburg. James Eads, the designer and builder of the jetties through the Mississippi River Delta, was not a military man, but he built the ironclads that help wrest control of the river from the South.

In 1863 U. S. Grant made war on the river almost as vigorously as he made war against Southern armies. He tried to bypass the stronghold of

Barges and faux gambling boat at Vicksburg, Mississippi, on the lower Mississippi River. The river here is lined with levees.

Vicksburg by digging a channel across a narrow isthmus of land. He failed, but not for want of effort. Thirteen years later the river, as though mocking Grant's feeble efforts, changed course and left Vicksburg in its wake. The people of Vicksburg protested mightily to the same government against which they had rebelled. Beseeching the federal government for help in controlling the Mississippi River soon became an honored southern tradition. In 1903 the Corps redeemed the city's fortunes by redirecting the Yazoo River from its confluence with the Mississippi so that it flowed past Vicksburg and thus reestablished the city's port.

That year—1903—the U.S. Army Corps of Engineers turned 100 years old and had spent much of the second half of that period attempting to control the river. After a fitful start, the Corps began building levees and did not stop until the entire lower river was lined with constricting levees. The Corps, anxious to redeem itself after early failures, was only too happy to do the bidding of southern senators and congressmen.

In times of flood, people along the Mississippi pray for the levees. During dry periods, they build more levees. Southern Louisiana's queen of culture, the "Crescent City" of New Orleans, is situated near the end of this gigantic levee system. It has existed in a precarious state since the first French settlers built on the only high ground around—today's French Quarter. The river that gives New Orleans life and enriches it also threatens the city with extinction. The thin line between those two extremes is the levee.

There is always a price for causing nature's posture to stoop. Louisiana is delta, and delta is river muck, moved over eons of time to the place where the river slows and what used to be topsoil in Indiana and Montana and Pennsylvania becomes fecund coastal Louisiana, splayed out over an area stretching from the Pearl River in Mississippi to the bayous of east Texas. The sediment carried with it a rich, nourishing load of organic material and native seeds. But the levees prevented this sediment and its organic mulch from building up the land and inseminating it. Instead, it is now channeled by the levees out to the edge of the continental shelf, where it sinks to great depths. Without this renewing earth, the coastline basically oozes slowly into the sea. In the 1930s, when the levee system was brought to fruition, southern Louisiana began to sink into the sea.

The levees were responsible for most of the land loss, but not all of it. Oil companies dug countless channels and canals to drill sites, which hastened the loss of soil and killed the groundcover with the infusion of seawater. Also, locks and dams upstream, on the Missouri, the upper Mississippi, and the Tennessee rivers, blocked half the historical flow of life-giving sediment. Then, in response to demands from the shipping industry, the Corps of Engineers dug a massive channel from the Mississippi River near New Orleans to the sea to create a straight-line shortcut for barges and oceangoing ships.[19] These alterations to the river made New Orleans and Baton Rouge the heart of a massive shipping industry. They protected hundreds of thousands of homes and businesses from floods. And they made possible the conversion of vast tracts of "worthless" swampland into farms and suburbs.

But what the waters give, the waters take away, and the southern Louisiana coast began to die without the basic building block of river-borne sediment. Most geologic change takes place over a period of time that is not readily comprehensible, but the Cajuns of southern Louisiana have seen their childhood haunts disappear into the sea in a single generation.

It is hard to imagine the rapidity of change in this area. East of New Orleans, squeezed between Lakes Ponchartrain and Saint Catherine, is a narrow spit of land that is home to the Bayou Sauvage National Wildlife Refuge and, before Katrina, a string of ramshackle second homes along the lakefront. These houses, made of wood and standing on stilts, had a temporary look to them when I visited there in 2003. The owners of these little cottage getaways had a tradition of attaching whimsical names to them: "Dats-a-No-No," "The Wet Spot," "The Passing Wind," "Camp *Camp*," and, prophetically, "The Hurricane Hut" and the "Gail Force." Such flippant monikers were a humorous way to flaunt their precariousness, but the cavalier attitude they reflect may be difficult to maintain. Nearly all these homes were destroyed by Katrina.[20] The thin stretch of land where they perched, strung out along Highway 90, is sloughing off into the Gulf of Mexico. Without the benefit of miles of wetland to buffer incoming hurricanes, these frail frame houses were like the houses built in nuclear-bomb test zones—felled by a single, catastrophic burst of rampaging force.

Across Lake Pontchartrain, to the northeast, is another national wildlife refuge called the Big Branch Marsh. In 2003 I walked along a boardwalk that extended into the swamp, through a mile of southern pine forest. At the end of the boardwalk was an information plaque that explained that the pine trees in the distance were dead because the land is subsiding, and pine trees cannot live in standing water. In the distance a thick growth of cypress trees formed a verdant horizon. Cypresses, unlike the pines, thrive in shallow water. The sign explained that subsidence is a natural process, implying that the changing topography is simply an artifact of geology and time. Although some natural subsidence has occurred, the great terrestrial retreat that is occurring all along the Louisiana coast is a result not of natural processes but of building levees, canals, and water diversions.

Today, a different sign can be found at the Big Branch refuge. It explains how the Katrina storm surge roared through here with a 6–8-foot wall of water, breaching the levee around the refuge. But it was not the storm surge that decimated the refuge; the saline storm waters could not return to the sea because they were blocked by the surrounding levee. It took several weeks to pump out the water, but by then the salt water had killed 80 percent of the trees and destroyed 1,700 acres of marsh.[21]

Both Bayou Sauvage and Big Branch Marsh are part of an estuarine complex that is rapidly losing ground. At the present rate of land loss, this section of the Louisiana coast will lose 45,000 acres of marsh and over 100,000 acres of swamp by 2050.[22] By then, homes with names like "The Hurricane Hut" and the wildlife refuges will be long gone unless dramatic restoration is undertaken.

Remember that the Mississippi River Delta has changed locations several times. At one time the river flowed to the sea to the east of the current river channel. It built a huge delta that extended into the Gulf for many miles. But the river, pregnant with sediment, built a terminal levee, effectively blocking its own passage to the sea. The Mississippi then formed another channel to the west where passage to the Gulf was still unobstructed. With the passage of time, the old delta melted into the sea, and today all that remains of this ancient extension of land is the Chandeleur Islands, a crescent of barrier islands 50 miles from the current shoreline. Hurricane Katrina dramatically reduced the area of the islands; some of them disappeared entirely. There is an important lesson in the fate of this former delta: when the river ceases to deliver its thick, fertile load of sediment, the sea reclaims its bed.

Coastal land loss is occurring all along the Louisiana coast at a stunning rate. Since the 1930s over 2,395 square miles of coastal land have been lost.[23] Hurricanes Katrina and Rita eliminated 217 square miles in a matter of days.[24] At the current rate of loss, 10 percent of Louisiana will slip into the sea by 2050.[25] The impact of these losses is of biblical proportions. A plan called Coast2050, published by the state of Louisiana, summarizes:

> The statistics are awesome; the ecosystem contributes nearly 30% by weight of the total commercial fisheries harvest in the lower 48 states and provides overwintering habitat for 70% of the migratory waterfowl using the Central and Mississippi Flyways; 18% of U.S. oil production and 24% of U.S. gas production come from coastal Louisiana and the adjacent Gulf of Mexico, with an annual value of $17 billion; Louisiana's ports rank first in the Nation in total shipping tonnage.[26]

The causes of the land loss have been known for forty years. Alarming testimony was provided to Congress in 1977 (described below). A report written by a panel of experts in 1994 pinpointed the most important sources of the problem: "Levees constructed for flood control and dredging for shipping channels reduce the supply of mineral sediment to wetlands. Restrictions of freshwater flow from the river, as well as other activities (including construction of canals and dikes), lead to changes in wetland hydrology and intrusions of saltwater in previously freshwater areas."[27]

Another major contributor to land loss is the Mississippi River Gulf Outlet, the shipping channel opened by the Corps in 1963. This channel grew to nearly three times its original width because of erosion and "contributed to the direct loss of more than 20,000 acres of wetlands and caused more than 36,000 acres of marshes and Lake Pontchartrain to become more brackish."[28] The channel also proved to be a very effective pathway for hurricanes to head directly into the heart of New Orleans.[29] The Gulf Outlet was deauthorized in 2006 in the wake of the Katrina disaster, and the area will be restored.[30]

In other words, some of the biggest components of Louisiana's economy—shipping, natural gas, oil, and new housing development—are destroying the southern part of the state and thus destroying the other major components of Louisiana's economy. This is a region at war with itself and its river. Land in this area appears to be divided into four starkly different land uses. The potential conflict among them is obvious to even a casual observer. There are impoverished neighborhoods—yesterday's suburbs—sprinkled liberally throughout the region. Numerous petrochemical plants, surrounded by ground the color of rust and used oil, belch out a brown-red haze that lies over the land like a carcinogenic shroud. New housing subdivisions—the faceless, cookie-cutter variety with the omnipresent strip-mall chain stores—have been added to the mix at an unprecedented rate. And interspersed among these are tracts of primeval swampland, full of amphibious wildlife, exotic vegetation, and some of the world's finest mosquito habitat. These four land uses—old modest housing for the poor, big new homes for the well-to-do, heavy industry, and dense, verdant wetlands—are intermixed as if they were all ingredients in a geosocial pot of stew, stirred frequently. One can encounter all four within a short drive in any direction.

Yet despite their close proximity, each exists as a threat to the others. Industrial pollution reduces real-estate values and threatens public health;

gentrification pushes out the poor and drains more swamps; and new levees built around burgeoning residential development push water elsewhere and contribute to subsidence. The only thing these uses all have in common is their vulnerability to land loss, subsidence, and hurricanes.

On the other hand, southern Louisiana's well-stirred pot of cultural stew has created one of the most diverse societies in the United States. There is more to lose here than just swampland and alligators; there is a unique mix of culture that has given the world everything from jazz music to Cajun cooking.[31] Looming over this rich cultural landscape like a bad dream is the steadily advancing sea, threatening to consume one of the oldest communities in the United States.

Blake Chaisson works for St. Charles Parish. He monitors pumps for the Corps of Engineers, ready to respond, as he put it, whenever rain and weather "make things too deep." As a small boy he hunted in cypress swamps and backwaters—in locations that no longer exist. "We had our places to go, favorite sites, but they're gone now. Whole swamps have just disappeared." Unless the management of the Mississippi River is altered dramatically, Evangeline may be expelled from yet another home.

Abuse Followed Ridicule

Admitting that a big problem exists in southern Louisiana has been a slow, arduous process that has followed a fairly predictable path: instigators were initially ridiculed by the powers that be, and slowly, very slowly, people began to listen to alternative points of view, fighting the establishment all the way. The first people to argue that the Corps' levees-only flood-control policy was a disaster were dismissed as fools. John Barry, in his book *Rising Tide*, describes how James Kemper, a young civilian from outside the circles of power and influence, argued during the years 1912–1927 that the Corps' policy of walling in the river was a recipe for disaster. He cited convincing statistics, his argument was logical and well founded, and his pleas for new studies of the river made eminent sense. But he was calling into question and therefore challenging the controlling dominance of the Corps and the Mississippi River Commission (which was controlled by the Corps). As he persisted, "abuse followed ridicule."[32] The Corps and its political allies ignored this pesky young outsider and issued countless reassurances that lining the river with levees and closing off all tributaries and outlets would handle a flood of any size. The heavens opened up in 1927 and immersed millions of acres in floodwater, but the levee building went on, and the basic elements for epic land loss—and catastrophic hurricanes—were set in motion.

By the 1960s appreciable wetland loss was becoming apparent to some people in southern Louisiana. In 1977 Senator Edmund Muskie's Senate Subcommittee on Environmental Pollution came to New Orleans and held hearings on the Clean Water Act. One of the people chosen to testify before this committee was Professor Barry Kohl, a young geologist from Tulane University. Professor Kohl was originally from Indiana, earned a bachelor's degree at Purdue University, and then went to Tulane for his doctorate. He quickly fell in love with New Orleans and decided to make it his permanent home. He began research in a field called biostratigraphy, which traces how sediment moves by tracking tiny fossils trapped in the sediment. Professor Kohl's research provided dramatic evidence that the coastal region was not being replenished, and he argued that the federal government, not local government, should regulate wetlands because they were "a national resource.[33] Kohl was challenging some of the most powerful entities in the state and arguing that the federal government should intervene to stop wetland losses.

This did not make the Yankee interloper popular with the Corps and local authorities. His testimony was immediately dismissed by the Corps and the oil and gas industry.[34] Then, as time proved Professor Kohl correct, the finger-pointing began in earnest. No one wanted to admit that his or her activities were causing land loss. In the meantime, Professor Kohl was joined by other instigators, many with ties to national environmental groups, and they began a campaign to restore coastal Louisiana. Additional studies were completed, and a host of powerful institutions began to admit that the problem was severe and a solution had to be found. In the 1980s a number of groups formed the Coalition to Restore Coastal Louisiana. Everyone knew that restoring the coast meant changing the way the Mississippi was managed.

Finally, the state of Louisiana adopted a formal position stressing the need to stop land loss. Even the oil and gas industry came on board; with thousands of wells and pipelines in the threatened coastal lands, it stood to lose millions in infrastructure damage. A publicity campaign, called

America's Wetlands, was funded in large part by Shell Oil Company; it produced a slick brochure and ran commercials on local television that showed the words "America's Wetlands" being washed away by incoming waves. Congress finally responded by passing the 1990 Coastal Wetlands Planning, Protection and Restoration Act, commonly referred to as the Breaux Act, which provided funding for small restoration projects administered by five federal agencies and the state of Louisiana.[35] This was an important step, but the act did not provide for a comprehensive all-out effort to stop the land loss.

The Corps of Engineers is the lead player in the Breaux Act's implementation. It responded with both long- and short-term solutions. The long-term goal was to study the causes of land loss and recommend projects designed to restore the wetlands. The more immediate goal was to build two projects to reduce saltwater intrusion, one at Caernarvon and one at Davis Pond. As Jim Johnson predicted, this was going to be a big one.

COON-ASS

In the early days of New Orleans, a massive break in the west-bank levee flooded an area larger than some European countries. Eventually the water receded, leaving only a small water-filled scar near the river. Many years later, it came to be called Davis Pond.

Jack Fredine had been with the Corps for over twenty years when I interviewed him in 2003. His "baby" was the Davis Pond Project. In a landscape dominated by impressive bodies of water, Davis Pond is a nondescript geographic feature. But serendipity brought it notoriety; it just happened to be located next to a proposed diversion from the Mississippi River just upstream from New Orleans. The Corps named this ambitious project after the little pond.

The Davis Pond Project proposed to cut through the levee on the west bank and direct a flow of water into the Bataria Estuary to improve oyster habitat with an infusion of fresh water. But in southern Louisiana, any time surgery is performed on a levee, it gets people's attention. In the minds of locals, slicing through a levee is a bit like cutting a hole in the bottom of a boat. Visceral reactions should be expected. In 1965, when the Davis Pond Project was first proposed, the locals did not like the idea. They said they needed to think about it. They thought about it for over twenty years. You just don't go messin' with the levee.

But there is more than one sacred cow in southern Louisiana. The oyster industry ranks with oil and gas as a source of local economic well-being, and that industry was ailing. Saltwater intrusion, caused by the levees, channels, and canals that had disrupted the natural flow of freshwater and sediment, was destroying the oyster beds. Oysters must have brackish water to survive; they die in fresh water. But they also need the protective cocoon of the still, shallow waters of an estuary. Open water invites predators, especially a homely, shelled slug known as an oyster drill. Because of human-induced changes in the coastal waters, oyster drills and other predators had replaced humans as the major connoisseur of oysters on the half shell. Saltwater intrusion, land loss, and open-water predation had narrowed the oyster habitat to a hair-thin crescent. So the Corps proposed to balance the saltwater inflow into the Bataria Estuary by diverting water out of the Mississippi River and into Lake Cataouatche, from there to Lake Salvador, and then to the estuary to the south. But that meant breaching the levee—something that has usually been the work of natural fiat or dynamite desperadoes who wanted to turn floodwaters onto someone else's land.

After twenty years of contemplation and watching the oyster beds disappear, the state of Louisiana finally decided to sign on to the Davis Pond Project. A similar Corps project, the Caernarvon Project, had experienced legal difficulties but appeared to be working, so anxious locals decided to support the Davis Pond diversion.[36] Work began in 1996 and was completed, under Jack Fredine's careful guidance, in 2002.[37] The $120-million project built a gated four-barrel box culvert through the levee with a 2.2-mile channel and 19 miles of levees, and relocated railroads, highways, and pipelines. Fredine watched over the project with paternal devotion, proud that the Corps was doing something that helped nature. He also had to reassure the new subdivisions being built just across the project's levee that the project did not present a threat to them.

Jack Fredine has strong opinions about past Corps projects that have damaged the coastal estuaries and gutted fertile wetlands. I asked him if it was common for Corps personnel to be so forthcoming about the Corps' past failures. "No, but I am almost 65. I'm not looking for a promotion, and I'll retire soon, so I say whatever I think."[38] Fredine is proud of his service with the Corps, but he is not afraid to admit that the agency has made some shortsighted decisions. For him, that is not important. "A lot of people talk about blame or who is at fault. I don't care about that; I only care about what we can do to solve the problems. That's why we built the Davis Pond Project."

Fredine invited me to accompany him on an inspection of the project and look at some new adjustments made at the lower end of the project. "We'll take an airboat out to the new breaks in the weir. We'll go with some guys from St. Charles Parish." At that point Jack touched my arm and grinned. "You need to understand that everyone in St. Charles Parish has a nickname. One of the parish council members is known as Snooky, and the assistant director of public works for the county is a guy everyone calls Poochy. Red LeBlanc will be our boat pilot. He's a regular 'ole coon-ass, and he really knows how to handle an airboat."

"What exactly is a coon-ass?" I could picture myself saying the wrong thing; this was a culture with which I was unfamiliar.

"It's a term of endearment, not an insult. That's what you call the Cajun boys around here. At one time the state of Louisiana tried to outlaw the term, but that just made it more popular. The coon-asses are good people."

Adrian "Red" LeBlanc and his sidekick "Squirt" were waiting for us at the edge of a swamp. Next to them was a 20-foot-long aluminum skiff with an engine and propeller mounted on the stern that looked like a crossbred Chevy muscle car and B-17. Red's Cajun accent was smooth and fluid, as though talking required no effort. Listening to him was a pleasure. Squirt handed out ear-protection muffs, and we were soon powering down a miles-long "borrow pit"—the source of the material used to build the levees (the borrow pit forms a channel). Red deftly maneuvered the boat through beds of water hyacinths—a pesky invasive species—and between thickets of willow and tall stands of cypress. When we approached a small footbridge that completely blocked the channel, Red simply stepped on the gas, threw the boat up on the bank, and powered around the bridge. When it became necessary to switch from one side of a 6-foot-tall levee to the other, he pointed the boat directly at the levee, accelerated, and turned the boat into a sled that hopped the levee and slid down the other side. I turned to look at him as he completed this amphibious maneuver, and saw a look of pure joy on his face. Red LeBlanc is a man who has spent his entire life in these wetlands. The life that he lives and the land that is disappearing are one and the same.

We inspected several channels cut in the weir, dug by Red a few weeks ago with barge-mounted heavy equipment to increase freshwater flows into Lake Cataouatche. On the return trip I saw small, fur-covered animals scurry out of our path. They looked like a cross between a beaver and a rat. Red made no effort to avoid them.

"Nutria," explained Jack. "They're a big pest, imported from Brazil for their fur. They have a voracious appetite. They eat the roots of young cypress tress and other plants—they can destroy a huge area. A female nutria can produce fifteen young every year. There's a $4 bounty on them in Louisiana—that's the only way we can get people to kill them."

There were plenty of native species too. Enormous white cranes rose from the water in front of our boat, seeming to levitate in slow motion. Wood ducks scurried madly across the water. Red-tailed hawks and a couple of bald eagles glided overhead, looking curiously at our noisy intrusion. An occasional blue heron flitted by. Red has a favorite alligator in this area, but the cool air temperature kept him out of sight.

After we finished our boat tour and were driving back to the Corps' district headquarters, Jack talked about the importance of the Davis Pond Project. "People need to understand that the Corps is doing some great things these days. This project won't solve the sediment loss problems, but it will help stabilize the salinity level in the estuary and bring back the oyster beds. That's the kind of thing the Corps should be doing."

The "new" Corps is clearly hoping to solve southern Louisiana's problems by building new restoration projects while at the same time protecting its old allies. It completed a draft plan in 2004, called the Louisiana Coastal Area Ecosystem Restoration Study, but Katrina struck before a final report was issued.[39] In 2007 Congress directed the Corps to take a much more comprehensive approach to coastal restoration and its relationship to protecting New Orleans from future storms.[40] That same year a master plan was approved that anticipates a $50-billion restoration effort.[41]

The effort to restore the coastline has been greatly complicated by Katrina. The original cost estimate of the restoration program was $15 billion.

The Bush administration cut funding for the program just months before Katrina.[42] Estimates to repair the damage caused by the storm range into the hundreds of billions of dollars.[43] However, all disasters create opportunities. In the case of Katrina, the Louisiana congressional delegation saw a chance to rake in federal dollars and requested $250 billion, including $40 billion for more of the Corps' flood-control projects that had caused the problem in the first place.[44]

In contrast, the disaster sparked the first serious discussion about the wisdom of building in vulnerable floodplains and along hurricane-swept coasts. Katrina was a man-made disaster, not a natural one.[45] The Corps is even exploring the idea of buying thousands of houses in sensitive areas and allowing nature to reclaim them.[46] The Breaux Act and subsequent legislation continue to funnel money to restoration programs, which have restored 120,000 acres.[47] And the Obama administration is considering a major revision to federal floodplain management to emphasize nonstructural solutions and minimize costly and dangerous development in flood-prone areas. The barge industry is fiercely opposed to that change because it may affect the depth of the river.[48]

Reversing the effects of 100 years of levees and dam building is impossible. Even if the entire Mississippi River were diverted to build up coastal sediment, the diversion would have limited success because half of the river's sediment load is trapped behind dams. Too many cities, including New Orleans, have been built in the floodplain to simply abandon levees, and no one wants to sacrifice the lucrative shipping channel. Rising sea levels due to anthropogenic climate change, combined with subsidence, will continue to submerge coastal lands.[49] The best we can hope for is remedial restoration and a powerful lesson about what happens when we try to force big rivers into narrow ditches.

THE HOUSE OF DEATH FLOATS BY

The phrase in the preceding heading is one of the chapter titles in Mark Twain's *Adventures of Huckleberry Finn*, but it reminds us all of the news footage of dramatic floods with houses, some with desperate people clinging to the roof, floating hurriedly downstream. It is always a disturbing image, given the sacred place of home in our psyche; "home" can literally be carried away. Our collective fear of floods has been a dependable source of support for the chimera of "flood control," even when it offers only a modicum of real security.

It is an illusion to think that humankind will ever truly control the Mississippi River—a river that has wandered laterally over hundreds of miles, flooded an area the size of Kansas, and created a delta that forms the heartland of the state of Louisiana. It is a river that Andrew Humphreys once described as a "turbid and boiling torrent, immense in volume and force."[50] It still is and may become even more so. Global warming is changing world weather patterns, with more dramatic and sometimes violent fluctuations in weather. In regard to weather, people tend to think in terms of the worst of this or the biggest of that. But the worst and the biggest have yet to occur; all climatic records exist only until they are surpassed by an even greater calamity. It is the greatest folly to assume that the world's worst weather has already occurred—that is an attitude that guarantees tragic underestimation.

As if to prove that point, the Mississippi once again broke records and levees in the spring of 2011. A Corps official in Vicksburg exclaimed, "There's never been a flood of this magnitude on the Mississippi."[51] One headline read, "Mississippi Flood Damage Could Reach Billions."[52] In a desperate effort to save cities and towns along the river, the Corps blew the "fuse plug" levee at the Bird's Point–New Madrid Floodway and flooded 130,000 acres in Missouri. This did not prevent flooding downstream—in Arkansas alone there was $500 million in damage. The flood surged around Vicksburg, Mississippi, and the Yazoo River flowed backward, flooding 2,000 homes. So the Corps also opened the Morganza and the Bonnet Carre spillways in Louisiana. The flood set a record for total volume of flow.[53] Critics once again blasted the Corps for relying on levees instead of opening up critical floodplains and relocating people rather than water.[54] Photos of drowned neighborhoods and inundated farmland once again dominated the news, but the Corps still claims that it has it all under control. The website for its Mississippi Valley Division confidently lists "control of floods of the Mississippi River" as its premier mission. It describes its efforts during the 2011 flood, which included distributing literally millions of sandbags, as a "flood fight."[55] The agency still does not admit that its own policies created many of the problems caused by the 2011 flood. The Mississippi will break free again, just as it has countless times in the past. Rather than assume that we have conquered it, we should respect the river for that eventuality. Those living in vulnerable locations should be encouraged to move; at the very least we should stop subsidizing their foolhardy enterprise. Municipalities lying in the path of an unrepentant river must be fortified like medieval towns and surrounded with levees. The remainder of the floodplain must be returned to its natural mission. The more latitude we give the river, the more forgiving it will be when it goes on a bender. The Mississippi is a volatile teenager blindly confident of purpose, not an aging retiree resigned to fate; it will not go meekly to a destination determined by others.

Throughout the nation we have spent enormous sums to build flood-control structures, and that investment has effectively protected property in some places, especially urban areas. But it has probably increased flood damage overall, in part by luring people into vulnerable areas.[56] A 2004 analysis of the cost of flood damage indicated that the total loss for that year was just as high as it was in 1927—after the expenditure of billions of dollars on "flood-control" levees.[57] As General Galloway put it recently, "Levees are the heart of the problem."[58] Would it not be prudent to move those houses, on our own terms, before Mother Nature does it for us in a much more abrupt manner? The Dutch, who know something about living low amidst high water, call this concept "room for the river."[59] We could learn a lot from them. In the meantime, the house of death just keeps floating by.

One of John Lee Hooker's most powerful songs, "Tupelo," is about a great flood that surged through that Mississippi town in 1935. He sang of painful loss as women and children screamed and cried amid the swirling waters, beseeching the Lord to save them.[60] It is a powerful, heartfelt, and tragic description of what happens when floods claim their victims. To the extent that it is possible, we should get women and children—everyone—out of harm's way. Then we can sing a very different song, this one by Bob Dylan, that advises us to sit contentedly on the bank, safely above the water, and just watch the river flow.[61]

8

DOWNSTREAM DILEMMA

Water Pollution

Yet pure its waters—its shallows are bright
With colored pebbles and sparkles of light
—FROM WILLIAM CULLEN BRYANT, "GREEN RIVER"

Americans tend to do things with panache and gusto; we even fought our Civil War that way. We slaughtered each other at a horrific rate for four years, and 620,000 Americans lost their lives. But most of them did not die from whizzing minié balls or blasts of canister. The majority of deaths in the Civil War were due to disease, in no small part because of tainted drinking water. No one, regardless of rank, escaped the foul scourge. General Robert E. Lee, at crucial points in the war, suffered from debilitating diarrhea. President Lincoln's 11-year-old son, Willie, died of typhoid in 1862, leaving his parents in agonizing grief. Willie probably caught the deadly bacteria by drinking unclean water; a few blocks from the White House was the Washington Canal—what is now Constitution Avenue—so fouled with sewage and offal that its miasma permeated the entire town. In 1862 the White House smelled like an outhouse.

But out of adversity comes innovation. The dead list of the Civil War gave rise to a realization that sanitation was critical when large numbers of people gathered, and Clara Barton, the volunteer Sanitation Commission, and others fought for improved medical care and sanitary improvements. From that tragic beginning, the public health movement gradually gained prominence. But it took another 100 years before the nation passed meaningful legislation that began the cleanup of many of our rivers.

For over two centuries we used rivers as a convenient dump to deposit our sewage, toxic waste, and agricultural runoff. This occurred despite the fact that two-thirds of our drinking water comes from rivers. The nation was slow to realize that we do indeed all live downstream.

That reality is more obvious to some than to others. Peoria, Illinois, became aware of its downstream status when Chicago built a canal in 1900 that funneled 274 million gallons of raw sewage into the Illinois River—the source of Peoria's drinking water.[1] The water in the Mississippi is used repeatedly by the 30 million people along the river. According to one old-time riverboat captain, there used to be signs in the restrooms of Quincy, Illinois, that said, "Be sure to flush the toilet. Hannibal needs the water."[2] The 25 million people who draw drinking water from the lower Colorado became aware of their downstream dilemma when it was revealed that 16 million tons of uranium tailings had been dumped next to the river just north of Moab, Utah. The thought of drinking radioactive water was apparently not too appealing to them.[3] These examples highlight the need for vigilance about what other people are putting in our water. One man's waste is another man's water—not an appealing image—unless we have truly effective water-quality control.

TOXICS RUN THROUGH IT

The effort to pass meaningful and enforceable national water-quality standards was a long, incremental struggle that finally bore fruit in 1972. Three years earlier much had been made of the Cuyahoga River catching fire, but in fact the Cuyahoga had caught fire many times before then. The 1969 fire just happened to coincide with a newfound awareness of the state of our rivers. It helped that troubadour Randy Newman penned a song about the river, mimicking Woody Guthrie's "Roll on, Columbia, Roll On," that went "Burn on, big river, burn on."

The culmination of the effort to clean up America's water was the 1972 Clean Water Act, to be administered by the newly created Environmental Protection Agency (EPA).[4] The Act set up a system of pollution-discharge permits that limited how much pollution could be poured into rivers and bays. Its lofty goal was to make the nation's waters "fishable and swimmable" by 1983. The law was amended in 1977 and again in 1987 to expand regulation beyond conventional pollutants, such as sewage, oil and grease, and biochemicals, to include toxics.[5]

The law proved to be fairly effective at limiting "point sources," meaning pollution that comes out of a pipe, such as that from a factory or municipal treatment plant. It has been much less effective in limiting "nonpoint" pollution from farms, feedlots, timber clear-cuts, mines, construction sites, city streets, and urban lawns.[6] The 1987 amendments established incentives for polluters to reduce their nonpoint releases.

In 1974 another important milestone was achieved when Congress passed the Safe Drinking Water Act, which controls pollutants in drinking water.[7] The Act was reauthorized and strengthened in 1996, with the added proviso that water suppliers had to inform their clientele of the amount of pollutants in their water.[8]

The politics of water quality lacks the sex appeal of fights over big dams and mighty rivers, but it is no less important, especially as we attempt to restore some rivers; it's no use saving rivers that are harmful to the touch. Unfortunately, much of the literature and law regarding water quality is virtually unintelligible to a lay audience. A full understanding of the issue requires expertise in chemistry, biology, medicine, public health, and civil engineering. It does not help that policy makers have become inured to countless acronyms and dense bureaucratese. It is not unusual to encounter obtuse perorations such as this: "Water quality-based effluent limits (WQBELs) in NPDES permits that implement WLAs in approved TMDLs must be 'consistent with the assumptions and requirements of any available WLA for the discharge.'"[9] Say what? Loosely translated, that means that permitted sewage levels must be measured consistently—I think. Reading an entire water-quality report is about as exciting as a history of plywood.

But even a cursory discussion of the topic requires a review of a few basic terms. The 1972 Act set up a permit system called the National Pollutant Discharge Elimination System (NPDES). This system basically gives point-source polluters a license to pollute, but it limits the amount of pollution to the *total maximum daily load* (TMDL). There is a vast and complex literature on how that magic limit is set, and how it can be reached, but suffice it to say that it is a difficult and controversial process.[10] Two additional terms that are necessary to understand municipal pollution are *combined sewer overflows* (CSOs) and *sanitary sewer overflows* (SSOs). These terms describe the source and makeup of the pollution that originates in cities and ends up in our waterways.

The Clean Water Act and related laws have been partially effective in cleaning up the nation's rivers. According to the EPA's 2000 Water Quality Inventory, 60 percent of our streams, lakes, and rivers were considered too polluted for safe swimming and fishing when the CWA became law in 1972; thirty years later, that figure was reduced to 40 percent.[11] I should note that the EPA survey included less than 20 percent of the nation's 3.7 million miles of rivers and streams. The data from this sample indicate that agricultural runoff from farms and feedlots is by far the greatest source of pollution, dirtying half the watercourses that were impaired. The next-greatest threat was "hydrological modification," meaning dams, dredging, and channelization.[12] The EPA's most recent update for the national inventory indicates that these figures have not improved since the 2000 report, and agriculture remains the biggest source of pollutants.[13]

A more disturbing picture was painted by a comprehensive survey of EPA data by the *New York Times* in 2009. The newspaper found that 40 percent of community water systems had violated the Safe Drinking Water Act in the past year, and 23 million people had ingested water from municipal systems that violated health standards. The *Times* survey cataloged 506,000 violations of the Clean Water Act since 2004; only 3 percent of the violators were fined.[14]

The U.S. Geological Survey also measures water quality from a broad sample of streams, rivers, and groundwater. The agency found fifty-two different pesticides in its samples; 94 percent of its water samples contained at least one pesticide, including half the wells it sampled. The report noted that these chemicals are usually present in small quantities that are considered "safe."[15] A subsequent analysis found that pesticides were present 90 percent of the time in the sampled water sources. This report also noted that the concentrations would "not necessarily cause adverse effects on aquatic ecosystems or humans," but then cautioned that concentrations "frequently exceeded aquatic-life benchmarks."[16] But the bottom line is that nearly everyone in America is ingesting small quantities of chemicals designed to kill other species, and, according to the Government Accountability Office, the EPA "lacks adequate scientific information on the toxicity of many chemicals that may be found in the environment."[17]

In recent years water pollution from agriculture has created new challenges. One of them is factory farms—massive livestock operations that produce 220 billion gallons of liquefied manure each year. In 2002 the EPA began regulating over 15,000 of these "Concentrated Animal Feedlot Operations" (CAFOs).[18] One of the results of the increases in agricultural pollution is the enlarging "dead zone" that has formed at the mouth of the Mississippi River. This hypoxic flow enters the ocean and then moves west along the coast.[19] It is now the size of New Jersey and threatens one of the richest fishing areas in the country.[20] It has grown larger in part because of the CAFOs and also because of federal subsidies for ethanol, which spurred intense production of corn in the Upper Mississippi Valley.[21] The massive BP oil spill of 2010 greatly amplified the damage.[22] Chesapeake Bay is facing similar problems as a result of both agricultural and urban pollution.[23]

There are several sources of pollution related to mining and drilling activities. Coal mining, including the type that removes entire mountaintops and deposits the refuse in streams, is a threat to water quality in places as diverse as Montana and West Virginia.[24] In Appalachia, coal mines inject a toxic stew called "slurry" into the ground, which then leaches into drinking water.[25] Hydraulic fracturing, or "fracking," has become a problem in numerous areas.[26] In western states, coal-bed methane drilling has produced large amounts of tainted water.[27] Mining operations have also created problems with cyanide used in gold mining, selenium, asbestos, and other toxics.[28] Water quality is also affected by the estimated 500,000 abandoned mines in the western United States.[29] Finally, a recent Supreme Court decision gave mines permission to dump millions of tons of "fill material," that is, waste excavated from mines, into lakes and rivers.[30]

Mercury concentrations are also reaching dangerous levels. A recent study by the U.S. Geological Survey found this neurotoxin in the fish of every stream it sampled across the United States; in one-third of them the level of mercury exceeded safe limits. Some mercury occurs naturally, but it tends to concentrate when wetlands are greatly reduced.[31] It is a waste by-product of mining operations, coal-fired power plants, and cement factories.[32]

Another threat to water quality is pharmaceuticals. Many communities are experiencing problems with drugs in their water supply that affect aquatic species, with unknown effects on humans.[33] For example, male fish in the Potomac River ingested so much estrogen—from birth-control pills—that they started morphing into females.[34] Could it have the same effect on humans? If so, we may finally get more women into the Congress—one way or the other.

Another growing problem concerns the increasing amount of turf grass in the United States, which has become the largest "crop" in the nation with the dramatic explosion of suburbs. About 40 million acres are now covered in lawns, which are routinely doused with heavy concentrations of chemical fertilizers and herbicides.[35] Home owners often think of a deep green monoculture lawn as healthy, but in fact it is a product of intense chemical interference with natural flora and pollutes our water supplies.[36]

Some threats to water quality are long-standing problems, such as high salinity caused by irrigated agriculture.[37] Others are holdovers from the Cold War, such as the radioactive contaminants leaching into groundwater at Oak Ridge, Tennessee, Hanford, Washington, Yucca Flat, Nevada, and old uranium-mining sites.[38] Other pollutants are from new chemicals of which we have little understanding; the Safe Drinking Water Act regulates only 91 of an estimated 60,000 chemicals in use.[39] An example is perchlorate—an ingredient in rocket fuel—that is now found in nearly every sample of women's breast milk.[40] Another is hexavalent chromium, which is found in municipal water supplies across the United States.[41]

No part of the country, however remote, is immune from water pollution. A 2008 EPA study found toxins in fish from every state, covering 43 percent of the nation's lake acreage and 39 percent of the nation's river miles.[42] Even the fabled Blackfoot River in Montana, the focus of Norman MacLean's novel *A River Runs Through It*, was inundated with toxic mining sludge when a holding dam broke in 1975. It killed everything in the river for miles. By the time Robert Redford's film was produced, the film crew had to find another river as a stand-in; the Blackfoot was too polluted.[43]

According to the river advocacy group American Rivers, we still dump 860 billion gallons of partially treated sewage into our waterways and costal areas every year. That is a lot of health-threatening stench, and it will take a large pot of money—estimated at nearly $1 trillion in the next twenty years—to clean it up.[44] But we have made progress, thanks to the Clean Water Act and advocacy groups that fought hard for cleaner waters. In this chapter we will visit three cities—Atlanta, Washington, D.C., and Seattle. Each city has committed grievous sins against a local watercourse, and each city has made dramatic efforts to correct past mistakes and restore poisoned rivers. The fact that these cities polluted a local waterway is not at all unique; most metro areas in America did the same thing. What is fascinating about them is their dramatic reversal of course, initiated by stalwart instigators and eventually championed by city leaders. In each case, citizens took matters into their own hands and changed the hydrological environment of their hometown.

SONG OF THE CHATTAHOOCHEE

Atlanta's water travails are partly due to the whims of history and weather, but they are also a result of myopic vision, irresponsible water policies, and a landscape that was as generous in space as it was parsimonious in water. The city was founded as a rail-transportation hub. The railroaders wisely built their tracks along ridgelines to avoid flooding and soggy ground. As a result, the great joining of these railroads, which became downtown Atlanta, was not near a major watercourse. Indeed, the meek but lovely Chattahoochee River was 10 miles away and often dried up completely in the late summer. The river drained a small river basin that rose in the hills of northern Georgia and flowed across a vast escarpment of impermeable granite. In short, while other cities were located so they could access water, Atlanta was located to avoid water.

Until after World War II there was not much fuss about Atlanta's water situation. It was thought that Peachtree Creek and a few other minor tributaries of the Chattahoochee would always have enough water for a modest rail crossroads. And modest it was. Founded in 1840, the "city" of Atlanta was just a small burg of 10,000 people at the beginning of the Civil War. When Sherman got through with it in September 1864, it wasn't even a small burg anymore. Although the city rebuilt, it remained a small southern outback for another 100 years.

In the late 1940s, amidst the postwar building boom, the Corps was requested to build a multidam, multipurpose project on the Chattahoochee River. The Corps drew up a plan, largely based on a study completed in 1928, that focused on navigation, hydropower, and flood control. The centerpiece was to be a dam on the upper Chattahoochee River, called Buford Dam. If the project happened to enhance urban water supply, that was fine, but that was not an authorized purpose of the project. During congressional hearings on the proposed project, several congressmen pointed out that a dam on the upper Chattahoochee would incidentally benefit Atlanta's growing need for drinking water, and therefore the city should help pay for it. One congressman in particular, a young upstart by the name of Gerald Ford, felt strongly that Atlanta should cover some of the cost of the dam, estimated at an impressive $47 million.[45]

But Atlanta, led by Mayor William Hartsfield, refused to pay anything. Hartsfield strongly rejected the idea that Atlanta should contribute to construction costs. He said that Atlanta had lots of other sources of water, and he didn't want the city to be "put in the category with such cities as are in arid places in the West or flat plain cities where there is one sole source of water."[46] So the dam would be built to meet other purposes and not for Atlanta's municipal water supply.

The problem is that Atlanta actually is like those other cities, given its unfortunate location away from any major water source. That did not matter much until Atlanta became one of the fastest-growing cities in the United States, with a metropolitan population that went from zero the day Sherman left to 5 million by the turn of the twenty-first century. This put extraordinary pressures on the Chattahoochee River.

In 1956 the Corps completed Buford Dam and named the reservoir after Sidney Lanier, a former Confederate soldier and poet who penned an ode to the river, "The Song of the Chattahoochee," that celebrated its idyllic but dutiful journey to the sea. The completion of the dam initiated a long period of conflict over the waters of the river. Everyone wanted something from the Chattahoochee, which joins the Flint River at the Florida border and becomes the Apalachicola River. The Corps turned the Apalachicola into a barge channel, so it needed water to float barges.[47] The dam also produced electricity, and the Southeastern Power Administration and other power producers wanted water to run their turbines.

The Corps also had to reserve considerable space in the reservoir for flood control.[48] All of these uses were specifically authorized by the act that built the dam.

In addition to the authorized purposes, Lake Lanier became a premier recreational resource; by the 1970s it had become one of the most heavily used Corps reservoirs in the nation, and today it sees 7.5 million visitors a year.[49] The other nonauthorized use of the reservoir is municipal water supply, and several towns and the city of Atlanta depend heavily on the outflows of the dam for their drinking water, even though they have no clear legal right to the large amounts of water they are consuming, according to a recent federal court decision.[50]

And that is only half the "use" of the Chattahoochee. What goes in must come out. Atlanta's antiquated sewage system, built between 1890 and 1930, used an assortment of CSOs and SSOs. The key word here is "overflow," which means that when the system lacks sufficient capacity to handle its load, it just dumps it untreated into the local river. In the case of Atlanta and other municipalities in north-central Georgia, that meant spilling it into the Chattahoochee. And spill they did. By the 1990s Atlanta was disgorging 300 spills a year, totaling 400 million gallons, and fouling the river with untreated human waste.[51] The Chattahoochee had become the source of drinking water for millions of people, the sewage dump for numerous municipalities, and a giant playground for flat-water recreationists. It was also expected to generate a lot of hydropower and float a bevy of barges. That is a lot to ask of a small river.

The Chattahoochee has one other attribute; it happens to be quite beautiful along the stretches that have not been lined with condos and strip malls—a thin thread of natural bliss among the endless sprawl of "greater" Atlanta. And that led to another "use" of the river.

The Way God Made It

"Can we help you carry that canoe?" My wife and I posed that question to two people carrying their craft up the bank of the Chattahoochee River along a path canopied by large trees.

"Oh, thanks, but we can get it."

It was one of those noes that actually sounded like a yes. The two people set the canoe on the ground to rest. It was jungle weather—midsummer Georgia—and they were perspiring heavily. While they were standing there, catching their breath, I figured I would exploit the moment and ask them a few questions about the Chattahoochee—one of the most fought-over rivers in America. "This stretch of river, in the National Recreation Area, seems really popular. Do a lot of people canoe this stretch?

The man standing at the end of the canoe nodded affirmatively as he wiped the sweat from his brow. "Yes, it's really popular. We figure that the Chattahoochee River National Recreation Area had about 3 million visitors last year."

I was surprised by the specificity of his answer. "How do you know that?" I asked with genuine surprise.

"Well, I'm the superintendent of the recreation area. My name is Dan Brown."

My wife and I looked at each other and smiled. Then I turned back to Brown. "I'm working on a book about river restoration. I've come to the Chattahoochee for the first time in my life, and I just happen to encounter the superintendent of the national recreation area; what luck!"

"I thought you were media—with your pen and notebook in hand— and I am always happy to talk to media." Brown seemed ready to take a break and chat about the river. Indeed, protecting the Chattahoochee— or at least the unspoiled parts of it—has been his full-time job since he became superintendent in 2007. It took just a little prompting from me to initiate an on-the-spot interview about the river. I learned that in addition to all the exploitive demands on the river, there is also a pressing need for the river to act as a calming device amidst the crowded fury of the city; that the river offers a healing balm to millions of Atlantans who want to escape into a verdant, soft-hued world for a few hours.

President Jimmy Carter recognized this need and in 1978 signed legislation to designate certain sections of the Chattahoochee as a national recreation area. At the signing ceremony, he noted that "it's a rare occasion when within the city limits of one of our major cities, one can find pure water and trout and free canoeing and rapids and the seclusion of the Earth the way God made it. But the Chattahoochee is that kind of place."[52]

The effort to protect the river faces formidable challenges. Nearly all of the river corridor is in private hands or owned by local governments.

Chattahoochee River National Recreation Area, Georgia.

Many of the riparian areas had been developed right to the waterline by the late 1970s. But the sprawl of Atlanta had yet to burgeon to its current stretch of over 100 miles across, and there were still stretches of the river that were relatively pristine, or at least agricultural. These isolated segments of the river formed the beginning of the recreation area. In 1984 Congress increased the authorized acreage area and then increased it again in 1999. In that year Congressman Newt Gingrich got an appropriation of $25 million to buy riparian acreage from willing sellers, which helped expand the area. Today the Chattahoochee River National Recreation Area includes 5,000 acres of land and 1,500 acres of river along a 48-mile stretch of river. Most recently the National Park Service, working with the Trust for Public Land, purchased the Hyde Farm property along the river, adding 53 acres. The protected segments of the river begin below Buford Dam and continue to the outskirts of the city of Atlanta.[53]

After chatting with Superintendent Brown, we wandered down to the river's edge just as a large family was "putting in" using a variety of

straight blondish hair and an angular face she still looks the part of a young and exuberant instigator with that fire-in-the-eye projection and the combative self-assurance of someone who has accomplished the impossible and plans to continue doing more of the same.

Riverkeepers have sprung up all over the United States, each dedicated to a particular river or bay. In the mid-1990s there was a riverkeeper for the lower Chattahoochee, but none on the upper river. Ted Turner's daughter, Laura Turner, was interested in protecting the upper Chattahoochee and met with Robert Kennedy Jr., who was deeply involved in the riverkeeper movement in the Hudson Valley (he became the first president of the Waterkeeper Alliance in 1999). The result of that meeting was a decision to start a riverkeeper organization for the upper Chattahoochee River. Funded primarily by the Turner Foundation— Ted Turner's fund for assisting river preservation—the new organization hired Sally Bethea as its first director in 1994.

It became immediately apparent to Bethea that if there was going to be a future for the river, the organization had to confront Atlanta regarding the enormous amount of sewage it was dumping. "We did not form the Upper Chattahoochee Riverkeeper to sue the city of Atlanta, but we knew that would be one of our greatest battles, and we knew it was going to be a political hot potato."

I asked her why it was necessary to start a new river group; were there no existing environmental groups that could protect the Chattahoochee?

"Back in the 1990s it was considered unseemly to sue the city over water pollution. There were not many staffed enviro groups in Georgia at that time and very few were willing to file lawsuits. Most of them had a policy to not get involved aggressively. But the Riverkeeper can take direct legal action; that's what is good about the riverkeeper model."

Bethea and her organization first tried meeting with Mayor Bill Campbell, a guy Sally described as a former inmate "in the Big House." They got nowhere. According to Sally, they discovered that the city was collecting sewage fees but not spending that money on maintaining the city's sewers. Campbell did not want to make fundamental changes. So they convinced the EPA and the state of Georgia to join them in a lawsuit alleging that the city was violating the Clean Water Act. It was a slam-dunk case; anyone with a pair of waders and a modicum of olfactory sensibility could tell what was going into the river. In 1998 a federal judge ruled against the city, which led to a consent decree that committed the city to clean up its CSOs. In a subsequent decree the city also agreed to solve the problems related to its SSOs. The decrees also committed the city to cleaning up garbage along 37 miles of tributaries—the city carted away 568 tons of it—and to purchasing land for a riparian greenway acquisition program.[55] Thus far 1,887 acres have been acquired for greenways.[56]

The consent decrees changed everything, but they were only the first step in a long process of bringing the Chattahoochee and its tributaries back to life. "We got a lot of great media coverage from the decrees," Bethea explained. "But first we had to figure out how bad the problem was. The city didn't even know where some of its pipes were, or how badly they were leaking. And people began to understand that this was going to be an expensive fix."

"Expensive" is a bit of an understatement. Big cities create big sewage problems. The city adopted a multipart strategy to meet the requirements of the decrees and rebuild its water-supply system, with a price tag hovering around $4 billion. The first task was to drill giant tunnels, 18 feet in diameter, through the solid granite under the city. These tunnels absorb the excess runoff and store it until it can be treated. The second strategy was to rebuild some of the streams' natural capacity through the greenway program by opening up floodplain and allowing some water to soak in through unpaved areas. Third, in some areas of the city where it was practical, sewage would be separated from storm runoff. The city's rebuild of its crumbling drinking-water system would also solve problems for a system that was plagued with leaks and so prone to contamination that the city had to issue repeated boil-water advisories.[57]

Funding and building such a large system take money, vision, leadership, and tenacity. In 2001, just as the city was beginning this herculean task, Shirley Franklin was elected mayor. She immediately recognized the problem, accepted responsibility for taking action, and dubbed herself "the sewer mayor." Her goal was ambitious: "I was determined that Atlanta not merely comply with the Consent Decrees, but actively embrace their requirements."[58]

Mayor Franklin immediately made two major changes to the city's water governance. First, she created the Department of Watershed Management, which combined all of the city's water-related functions into one agency. Second, she reacquired the drinking-water system, which had been sold to a private entity, with disastrous results. Her next task was to raise additional revenue—always a difficult task. But in 2004 she pushed through an increase in sewer fees, and that same year voters approved an increase in the sales tax to cover the water system's overhaul.[59]

The effort has paid off. The CSO improvements were completed in 2008, right on schedule, prompting the federal judge who oversaw the consent decree to call it "a remarkable accomplishment."[60] The new SSO system will come online in 2014, as required by the consent decree.

This does not mean that the Chattahoochee is free of danger. Other municipalities have not cleaned up their sewage systems, and perhaps the greatest threat to the river concerns water quantity. Georgia emerged from a long drought in 2008, which made it abundantly clear that the Chattahoochee cannot provide enough water for endless growth—or even for the existing population in dry years. And the recent district-court decision made it clear that other water users downstream are not going to play dead and let Atlanta suck up all the water. The Georgia legislature finally got that message and passed one of the most aggressive water-conservation laws in the country in 2010.[61]

The goal of the Waterkeeper is simple: "Our vision is that communities of people and wildlife have enough clean water throughout the Chattahoochee–Apalachicola River system."[62] Achieving that goal will require an unprecedented level of cooperation and vision—and Sally Bethea.

On our last day in Atlanta, my wife Jan and I went to breakfast at Ray's on the River, an upscale restaurant sandwiched between the Chattahoochee and an interstate highway that was packed with speeding motorists even on a Sunday morning. After our meal I wandered down to the river's edge and marveled at the contrast between the languid flow of the water and the hurried crush of modern Atlanta. Sidney Lanier's poem came to mind:

Run the rapid and leap the fall
Split at the rock and together again
Accept my bed, or narrow or wide
And flee from folly on every side

The river is still trying to flee from folly on every side.[63]

TOO MANY FEDERAL TOILETS,
NOT ENOUGH SENATORS

We were in the "bad" part of town on a "forgotten river." Lee Cain, of the Anacostia Watershed Society, piloted the boat slowly downstream. "See that little creek over there? That's Dueling Creek; it's just outside the Washington, D.C., line. Dueling was illegal in the District, so men came out here to duel. Commodore Stephen Decatur ['My country right or wrong'] was killed there in 1820. There's so much history on this river, it's just amazing."[64]

The Anacostia River is one of those waterways that helped develop the new nation but then slowly slipped into neglect and abuse. When the British invaded in 1812, they had to get through Bladensburg, Maryland, to get to Washington so they could burn the White House. Bladensburg at that time was a bustling port on the Anacostia River, handling tons of tobacco and other goods. British troops took over one of the larger buildings in the town, a tavern called the Indian Queen, built in 1752, that had been frequented by George Washington. Local militia took shots at them while First Lady Dolly Madison, staying in a nearby house, took flight back to Washington.

Today Bladensburg is no longer a port—the upper river is filled with silt because of poor agricultural practices and no longer has sufficient depth. But a small marina and river park here mark the beginning of the 8.5-mile-long intertidal run of the Anacostia River. From here, the river enters the federal district and flows through a part of town that is rarely mentioned in glossy travel brochures about Washington. Here, along the banks of the Anacostia, live some of the poorest people in the region, nearly all of them people of color. And while the Potomac River has received much attention and federal clean-up funds, the Anacostia has struggled with its status as a "forgotten river" in a poor section of town.

All of that is about to change, and in a big way, thanks in large part to the Anacostia Watershed Society. Its headquarters is in the historic structure in Bladensburg that once housed the Indian Queen Tavern; where George Washington once walked, river activists like Lee Cain now tread, anxious to dispel the notion that the Anacostia is a forgotten river without a future.

Lee grew up on Chesapeake Bay. As a boy he could easily catch a bushel of crabs, but as the years went by, Lee watched the Bay decline in dramatic fashion, burdened with toxic runoff, silt, overfishing, and overdevelopment. After earning a college degree in physics, he decided to move to the District and devote his time to reversing the damage done to the area's rivers, which flow into the Chesapeake. He speaks of the Anacostia as though it were an honored but aging family member. "Captain John Smith, after his life was spared by Pocahontas, sailed up the Anacostia. You know, that was a long time ago. This became a really busy river, but by 1850 the silt had clogged it up so much that big ships could no longer get up here. That was the beginning of the problems for the river."

In the 1930s the Civilian Conservation Corps (CCC) built retaining walls along the river to channel the flow and drain riparian wetlands— these walls are still in evidence along the lower river—and this hastened the decline of the riverine wetlands. As the river basin developed, the Anacostia became a handy dumping ground for sewage and garbage. The city of Washington's antiquated sewage system routinely spilled its stinking offal into the river. As the neighborhoods in the area declined, the river went with them. In 1938 a child wading in the river contracted typhoid fever.[65] It became hazardous to eat the fish from the river, many of which displayed tumors and deformities due to the poor water quality. The Washington Navy Yard, as old as the nation itself, routinely discharged pollutants into the river, including polychlorinated biphenyls (PCBs). By the late twentieth century the District had no less than seventeen CSOs dumping untreated sewage and storm runoff—2 billion gallons of it every year—directly into the river.

But that repulsive image was in stark contrast to what I saw when Lee and I motored quietly down the river. "This is not what I expected," I told him. "I anticipated a river lined with tenements, warehouses, and abandoned cars, but this is beautiful." I had read the sordid history of the river, with its overtones of classism and racism—who cares about a river in this part of town? But what I saw as we quietly worked our way down the river was a verdant corridor of peaceful water and dense riparian vegetation, as though we were miles away from any human habitation.

About halfway down the river we cruised by Kingman Island, a man-made strip of land that is dense with towering trees and thick under-brush. On the southern tip of the island is a literal bridge to nowhere, started just before the outbreak of World War II but never completed because of lack of funds. This lack of access helped prevent development of the island. Recent efforts to build an amusement park on the island were stopped by river advocates, and the island has been allowed to evolve into a natural state.[66]

"Are you sure we are on Washington's southeast side?" I asked incredulously.

"We are," Lee assured me, smiling. "See that area over there?" He pointed to a densely vegetated wetland beside the river. "That's called the ANA11. It used to be an unofficial landfill—people just brought stuff down here and dumped it. But when the government rebuilt the Wood-row Wilson Bridge, downstream, they destroyed some wetlands, so they had to replace them with wetlands elsewhere. So the state of Maryland and NOAA [the National Oceanic and Atmospheric Administration] converted the old dump into a thriving wetland—remember that 96 per-cent of the Anacostia's wetlands have been drained. The Society has also done a lot of work restoring riparian areas and replanting native vegeta-tion such as wild rice and spatterdock."

On the day that Lee and I were on the river, we passed numerous sculls being oared by schoolkids. Each sculling crew was a collection of children of all races—a visible embodiment of how a river can bring people together, pulling in unison. Lee mentioned that the Society hoped to increase recreation on the river, which prompted a question from me regarding the trash that was still visible on the water. Lee explained that the Society's volunteers had gathered literally tons of garbage from the river, and then he pointed to a small side stream.[67] "We pulled 6,500 tires out of that stream a few years ago."

"It's still a bad idea to get in the water, right? Fecal coliform has a nasty ring to it."

"I've fallen in a few times; I just go home and shower really well. But yeah, you don't want to get in this water." Lee grinned, but we both knew that the health threats in this river are very serious.

As we motored along, we saw about a dozen people fishing from the banks; these were not upscale fly fishermen casting $400 Scott fly rods but humble local folk. I asked Lee if they had any idea how risky it was to eat fish from this river.

"I'm an environmental educator, and it's part of our job to inform people about those dangers. We take hundreds of schoolchildren down the river and teach them what the river has to offer and what the problems are. Those people fishing—well, we're working on educating everyone about the river. But the goal is to create a river where everyone can fish and swim here without worrying about a health risk. We've worked hard to educate people about the value of this river."

The hard work has paid off. As we moved down the river and sailed under the Metro tracks and Pennsylvania Avenue, we saw numerous cormorants, egrets, and blue herons. Kingfishers darted among the trees, and beaver-felled trees could be seen along the shore. Half a dozen red-eared slider turtles basked in the sun on a log. Mallards and Canada geese competed for space and chattered noisily, reminding me of Congress.

"We've seen lots of osprey along the river—they come here to eat the shad, which are becoming more numerous. We've even seen a few bald eagles," Lee said.

"Amazing; bald eagles in the nation's capital." I savored the irony that the nation's symbol had finally returned to Washington. "The only eagles I've ever seen here were on lapel pins and flagpoles."

This description of the Anacostia sounds more like a wildlife refuge than a trashed river. Indeed, the river is both, and there has been a long, uphill battle to make it the former rather than the latter. Restoration efforts go back thirty years. A consortium of local governments signed an agreement in 1987 and formed the Anacostia Watershed Restoration Partnership, but it was not at all clear that the participating governments would pony up the money and commitment necessary to make it happen. So in 1989 a local river walker named Robert Boone started the Anacostia Watershed Society. He was determined to make the restoration of the river a reality, not just a dream. I knew that I would never understand the saga of the Anacostia without having a long talk with Boone. We met at a local brew pub for lunch and talked for nearly three hours.[68]

"Oh, they used to call me a loony tree hugger," he said wistfully. His slight accent hinted at his upbringing in the Piedmont country of North Carolina. "I used to walk the Potomac with my boy when he was little. We walked the whole length of the Anacostia. It was beautiful, but corrupted and abused."

Boone spoke with a passion that transcended politics; he saw his work on behalf of the Anacostia as a calling, the inevitable path of a spiritual quest to, as he put it, "understand the miracle of the elegant design of living things." On one of his walks with his son, he noticed an oil slick—he called it a "dead rainbow"—seeping into Hickey Run, a tributary of the Anacostia River. With a little sleuthing he discovered that the pollution came from the D.C. Metro bus maintenance yard. For years the yard's leaking underground storage tanks and surface runoff had been oozing oil and other contaminants into the river. He decided to do something about it.

Robert Boone was not an accomplished lobbyist at that time. For most of his life he had worked a great variety of jobs, ranging from psychology professor to cabinet maker. But he had worked briefly with the Interstate Commission on the Potomac River, so he had an idea of what worked, and what didn't work, when it came to restoring rivers. What did not work, in his view, were government commissions, which he described as "hogtied and blindfolded." So, with the help of a donor, he formed the Anacostia River Society with a staff of one—himself—and went to work convincing the District that it had an obligation to clean up its mess. Using a combination of embarrassing media coverage and the threat of legal action, he cajoled the District into cleaning up the bus facility.

This early victory convinced Robert that the Anacostia could have a bright future, but early on he realized that the river's condition was not just an environmental problem but was also a question of social justice. "Remember that most of the pollution originates in D.C. and upstream in white communities. But the feds, the state of Maryland, and the counties just ignored it; they'd say, 'Oh, that's a D.C. problem down in the Black community.'"

It was obvious to Robert that the Anacostia Watershed Society would not be effective unless it forged a partnership with that community. He described going to a local neighborhood meeting, the only white person in the room, and trying to convince the attendees that they should help restore the river. "They told me that environmental issues were for little old white ladies and the Sierra Club, and they had to deal with other issues such as crime and keeping their kids in school. I was told that over and over. And I was faced with distrust, not because I'm an environmentalist, but because I'm white. I told them I was there to represent the river, and the river was in their neighborhood, so we were natural allies. But it took a long time to get credibility with the community."

The turning point came when Robert began working with a local social worker named Brenda Richardson; they both understood that the way to effect change was to start with children. "We started this project called Eco Patrol, where we'd take kids out and pick up trash in the community and plant trees. Word got out about that. I started being perceived as a relevant player."

At about the same time, a new threat confronted the Anacostia River. The wealthy owner of a professional football team[69] announced that he wanted to build a new stadium on the banks of the Anacostia. The stadium and its attendant roads and parking lots would have had an enormous impact on both the river and nearby communities. The Watershed Society, working closely with black community leaders, managed to stop the stadium.

"We held demonstrations, we put on our ties and lobbied Congress. Together, we stopped a wealthy, well-connected white man from despoiling the community and the river. This gave the Society 'street cred' in the black community."

What the Society lacked in money and size, it made up in sheer chutzpah. In 1996 it decided to sue the U.S. Navy. Jim Connolly, a staff member at that time, has vivid memories of when it decided to take on the Navy. He had been with the Watershed Society for only a few years, but before that he had worked with impoverished kids in Washington, D.C., so he was familiar with difficult challenges. Connolly's work with kids from, as locals call it, "the east side of the river" (meaning the poor side of town) prepared him well for his work on the Anacostia and shaped his view of the river as an environmental justice issue.

"One of the things embedded in this river is the racial dynamic. It's like the big elephant in the room that no one wanted to talk about."[70] The lawsuit against the Navy pitted the tiny environmental group, allied with poor people, against a behemoth. "We only had a staff of three or four, and we knew it would be a David vs. Goliath battle, but we went ahead with the lawsuit because we knew we had a strong case. Even the commander of the Navy Yard saw the need to clean up their base, and said he wanted the Navy to be part of the solution rather than the problem. So we signed a settlement agreement with them, and they have become one of our best allies."[71]

After its victory in the Navy case, the Society sued the District of Columbia in 1999. Connolly explained why it took so long to get the District to meet its obligations:

> There is federal money available to states to clean up their rivers because they have senators and congressmen that get money for their home states and districts. But D.C. has no voting members of Congress, so we had a hard time convincing Congress that they needed to spend money on cleaning up this river. And the irony is that the feds created most of the problem; there are literally thousands of federal toilets flushing into the Anacostia watershed, but they didn't want to put up any money. There were just too many federal toilets, and not enough federal senators who wanted to help.

So the Society sued, working in conjunction with a local neighborhood association called the Kingman Park Association and other stakeholders. Again, it won big-time and forced the District's Water and Sewer Authority to agree to build an up-to-date sewage-treatment system. It will be expensive—something like $2.5 billion—but it is a necessary step in returning the Anacostia, the Potomac, and Chesapeake Bay to an acceptable level of health and productivity. The Society also sued upstream polluters in Maryland, and again it won.

To complement its courtroom victories, the Society has also covered its political bases. In 1991, in the early days of the Society's existence, it invited two dozen influential politicians to see the river firsthand. These people were used to being feted at posh locations, but on a cold, rainy April day they huddled together on the Society's boat to show their support for cleaning up what the *Washington Post* called "D.C.'s Dirty Little Secret." The group included a senator, two congressmen, county officials from Maryland, and two members of the District of Columbia City Council.[72] These trips became a valuable way for the Society to acquaint local people, from schoolchildren to powerful elected officials, with the river. Senator Benjamin Cardin of Maryland took one of these trips and later introduced a bill that directed the federal government to work with Maryland and Washington, D.C., to develop a 10-year plan to restore and protect the Anacostia River.[73] Congressman Steny Hoyer was so impressed that he helped the Society organize a special river trip for the congressman's friends in 2007. I'm sure many of them experienced the same startling surprise that I did when I visited the river. The latest victory is the passage of a new bag fee in the District that charges five cents for each plastic bag handed out by retailers. It has already had an impact on the river.[74]

When Robert Boone started the Anacostia Watershed Society over twenty years ago, he knew he was up against difficult odds. But he is a classic instigator, unfazed by controversy, adversarial encounters, or biting criticisms. His perseverance has been due to his spiritual faith and a belief that the restoration of the river is an important part of a larger movement to sustain the planet. In the Society's very first newsletter, he wrote: "Parts of the Anacostia remain green, verdant, full of life and beauty. The Society exists because the Anacostia is the little corner of spaceship Earth that we are going to care about."[75] The Anacostia is no longer the forgotten river.

SUPERFUND RIVER

Elliott Bay, just south of the glittering skyline of the Seattle waterfront, has seen years of hard use. It is at the center of Seattle's industrial economy, producing and shipping goods to all corners of the world. But a heavy price has been paid for this economic success story. A string of superfund sites mars the south end of the bay, where the Duwamish River is split by Harbor Island—a place so polluted that part of it had to be capped with asphalt, fenced, and declared too toxic for human visitation. The Duwamish River itself is a superfund site because of heavy concentrations of PCBs and other poisonous chemicals in the sediment.[76] It is an unlikely site for a river-restoration project, but the river is making a very gradual comeback thanks to years of carefully selected restoration activities.

I was invited to kayak the river with two employees of the city of Seattle, Jennie Goldberg and Judith Noble, who have been leading the effort to clean up the river.[77] We began our kayak tour of the Duwamish at a point that could only be described, charitably, as scenically challenged. The put-in was under an elevated highway. On one side a surface highway was jammed with an endless line of trucks; on the other side a freight train belched smoke. The flyway for the Sea-Tac airport and Boeing, directly overhead, added to the cacophony. Some homeless people, camping in the weeds, watched us with great curiosity. We walked past scattered trash and over a discarded carpet to reach the oily banks of the Duwamish. All the senses were affronted. I made an effort to get into my kayak without letting my feet sink into the sediment, trying to remember what PCB stood for.[78]

But the absence of an attractive context stood in stark contrast to the ebullience of Jennie and Judith. They were absolutely delighted to be on the Duwamish again and obviously felt a sense of kinship for this river that has absorbed so much of their time and energy. This tour gave them the opportunity to showcase the progress that has been made in partially restoring the river and its riparian area despite enormous challenges and difficulties.

The effort to restore the Duwamish and its main tributary, the Green River, began very gradually. There has long been a sensitivity in the Northwest to the health of local rivers; this is, after all, the home turf of the famous Chief Seattle. The successful cleanup of Lake Washington in the 1960s —once heavily polluted with sewage—gave people a sense of what could be accomplished if various political jurisdictions, prodded by citizen groups, worked together. In the early 1980s the Port of Seattle proposed an industrial development for an island in the river, Kellogg Island. At that time the island was about the only green natural space on the entire lower Duwamish. On a river where 98 percent of the intertidal wetlands and marshes have been destroyed, the remaining 2 percent become quite important. Citizens demanded that the island be spared, and the city responded by canceling its development plans. As in so many river restorations, a few small early victories served as a catalyst for more ambitious projects.

The idea of restoring parts of the Green and Duwamish rivers received another boost in the early 1980s when a feisty Vietnam veteran, John Beal, began agitating to clean up Hamm Creek, a tributary of the Duwamish. Beal started a nonprofit group called "I'M A PAL" and won an environmental award from the Environmental Law Institute in 2003 for his restoration efforts.[79] Beal passed away in 2007, hailed as the "hero of Hamm Creek."[80] By that time there were many people active in the campaign to restore the Duwamish; it takes an army of John Beals to successfully restore a river.

In addition to citizen action, cleanup efforts were greatly aided by a successful CERCLA (the "superfund" law) suit brought against the city of Seattle and King County by the federal government in 1990.[81] A 1991 consent decree set in motion official efforts to remediate contaminated sediment and restore the natural habitat.[82] This did not stop the EPA from declaring in 2001 that the river itself was a superfund site because of sediment contamination. Restoration efforts were also boosted when the chinook (king) salmon and bullhead trout were declared "threatened" under the Endangered Species Act in 1999. At one time the Duwamish/Green watershed supported large salmon and trout runs, but only a remnant of those runs remains.[83]

The challenge of restoring the river is formidable. At the put-in, Judith provided several maps of the river, one of which showed the historic meanders of the river before it was straightened and channelized. Other rivers—the Black, the Cedar, and the White—used to flow into the Duwamish but were diverted into other river basins, greatly reducing the flow. The current channel is heavily dredged, the banks consist of steep riprap, and the watercourse is filled with a great variety of ships, wharves, barges, and decaying piles.

To the untrained eye, it was difficult to discern how the river had changed in response to the restoration activities. Judith pointed to a new wharf built by the Muckleshoot Tribe. "They used a metal screen rather than planks for the decking on the wharf; the screen allows sunlight to pass through to the intertidal land below, which improves fish habitat." At another point along the river she pointed out that the rock used for the riprap was smaller than usual. "We got them to use smaller rocks along the banks below the high-water line; it is more fish-friendly." In other places the steepness of the bank has been reduced to improve habitat.

As we kayaked upriver with the tide, the first real sign of significant restoration progress was Kellogg Island. The effort to save it from development many years ago has paid off, and today Kellogg is literally and figuratively an island of verdant nature in a sea of industrial development. Blue herons like to frequent the island, and the northern end offers a large area of intertidal wetland. Further upriver we began seeing pocket parks that have been built in several places. Jennie noted how popular these were with the local neighborhoods.

Nearly all the restoration on the Duwamish must take place within the confines of existing infrastructure. Just as the Charles River and Los Angeles River restoration projects (see chapter 9) are severely limited by their surroundings, the Duwamish restoration must take place in a manner that does not greatly diminish the economic activity in the area or significantly disturb transportation and service delivery. In such a context, a great deal of careful planning and innovation, tempered by political reality, is required. That does not mean, however, that not much is going on. A coalition of the City and Port of Seattle, King County, and the Boeing Company (which owns land along the river) has worked diligently since 2000, spending $70 million to clean up the contaminated sediment. And over the past twenty years the City and Port of Seattle have spent $28 million to restore habitat along the river.[84] There are dozens of additional projects upstream on the Green River.[85]

The fruits of all this labor were not often evident as we kayaked the 5 miles of the shipping channel, in part because much of the progress concerns what is not present. Judith and Jennie pointed to the outlet of the Diagonal Combined Sewer Overflow (a CSO) and noted that the city no longer discharges untreated runoff from that pipe. Much of the contaminated sediment in the river has been removed or capped. But not all progress is invisible. There is beauty in this river, which becomes apparent at a wide spot in the channel with the mundane name of the Turning Basin. Once the site of sunken wrecks and detritus, it now has an almost parklike feel to it. Kingfishers dart along the surface of the water, and a pair of nesting osprey keep vigil from a man-made platform far above the water.

There is a lesson in the long, difficult effort to restore parts of the Duwamish: even the most degraded and developed rivers can regain a semblance of life and vitality with sufficient courage, commitment, and dialogue. It is tempting to think of river restoration as a process where rivers are returned to nature, and in some cases that is possible. But society can also benefit from simply making rivers a little more natural, a little more appealing to humans and other forms of life, and a lot less of a threat to our health. River restoration comes in many different shapes and dimensions, just like rivers.

PURE RIVER OF WATER OF LIFE

The nation's effort to cleanse its rivers has been only partially successful, for three reasons. First, forcing polluters to clean up their act increases their costs; it is cheaper to send your waste downstream and make it someone else's problem than to accept responsibility for the problem and solve it locally. Second, in regard to nonpoint sources, water-quality policy has tended to focus on cleaning up the mess after it has occurred rather than solving the problem at its source. We have spent hundreds of millions of dollars on water-treatment plants, but precious little on preventing nonpoint pollution in the first place.

Third, we still have a mind-set that allows a disconnect between the quality of water we want and what we put into the water. We will never achieve the lofty goals of the Clean Water Act until we adopt a new ethic that focuses on individual, municipal, and corporate responsibility for preventing water pollution, rather than depending on government to clean it up after the fact. This requires a cognizance on the individual level about everything from the chemicals used on lawns to the things we pour down the drain. It also entails a consciousness about what we eat; organically grown food may appear more expensive, but what are the "costs" associated with a whole nation, especially children, ingesting agricultural chemicals?

At the municipal level, the cleanup requires an attitude like that of Mayor Shirley Franklin of Atlanta—a commitment to the spirit as well as the letter of the law. That means a willingness on the part of city leaders to charge sufficient fees to enable their city to adequately treat their sewage and storm water. That is a tall order; the EPA estimates that there is a gap of $500 billion between what is being spent and what should be spent on urban systems.[86]

At the corporate level, it requires a sense of civic responsibility to prevent pollution from reaching the watercourse and not relying on downstream utilities to clean up after them. When my son was growing up, we had a basic rule about caring for our home: you mess it up, you clean it up. This should be the heart and soul of our laws protecting water quality.

Finally, we must begin to view water holistically. The quality of water is related to all the other uses of water. Water diversions concentrate pollutants; dredging for barge channels destroys natural filtering processes; dams concentrate pollutants in slow-moving water; flood control cuts off rivers from the natural filtering capacity of floodplains, as does the destruction of wetlands. Meeting the lofty goals of the Clean Water Act will require a reappraisal of all aspects of water policy.

The Bible both begins and ends with references to the quality of water in rivers. In Exodus, God punished the Pharaoh by turning the Nile River into blood, "and the fish in the river shall die, and the river shall stink, and the Egyptians shall loathe to drink of the water of the river."[87] And the last chapter of Revelation begins, "And he shewed me a pure river of water of life, clear as crystal, proceeding out of the throne of God and of the Lamb."[88] The purity of water and its healing power play an enormous symbolic role in Judeo-Christian tradition. This is perhaps best illustrated by the custom of baptism by immersion. The purpose of this ritual is to cleanse the soul and the body and rid them both of unwashed impurity. The entire concept is based on the notion that water heals the soul by washing away our sins. To accomplish that, we must have clean water in our rivers.

III

RESURRECTION

In the early years of our country we lived next to rivers, drank from them, and ate from them. We lived in harmony with our rivers. But the sheer weight of our numbers and a focus on exploitation and extraction sundered that relationship. Rivers were viewed as a limitless resource, as were the prairies, the game, and the forests. But when we realized that the rivers and the game were almost gone, we entered an age of limits, an age of understanding that there must be a balance between what we take and what we leave intact. This section of the book is about that newly acquired sense of balance, where we appreciate the role of rivers as both human and nonhuman habitat, play raucously in or by rivers, and rediscover the value of a clean, healthy, flowing watercourse.

9

RIVER CITY

Urban Riverscapes

Yet, fair as thou art, thou shunnest to glide,
Beautiful stream! by the village side
—FROM WILLIAM CULLEN BRYANT, "GREEN RIVER"

In 1969, when Neil Young sang about shooting his girlfriend down by the river on his best-selling album *Everybody Knows This Is Nowhere*, there was a twisted logic to the lyrics. In those days, in most American cities, the riverfront was a good place to go if you wanted to commit some heinous crime. The river itself was polluted and unfit for decent human activity, and urban riparian zones often consisted of abandoned warehouses and dilapidated buildings, crowded with failed dreams and moral lapses. When Tom Murphy, former mayor of Pittsburgh, was growing up, his mother would warn him, "Make sure you're home before dark and don't go near the rivers."[1] Down by the river was the no-good part of town. Think of Jimmy Hoffa and concrete shoes.

It had not always been that way. Before modern roads, rivers were the only viable mode of transportation. And without modern pumps and reservoirs, citizens drew their water directly from rivers. So cities faced the waterfront—that was where the action was. All of America's greatest nineteenth-century cities were located next to a major watercourse. The illogic of placing megacities in the middle of a desert with no discernible source of water had yet to become fashionable.

But our river orientation began to change with three significant developments. First, we developed an extensive network of highways and railroads. Second, the U.S. Army Corps of Engineers and other government agencies literally built walls between cities and their rivers for flood control. Third, we abandoned many of our inner cities and moved to the suburbs. This was due to racial animus, an ethic of consumption, and a sense that we would rather be walled off from our neighbors than part of a neighborhood. America secured itself behind sliding-glass patio doors and left its historic downtowns, which were always along rivers, to those who could not afford a split-level with a two-car garage.

In the 1970s American cities gradually began to see the benefits of a revitalized waterfront, for several reasons. In the industrial age factories were near rivers to exploit their transportation, their energy, and easy access to water for cooling and lubrication. As factories closed, riverfronts became available for other uses, often at bargain prices. These former industrial heartlands could be transformed into desirable places to work, play, and live.[2] River restoration is also a critical part of the "new urbanism" that is transforming downtown areas into sustainable, walkable, carbon-efficient communities. Riverfront property can be redeveloped with an emphasis on these themes while also incorporating the recreational and scenic attributes of riparian corridors and historic buildings.[3] This dramatic return of the urban waterfront is happening all over the developed world.[4]

Most major cities in the United States today have some form of riverfront-restoration program. These programs have required an expansion of our concept of "urban water supply" to include also the amenity value of urban rivers. Even in arid regions, the amenity demand for rivers can be met while also meeting reasonable demands for consumptive use. To get rivers to meet both consumptive needs and amenity demands requires innovative water management, creative efficiency and conservation measures, and increased attention to water quality. In this chapter we will look at a small sample of the urban river-restoration projects in the United States, including the history-based project in Richmond, Virginia, the ambitious plans for Boston's Charles River, and the seemingly impossible task of bringing a little life back to the Los Angeles River. We will also look at the removal of Embry Dam on the Rappahannock River.

BELLE ISLE

Richmond, Virginia, had a unique reason to rehabilitate its waterfront on the James River. Richmond was the capital of the Confederacy during the Civil War—ground zero for four bloody years of conflict. Much of its history from that wrenching era is scattered along its waterfront. The city, the National Park Service, and a private donor have created a unique blend of historic structures and natural areas along the river.

Just above Richmond the James River begins to churn into white water at the fall line, where the water begins its descent from a rocky shelf to Chesapeake Bay. The 7-mile-long Hollywood Rapids are both a challenge for kayakers and a lovely natural feature—something that has become quite rare in sprawl-tainted northern Virginia. At one time much of this rapid was submerged behind power dams, but spring floods, and a hurricane in 1969, demolished sections of the dams and freed this river to flow once again over a jumble of rocks and ledges.

On one side of Hollywood Rapids is the eponymous cemetery, home to the remains of many Confederate heroes. Just downstream is the Tredegar Iron Works. During the war it produced cannon and ammunition for the Confederacy. Much of Richmond's waterfront burned before the city was surrendered to the Yankees, but most of Tredegar, which predated the war by decades, survived. Today it is administered by the National Park Service and provides a rich historical background to the James River riparian corridor, much of which has been preserved in the 550-acre James River Park.[5] Just downstream from Tredegar, the city has preserved the eighteenth-century canal that runs along the river; today it is a park and mile-long walkway that offers stunning views of the river and the green hills beyond.

But perhaps the most dramatic transformation is Belle Isle. This half-mile-long island is just opposite Tredegar and Hollywood Cemetery. During the Civil War it was a prisoner-of-war camp for captured Northern soldiers. As in all prison camps of that era, the conditions were abominable, and men died by the thousands. Their bodies were tossed into hastily dug mass graves just beyond the deadline—the demarcated zone around the prison camp that meant instant death for anyone who ventured too close. Today Belle Isle is a densely vegetated natural area, popular with joggers, sunbathers, and picnickers. Not far from where men collapsed in the painful agony of starvation, people in bathing suits now bask in the warm rays of the sun and frolic in quiet pools beside the river. The contrast between the tortured suffering of 150 years ago and the family-friendly park of today is difficult to imagine. It does, however, demonstrate in stark visual terms how far we can advance when we focus on livable environments rather than intransigent hostility to change.

The managers of the modern Belle Isle did not turn their backs on the island's difficult history; rather, they made it part of the experience. The only part of the island that is not heavy with trees and undergrowth is the former prison camp. Signs explain in graphic detail what happened. The soldiers buried here were later disinterred and reburied in a national cemetery, but the dismal history of this place hangs in the air like a thick mist. At the other end of the island are the remains of an early hydroelectric operation. The generators and canals finally succumbed to nature and economics, but the remnants of the stone races and rusting iron intake gates are now part of the island's rich history. A trail that circles the island has interpretive signs that explain the meaning and significance of these relics. Richmond's waterfront today is an innovative blend of natural riparian areas, urban parks, and turbulent waters coursing through a turbulent history.

RIVER OF THE ANGELS

Councilmember Ed Reyes represents what he calls the "original" suburbs of Los Angeles. His district includes some of the toughest gang-plagued areas of the city and lots of working poor. Ed grew up in those neighborhoods and knows the sense of alienation and hopelessness that overwhelms many of its residents. There are almost no city parks, no natural areas, and virtually no open space. But there is the LA River (no one calls it the Los Angeles River), and Reyes wants to make it the catalyst for a new vision of this hardscrabble section of the city. I met with Councilmember Reyes in his office in the imposing tower of Los Angeles City Hall in 2006. He spoke firmly, yet with a slight smile. Even though his ideas are controversial—some would say outrageous—he exuded a quiet confidence.

"It's an issue of social justice. We have 30,000 to 40,000 people per square mile in the area along the river, but we are one of the most park-poor cities in the nation," he explained. "We need to change the image of the river from being your back yard to your front yard."[6]

It is a daunting task. Most people think of the LA River as a sterile, concrete-lined sewer. It has appeared in numerous Hollywood films, which typically portray it as a forbidding death trap fit only for madcap car chases and dead bodies. Indeed, most of the 51 miles of the river was totally encased in concrete between 1938 and 1960 by the Corps, following a series of devastating floods. But there are three sections of river bottom that were not lined, and the county of Los Angeles established an easement along each side of the river to maintain access to the channel. The easement and the absence of concrete in some stretches were not due to some farsighted vision concerning the environment. Rather, they were necessary for engineering reasons, but they inadvertently created the possibility that some day, with sufficient vision and panache, city planners could reclaim some of the river and convert it from an eyesore into a neighborhood resource.

The change in image for the river began with an eccentric poet, Lewis MacAdams, who was able to look at the river and see not what it was, but what it could be. He and a small group of friends (Councilmember Reyes described them as "bohemians") began spending time in the river corridor. It wasn't easy; the Corps and Los Angeles County had essentially blocked public access, fearing that the occasional flash floods that roared down the straightened channel would drown anyone foolish enough to slip a toe into the current. But the flash floods are rare; most of the time the river consists of a slow-moving stream coursing along in the sun. The areas where the river bottom was not covered with concrete—because of spring water bubbling up from below—became lush with vegetation. The Corps and the county routinely bulldozed this vegetation until the 1980s, but then quit. In the ensuing two decades, eucalypti, cottonwoods, palms, and willows grew into dense stands of jungle green. This attracted a variety of waterfowl and other species.

In a pattern that is quite typical of river-restoration efforts, Lewis MacAdams and his bohemian friends were not taken seriously, at least initially. MacAdams published books of poetry about the river, and he and his colleagues attended meetings and generated media coverage. In 1986 they cut a hole in the fence that isolates the river, gathered together at the water's edge, and formed Friends of the LA River (FoLAR). Gradually, people started paying attention to their message. Ed Reyes, then working as a planner for the Los Angeles City Council, listened to the ideas of FoLAR and took the vision even further; to him, it was not just a river restoration but the rejuvenation of an entire community.[7]

The river activists got some help when Mayor Tom Bradley, toward the end of his last term in office, established a committee to explore the possibility of a greenway river corridor project. By then a growing number of environmental groups began to sign on to the effort to re-green the river, including Northeast Trees, Treepeople, and traditional environmental groups. FoLAR continued to grow and advocate aggressively for a living river. In 2001 Ed Reyes was elected to the city council and became an essential catalyst for changing attitudes toward the river.

He was initially greeted with skepticism by the city engineers, who told him that nothing could be done with the river without compromising the flood-control capacity of the channel. But he persisted and began to put together a coalition of stakeholders. "We began a dialogue, not just about the river, but about the whole river system, and about how it could contribute to our communities and bring us together. Up until then, the river corridor was seen as single-purpose—flood control. We wanted a multipurpose river, for recreation, a park, and improved opportunities for economic development. I wanted a river that enhanced the quality of life for people living in that area."

Councilmember Reyes told me a story of how, a few years ago, he was talking with some of his friends from boyhood. They asked him why he was wasting his time trying to revive a dead river. They had already written it off as a lost cause because throughout their lives it had been nothing more than an eyesore. So he invited them to spend a day in the river channel and reminded them that when they were boys, they had caught catfish and tadpoles there. "We were Tom Sawyers on a concrete river."[8] He convinced them that the river, no matter how compromised, was the only thing they had that even approximated a natural environment. "I want to give every child a chance to experience a real river. In this part of the city the LA River is all they have."

The lower river—the part in Reyes's District One—runs through mixed industrial and residential areas. There are many warehouses along the river, and a number of brownfields—sites polluted by toxic wastes. Railroads run along both sides of the river. It is devoid of greenery, and the river channel is a broad, straight concrete chute that covers an enormous amount of surface area. With sufficient innovation, the river could become part of an area redevelopment project with multiple objectives:

vegetate the channel; maintain flood-control functions; clean up the toxic mess; and replace some of the defunct industrial areas with mixed-income residential, riverside parks, and floodplain.

But not all of the LA River is in distressed neighborhoods. It rises in the Santa Susana Mountains and begins its course to the sea by flowing due east through the ritzy San Fernando Valley. In this region talk of creating a river park stirs the wrath of residents who do not want the public "riffraff" parking on their streets and hanging out in their posh neighborhoods. Restoring the LA River will require that rich and poor alike agree on some sort of plan.

As the river leaves the San Fernando Valley, it turns south and enters a section called the Glendale Narrows, which is squeezed between the Santa Monica and the San Gabriel mountains. This stretch is devoid of concrete at the bottom of the channel and, except for the sloping concrete banks, looks a lot like a real river. One can imagine the potential for restoration if more concrete were removed. Along much of this section of the river there are bike paths and newly developed pocket parks, so many, in fact, that the corridor is now a complex web of disconnected trails, bike paths, and parks. These parks reflect the contemporary culture of Los Angeles; there is a park just for yoga, a park modeled after a rattlesnake, one with steelhead fish sculptures, and a small park that used to be a crack house.

The river has become sufficiently attractive that an enterprising young man named Joe Linton wrote a guidebook to it.[9] Joe agreed to meet me on a warm January day in 2006 and to lead me on a biking tour of the river, starting at the Glendale Narrows.

By profession he is an artist and an author, and he illustrated his guidebook with his own watercolor and pen-and-ink drawings. He is a native of Los Angeles and has been working on river issues since the early 1990s, when he got to know Lewis MacAdams and adopted MacAdams's vision of a living river running through downtown Los Angeles. Joe looks like he could be a former linebacker, but he is soft-spoken and mild-mannered. He does become exercised, however, when he talks about the river and its promise of future greatness.

"I agree with Lewis MacAdams; the LA River is a 100-year art project. For this river restoration, artists, poets and singers have led the way. We are changing the script from gang violence and sewers to nature and wildlife."

Joe first became involved in river issues when the Corps of Engineers developed a proposal to build massive flood-retention walls along both sides of the lower river in the 1980s. FoLAR sued the Corps in an effort to stop the project. It lost, but this occurred at the time when the Corps was evolving and looking for new, nontraditional projects. It modified the retention walls in response to public input and began a dialogue with FoLAR and other river-restoration groups that eventually blossomed into something of a partnership (Councilmember Reyes remembers flying over the river in a helicopter with Corps officials and seeing them get excited "like kids in a candy store" over the prospect of a new restoration project). Joe Linton also views the Corps in a positive light and agrees that it has an essential role in the river's restoration.

Some of the river's problems are jurisdictional; the Corps controls and maintains some of the river, while Los Angeles County controls other sections. To complicate matters further, the city of Los Angeles controls the river bottom for the 31 miles where the river flows through downtown Los Angeles, while the upper river cuts through numerous towns and bedroom communities. At one of the bridges on our bike tour, Joe explained why some concrete banks were covered with graffiti while others were not: "The Corps just does flood control; they feel that graffiti is a local law enforcement problem, so they don't remove it on their sections of the river. But the County has an aggressive graffiti-control program. So that's why some stretches have graffiti and some do not."

About noon Joe and I stopped at a ramshackle sidewalk Mexican restaurant not far from FoLAR's office, got some takeout, and rode our bikes to a spot near the river. Joe shared with me a comprehensive vision of the river, including appropriate river-friendly development, the renewal and rebirth of local neighborhoods, and a living river channel free of concrete. As we talked, four locals, so inebriated they could barely walk, staggered past us on their way to the river. Joe spoke to them as if they were old friends, and we gave them the remainder of our lunches. At that moment it took a great deal of imagination to envision this area as a green, healthy, inviting river corridor. But for Joe Linton, that act of imagination is not difficult. "The river should be a place for everyone—

all the people of L.A. Everyone has a right to be happy here. It's all about art, it's all about people, and it's all about nature's river."

The rehabilitation of the river will be guided by the Los Angeles River Revitalization Corporation, which met for the first time in December 2009.[10] Its work got a bit easier when, in July 2010, the EPA reversed an earlier decision and declared the LA River to be navigable, which makes the Clean Water Act applicable.[11] It may be difficult to imagine, but some day the LA River may be more than just a setting for Terminator movies; this river will be back.

RIVER OF DEMOCRACY AND THE EMERALD NECKLACE

Renata von Tscharner, the president of the Charles River Conservancy, looks a bit like a staid librarian, with slightly graying hair in a bun, low-profile glasses, and a thick scarf around her neck. But in her case, looks are quite deceiving. Von Tscharner's favorite hobby is wind surfing, and her favorite place to do it is on the cold, forbidding waters of the Charles River in downtown Boston. And she's not afraid to go for a swim in the river, whose waters became so polluted in the 1950s that the state legislature outlawed swimming. Renata routinely swims and surfs the river but makes a point of never touching the bottom. "The bottom sediment is full of toxins; you name it, and it's there," she says in her clipped German accent.[12] Renata came to the Boston area twenty-seven years ago from northern Switzerland and was immediately drawn to the beauty of the river, despite its foul water and nearby urban blight. Where other people saw devastation, she saw hope. As a professional planner and architect, she could imagine a restored, healthy Charles River that could be the centerpiece of Boston's urban parks—a place where everyone could enjoy a quasi-natural setting amidst the skyscrapers and historic buildings of Boston. Her goal was to "struggle against negligence and overcome the detritus of the industrial age." To renew the river and its parks, she had to "rock the status quo, and make people fall in love with the river again."

No one is talking about re-naturing the Charles River; the opportunity to do that passed when the Puritans first began filling in the bay to increase the land available for the city's growth. Today, the lower river, once several miles wide, is much reduced by landfill created when horse-drawn wagons began removing dirt from Beacon Hill and dropping it along the south bank of the river to create an area that became known as the Back Bay, a pleasant part of the city that in years past was zoned to keep out dirty Irish scruff.[13]

The Charles River has one unique feature that has kept it from being totally abandoned and abused; nearly the entire lower river is a state park, thanks to a visionary planner named Charles Eliot.[14] Eliot's dream was to create a linear urban park that would welcome all classes of Bostonians—a great democratizing gathering place where everyone, regardless of class (except perhaps the Irish) could enjoy swimming, beaches, and strolls along the banks. His vision of a "democratizing common ground" helped persuade community leaders to save the riparian zones along the lower river as a series of state parks. His plan became the envy of the world, presented at the World's Fair of 1883 to much acclaim. Much of Eliot's plan became a reality. Parks along the river became popular among Bostonians. In the latter part of the nineteenth century, shoreline beaches along the Charles River were thronged with bathers; photographs from that era show crowds of people venturing into the water, like a busy day at Miami Beach. Parks were created from the landfill along the south side of the river. Frederick Law Olmsted, of Central Park fame, designed and built a bathhouse at Magazine Beach, so named because it was the site of a military armory.[15]

But Eliot's dream was never totally fulfilled. As the city grew, streets and highways crowded the river, Boston and upstream towns dumped raw sewage into the water, and much of the public infrastructure along the river began to decay. It took less than fifty years to turn the river into a cesspool, and the city turned its back on the river. When Michael Dukakis's problems with pollution in Boston Harbor garnered national attention in 1988, people began to rethink how they treated the Charles River. One of those people was Renata von Tscharner.

She was teaching at Radcliffe when she realized that no one was paying attention to the lower Charles River, even the most educated elite in the city. Relying on her expertise and experience in planning and architecture, she began by writing a mission statement for a new organization, the Charles River Conservancy. She wanted to complement, but not overlap, the objectives of the Charles River Watershed Association, which had been in existence since 1965. So she focused on the riparian parkland on the lower river—the heart of what Eliot had envisioned as a democratic park for the people. She then sent a letter to 100 friends and influential individuals, asking for advice and support. With their encouragement, she formed a board of directors, starting writing grant proposals to foundations, and challenged the city of Boston to live up to Eliot's dream. "My greatest opposition was the lack of money and awareness," she said, describing the early days of the Conservancy.

Efforts to restore the parklands along the river received a boost when the "Big Dig," the nickname for a massive urban transportation project in downtown Boston, allocated funds for environmental remediation. The Massachusetts Turnpike Authority developed a plan that included environmental remediation, and thus began a coordinated effort to give life to the lower river. Things began to happen.

The Charles River Conservancy focused on bringing art, culture, and tourism to the river. Each year the Conservancy sponsors a "River Sing" where groups of people line both banks and several bridges and sing river songs. It is an effective way to make a link between culture and water. Another unique use of the river concerns an odd vehicle known as a DUKW, pronounced "duck." These are amphibious vehicles produced in great numbers during World War II. The Museum of Science in Boston employs about a dozen of these boat-buses to give visitors a tour of the city that includes a jaunt up the Charles River. About 500,000 people take the "duck" ride every year, getting a perspective of the city from the river. The ducks must maneuver among an often chaotic fleet of sailboats and sculling craft.

To improve public access to the river, the Conservancy has been working with state and local agencies to create a 40-acre park at North Point near the 1910 dam that bridges the river between Boston and Charlestown. The land was formerly a run-down area of warehouses, railroad tracks, and abandoned buildings. And in a very unusual transformation, the Charles River Conservancy has taken control of a piece of land just below the new Leonard Zakim Bridge (part of the Big Dig), and is building a skateboard park. This kind of river-friendly development draws people to the area and gives them reason to enjoy the river and protect it. Von Tscharner showed me the site of the future skate park—at the time of my visit it was still just a pile of debris and construction materials—and asked me to envision a day when many families would come here and enjoy this riverside amenity. It takes a special kind of person to create and implement such a vision for the future.[16]

Not everyone is happy with such developments, noted Renata. "There are people who talk about returning the river to a natural state and making it a wildlife area. But this just isn't feasible. The best way to protect the river is to encourage development that is compatible with the river. If people use the river, they will learn to love the river."

The linear system of parks envisioned by Charles Eliot is just one part of an extensive river-park system in the Boston area. Another component is known as the Emerald Necklace, designed by Frederick Law Olmsted. It consists of a series of parks connected largely by the Muddy River and its tributaries. The course of the river channel is Olmsted's creation; it is urban landscaping at its finest, and over the years the oak trees he planted, and other vegetation, have grown to the extent that the river looks quite natural. It offers an attractive haven that contrasts vividly with the surrounding welter of high-rises and highways. The Necklace carves a graceful arch through a series of colleges and universities in Brookline and the Back Bay area.

But over the years the Emerald Necklace suffered from encroachment, inadequate maintenance, and overuse. In 1996 the Muddy River experienced a 100-year flood that destroyed millions of dollars of property and caught virtually everyone by surprise. The Corps initially denied that a flood had occurred and instead blamed the problems on the city's antiquated combined sewer overflow (CSO) system. But after assessing the damage, even the Corps acknowledged that a major flood event had occurred, and it quickly suggested building levees and dredging. Olmsted's dream child was about to be demolished. That was when the Charles River Watershed Association went into action.

The Charles River Watershed Association was formed in 1965 as an advocacy organization, primarily in response to the tremendous levels of pollution that was borne by the river. Many of the towns along the 80-mile length of the river rely on the old-style CSOs that combine storm water with sewage. Even though the 1972 Clean Water Act forbids piping

The Muddy River and the Emerald Necklace, downtown Boston, Massachusetts.

sewage directly into rivers, it still occurred during extreme rain events. The Association was formed to pressure local governments and industries all along the river to clean up their act.

With the passage of time and some notable legislative victories, the Association began to change its orientation; it realized that to remain effective, it had to examine all the various threats to the river, not just water quality. When Robert Zimmerman became director in 1991, he shifted the focus of the group to emphasize scientific and legal expertise. The first scientist it hired was Kate Bowditch, a hydrologist. She took me on a tour of the Muddy River and explained how the Association and its allies were working to restore and protect it.

Kate parked the car and suggested that we walk. "Forget Central Park; Olmsted's finest accomplishment was the Emerald Necklace. It's a beautiful public amenity, on the national register of historic places."[17]

When the Corps proposed to channelize and levee the Muddy River, the Association and its allies immediately began pushing for alternatives.

They were hopeful that the Corps would adopt, not its usual structural approach, but a different technique the Association had pioneered on the upper Charles River. The Corps had constructed the dams on the lower river and initially had planned a similar structural approach to flood control on the upper Charles River, but the Association and other groups had worked with the Corps to adopt a floodplain-management approach instead. The Corps had agreed and in 1977 had begun to purchase 8,000 acres of flood-prone lands along the river for use as parks and public areas. This project became known as the Natural Valley Storage Area and became a model for the development of a flood-control plan for the Emerald Necklace. When the Corps expressed a willingness to apply that concept to the Muddy River restoration, the Association heartily approved.

"Using that approach, the Association and the Corps have been allies on the Muddy River restoration," Kate explained. "We all see the benefits of solving the flood-control problems with techniques that also solve the problems with toxic sediments, habitat protection, invasive plants, erosion, and inadequate park maintenance."

As we walked along the Muddy, the nature of both the problems and the promise of this river became obvious. To a great extent, the Muddy River was being loved to death. A thin strip of natural beauty, it was crowded with joggers, walkers, picnickers, and those who were in need of a quiet, contemplative moment. It was easy to see why many of the high-stressed students from the surrounding universities would come here to seek everything from solace to exercise or a romantic respite. People flock to this island of peace in a sea of noise and motion. The riparian area along the creek still supports "victory gardens," started during World War II and popular among the locals. They line certain sections of the bank and lend a unique beauty to the riparian landscape.

But the problems are also evident. In some places the river disappears entirely, pushed into a culvert despite its obvious attraction to people. There is so much foot traffic along the river that the plant life is trampled and erosion results. And an invasive plant, phragmites, grows up to 20 feet tall and creates a wall of bamboolike vegetation between people and river. Maintenance of the park is so underfunded that prisoners from the county jail are used to maintain the grounds. In some places the water gathers into fetid, trash-filled pools. These are all solvable problems, but they will require a comprehensive vision of renewal and maintenance. Emeralds cannot shimmer without an occasional polishing.

It's hard to find a river in America that gently courses past redbrick buildings built by English colonists in the 1600s. The "restoration" of the Charles River is a relative concept, not re-naturing, but a goal that encompasses social, cultural, and natural renewal to create an enhanced quality of life. There are many lessons to be learned from the Charles River. This watershed witnessed the dawn of the American industrial age, with all of its attendant material advantages and environmental costs. The sediment at the bottom of the Charles River includes America's oldest industrial toxics. These poisons were dumped into the river hundreds of years before the word "toxics" entered our lexicon. By the time the concept of environmentalism became part of the political landscape, the Charles River had been dammed and slammed for over 300 years. When compared with the troubles experienced by other rivers in America, the Charles is the granddaddy of both abuse and renewal. That long history makes its revival even more impressive. A 2005 settlement between the EPA and the Massachusetts Water Resources Authority led to dramatic reductions in sewage discharges into the river; by the time that process is completed in 2013, sewage inflow from CSOs will have been virtually eliminated.[18] And the American shad, a pollution-sensitive fish, has returned to the river.[19] In 2011 the river was named as a finalist for the international Riverprize awarded by the International River Foundation.[20] That is quite a change from the days when it was illegal to dip a toe in the river.

On my last day in Boston, I grudgingly coughed up the $27 to take the DUKW "duck" ride. I'm not much for canned guided tours, but this one held the attraction of actually spending a little time on the river, so I signed up. The tour guide knew a lot of Boston history, but his finest moment came when he gave the wheel of the duck to a passenger while we were cruising upriver between the old dam and the Longfellow Bridge. This gave him a chance to wax poetic, in heavy Boston accent, from his vantage point in mid-river. "You can see Fenway Park from here, you can see Bunker Hill, the capitol dome, the Back Bay and Cambridge. You can see downriver to the museum, you can see upriver past Harvard and Boston University. Ain't it great, ain't it great, ain't it great!"

A GREAT MULTITUDE OF FISH

Imagine 3,000 dead Americans laid out in heaps and windrows in a field. Today, even in this age of terrorism, it is hard to fathom such bloodletting. Yet that was the scene in front of a stone wall just outside the prosperous Virginia community of Fredericksburg on December 13, 1862. The fallen and those who felled them were all Americans, immersed in the great tragedy of the Civil War. The town of Fredericksburg was fundamentally changed by the battles that took place there, both physically and psychically, and preserving its tumultuous history has been a part of Fredericksburg ever since. Change does not come easy to this community, and it may seem surprising that such a town could support the removal of a local dam. But on February 23, 2004, a great crowd of locals, dignitaries, and the press gathered to watch plastic explosives make a different kind of history when they shattered Embry Dam on the Rappahannock River just upstream from Fredericksburg.[21] This was not the work of crazed monkey-wrenchers or heinous terrorists. Rather, the charges were placed by demolition experts from a joint U.S. Army/Air Force team.

Squeezed among the verdant hills of northern Virginia, the 184-mile-long Rappahannock River was once home to vast runs of American shad, striped bass, and alewives. Shad has long been a staple of southern cuisine, and a shad bake was always a social event worthy of attention. Indeed, a shad bake may have hastened the fall of the Southern Confederacy. At a pivotal moment in the Battle of Five Forks, General George Pickett, who lent his name to the most famous charge of the war, took leave of his post to attend a shad bake. While the general was enjoying his fish dinner, his troops were being overrun; General Lee surrendered eight days later. Pickett was the only general officer sacked by Lee during the war, all because the allure of a well-cooked shad was just too tempting.

The construction of Embry Dam, and an earlier crib dam built in 1855, stopped the shad from running upstream to spawn, and the fish declined precipitously. Embry Dam, completed in 1910, was built to supply newfangled electricity to Fredericksburg. It also helped funnel water into a canal that became the source of Fredericksburg's drinking water.

Embry Dam, with holes blown through it, on the Rappahannock River, Virginia. It was later removed.

But the town eventually found other sources of water and electricity, and the dam became a liability. A small group of people began talking about removing it.[22]

Instigators are usually local people, common citizens with little or no experience in politics. In the case of Embry Dam, there were plenty of those. In 1985 a small group of locals formed Friends of the Rappahannock and began quietly agitating to remove the dam. But it was not until Virginia senator John Warner got on board that things started to happen.[23] The senator was quite fond of casting a line, so the president of Friends of the Rappahannock invited him to fish near Fredericksburg, which is only 35 miles from Washington. The senator made many trips to the area, quietly fishing in the rocky current below the dam. It was the perfect refuge from the intense pressure of Washington. It did not take Senator Warner long to see the benefits of removing the dam. Local support for removal had grown quite strong, but the estimated $10-million price tag was beyond the community's means. Warner got a federal appropriation for the removal and was honored with the privilege of operating the plunger that activated the blast. He then quoted the Bible, at Ezekiel 47: "There will be a very great multitude of fish, because these waters go there; for they will be healed and everything will live where the river goes."[24]

On the day I visited Embry Dam in 2004, I stood beneath the superstructure of the dam, marveling at the giant holes blown in the concrete a few months earlier. Young boys fished from the top of the remaining parts of the dam while a family fished in the current that rushed through the newly opened fissures. I wandered up a narrow dirt lane and encountered twenty-five people sitting in a shade shelter, all of them wearing life jackets as though they expected high water at any moment. In front of them was Bill Mix, a stout, ruddy-faced man of about 50. I introduced myself and asked if he could tell me anything about the dam.

"I can do better than that, but first I have to talk to these people, excuse me," he replied as he waved his arm at the group impatiently waiting for high water. He asked for their attention and explained apologetically that sediment buildup in the river would prevent them from paddling their rafts to a nearby island. Instead, as a sort of consolation prize, he would give them a personal tour of the remnants of Embry Dam, despite all the no-trespassing signs. He motioned for me to join them. It was a stroke of fortuitous luck for me, because Bill knew well the story of Embry Dam and how the idea of removing it had gone from an absurd dream to planting explosives. Bill had taught in the local public school for thirty years before retiring and devoting much of his time to the restoration of the Rappahannock. Bill, along with other members of Friends of the Rappahannock, explained the history of Embry Dam and how it had evolved from a cutting-edge hydropower project to a useless wall of concrete. Bill became quite animated when he described the massive crowds that lined the banks to view the detonations. "How often do you get to see a dam blown up? We worked hard for that."

Since my visit the dam has been completely removed, and shad are once again swimming up the Rappahannock.[25] I returned to the dam site in 2010 and was struck by the vast expanse of uninterrupted water. It's enough to make General Pickett smile in his grave.

THE PLEASANTNESS OF THE RIVER

In 1690 William Penn petitioned the king of England for permission to build an entirely new city in the English colony he had founded a few years earlier. His petition for the site he had in mind described a cornucopia landscape embraced by languid rivers:

> That which particularly recommends this settlement, is the known goodness of the soil and situation of the land, which is high and not mountainous, also the pleasantness and largeness of the river being clean and not rapid. . . . The sorts of timber that grow there are chiefly oak, ash, chestnut, walnut, cedar, and poplar. The native fruits are pawpaws, grapes, mulberrys, chestnuts, and several sorts of walnuts. There are likewise great quantities of deer, and especially elk [that] use that river in herds. And fish there is of divers sorts, and very large and good, and in great plenty.

Penn got permission to build his city at this idyllic location, which he named Pittsburgh in honor of the British prime minister.[26]

But 178 years later a visitor described that same location as a filthy and fetid "hell with the lid off." Another observer, the English writer Anthony Trollope, was even more caustic: "Pittsburgh, without exception is the blackest place I ever saw."[27] The once-pure rivers that surround the heart of Pittsburgh had become cesspools, and the air was so dirty that even a modern-day Los Angeleno would gasp for air. The heart of the city was an area called the Golden Triangle, a slice of land between the Monongahela River and the "mournful" Susquehanna River where they join to form the Ohio River.[28] By the 1950s that area had, in the words of the state of Pennsylvania, "deteriorated into a commercial slum."[29]

Fast-forward to 2009; the city of Pittsburgh was the top-ranked U.S. city in the *Economist* magazine's annual poll of the world's most livable cities.[30] The waterfront is being restored, re-natured, and preserved, thanks in large part to the high-energy efforts of Friends of the Waterfront and an ambitious blueprint called the 'Riverfront Development Plan."[31] In short, Pittsburgh went from an idyllic setting to the grime capital of America and then to a place that values its rivers, its waterfront, and its natural setting. The city's plan to revitalize the Golden Triangle was aptly named "Renaissance."

Pittsburgh has helped set an example of how any urban riverfront, no matter how tarnished, can be made to shine. In addition to the urban rivers described in this chapter, there are many other examples all over the country as America turns to face its rivers. Chattanooga, in an aggressive program, completely rehabilitated its interface with the Tennessee River. The Walnut Street Bridge was made into a pedestrian walkway, a spectacular aquarium and art museum now overlook the river, and the south bank is lined with a promenade and bike lane. Columbus, Ohio, has its "Riverfront Vision Plan" that encompasses a variety of innovative strategies along the Scioto River. Reno, Nevada, turned its stretch of the Truckee River into a downtown kayak park; Denver, Colorado, has a similar facility on the South Platte. Riverfront Park in Spokane, Washington, is the anchor of that city's redevelopment. San Antonio has its famous River Walk on a stretch of river that was once slated for destruction.[32] Salt Lake County in Utah is working to bring the Jordan River back to life as a linear park.[33] The much-maligned Cuyahoga River in Cleveland is now a national park.[34] Portland, Oregon, has the Tom McCall Waterfront Park on one side of the Willamette River and the Vera Katz Esplanade on the other. Sacramento has its Promenade and Old Town. On the other side of the country, Portland, Maine, boasts of the Westbrook River Walk along the Presumpscot River. New York City wants to spend $3 billion on its waterfront.[35] Even the Chicago River, once a fetid sewer lined with tenements, is being resuscitated.[36] There are dozens of other examples, ranging from modest cleanups to complete makeovers.[37] I can envision a new TV show. As William Least Heat-Moon put it, "City rivers everywhere across America still offer opportune avenues to enchant wayfarers."[38]

Urban river restoration must take place within the larger context of river uses. Public water suppliers deliver 43 million gallons of water each day, and two-thirds of it comes from surface water sources, so we must meet drinking-water needs while also recognizing the value of intact urban rivers. That is not an impossible task. Only about 13 percent of all water withdrawals are actually for human consumption, so much of our water-supply "problem" concerns uses other than domestic use.[39] This means that the potential for water conservation and enhanced efficiency is enormous.[40] If we begin to treat water as a precious and scarce resource while at the same time recognizing that an intact urban river is a tremendous asset to a community, we can meet both amenity and consumptive needs for water in urban areas.

Urban river restoration and preservation are not about being antidevelopment. On the contrary, river restoration in urban areas restores property values and attracts development. Most urban riverfront-restoration projects involve recycling abandoned and abused areas of town and remaking them as trendy new areas in which to live, work, and enjoy life. This requires an unprecedented level of imagination and cooperation and an orientation to the future. As one river plan puts it, "All good things begin with a vision of what may come."[41]

Last, rivers are often the only open, quiet space in crowded urban areas. They are a way to enrich the lives of city dwellers and provide a relaxing contrast to concrete and steel. In a nation that at times appears to be collectively manic, there is immeasurable value in providing open space, solace, and relative quiet. The use of antidepressant drugs in the United States has increased 400 percent since the late 1980s.[42] Perhaps we just need a bigger dose of river.

10

NET LOSSES

Habitat and Endangered Species

Oh, loveliest there the spring days come,
With blossoms, and birds, and wild-bees' hum
—FROM WILLIAM CULLEN BRYANT, "GREEN RIVER"

When I first began graduate school in 1974, I was desperate for money. Fortunately, through a series of connections, I was able to land a job on a salmon-processing ship off the coast of Alaska. The ship had left Bellingham, Washington, that spring, but the ship's owner said that if I could catch up with it, I could have a job cleaning, freezing, and stacking salmon. So, at the close of the semester, I hitchhiked from Tucson, Arizona, up the Alcan Highway to Anchorage, Alaska. The trip took nearly a month, but hitching was the only mode of transportation I could afford.

From Anchorage I took a commercial flight to Dillingham, then hired a private plane with the last of my cash reserve to fly me to the coastal village of Naknek. There I discovered that a small boat from the ship had come into the harbor for supplies; I bummed a ride when the boat returned to the ship in Kvichak Bay, just a few miles offshore. As soon as I arrived, I discovered two immutable facts: one, I was in a natural environment that was strangely alien and awesomely beautiful; two, the captain of the ship packed a sidearm and a pathological hatred of anything he did not understand, which included the entire human race. He immediately noticed I was a member of that disreputable mob.

My worst experience of the summer came early in the season when ice still came down into the bay from the Naknek River and the upper bay. We would occasionally see walruses riding along on chunks of ice. One day when the salmon weren't running and we had plenty of time on our hands, several crew members and I were loitering on the deck, marveling at the dramatic scenery. Someone told us to lower our voices and pointed up-bay. An enormous bull walrus was riding a chunk of ice as it flowed silently passed the ship. It had tusks the size of bowie knives, and it just stared at us as it glided by, as though we were the weird animal. Suddenly there was the loud crack of a rifle, and the walrus's head exploded in a cloud of pink froth. The massive animal instantly went limp and slid from its perch into the water, surrounded by a reddish tint. We heard the captain laugh, the gun in his hands. My rage was tempered only by my fear as I faced him. "Why the hell did you do that?"

"You idiot college kids," he said with pure unalloyed hatred. "It's a damn walrus—they're useless!"

People feel strongly about wild animals. Whether it is arguing over the fate of our national symbol or having a passion for or against wolves, we speak our minds on wildlife. Our world would be incomplete without wildlife in some form. Yet despite our love of wild animals and plants, we have decimated habitats, caused the extinction of hundreds, perhaps thousands, of species, and pushed wildlife to a far corner of our existence. This is especially true for riverine and riparian species. As wild rivers were tamed, they lost their status as prime habitat.

There was a time when the American continent teamed with wildlife, and our rivers and lakes offered an unbelievable cornucopia of fish. It is difficult to imagine today just how bounteous America's waters were. An early settler of the Jamestown colony noted that the rivers "abound with fish both great and small."[1] The pilgrims at Plymouth Rock discovered that "the greate smelts passe up to spawne likewise in troupes innumerable," and they watched the local Indian people pull eels out of the mud by the hundreds.[2] In 1705 a writer noted that the "herrings and shad come in such abundance in the brooks of Virginia that it is almost impossible to ride without touching them."[3] Early settlers found the Delaware River to have "a great plenty of fish," and Pennsylvania rivers were "full of fine trout."[4] The shore of Lake Erie was so choked with bass in the early 1800s that "a gig maybe [sic] thrown into the water at random and it will rarely miss killing one."[5] The Mississippi River used to be filled with "monstrous fish."[6] On the upper Missouri River, Lewis and Clarke caught "great quantities of trout, and a kind of mussel, flat backs and a soft fish resembling a shad and a few catfish," and at the Dalles on the Columbia they sat in astonishment as the salmon, which "abound in such great numbers," leaped Celilo Falls.[7] A traveler to the Willamette River in 1841 saw a similar phenomenon and marveled, "I never saw so many fish collected together before."[8]

In this chapter we will look at the "use" of rivers as valuable habitat. In a prior era that use dominated our culture, then receded; recently it has made a startling comeback. The reemergence of riverine habitats not only protects endangered species but also plays a crucial role in the economy, lifestyle, and culture of most regions of the country.

ALPHABET GUMBO

In a sense, the United States does not have "wildlife management." Rather, we have a complex farrago of state and federal agencies that sometimes work together but often work at odds with one another. For aquatic species, we have an additional layer of water agencies that affect aquatic wildlife.[9] This might make some sense if there was an overall plan of coordination for these myriad government entities, but there is none. Instead, we have the Commerce Department (the National Oceanic and Atmospheric Administration [NOAA]), the U.S. Army, the Department of the Interior, the Department of Agriculture, the Environmental Protection Agency (EPA), the Tennessee Valley Authority (TVA), and the Federal Energy Regulatory Commission (FERC) all trying to control aquatic species and their habitats. Toss in fifty state wildlife departments, another fifty state water divisions, and several thousand special water districts, and you have the ingredients for a first-class alphabet gumbo.

To make matters worse, aquatic species have a disconcerting tendency to wander all over the map, from jurisdiction to jurisdiction. Some of them even venture beyond our shores. They also share their riverine habitat with many nonaquatic species; more than 70 percent of land animals use riparian areas.[10] In other words, life concentrates at the river, and so do government agencies.

So do people. We love to be near water, probably because of evolution; venture too far from fresh water, and you're history. It is also because we are attracted to riverine flora and fauna, initially as a source of food and increasingly for more complex reasons as we climb Maslow's hierarchy.[11] And we haven't given up the food sources found in water; fish and seafood are the only wild animal species that are still widely consumed. Thus rivers are far more than just the source of drinking water for two-thirds of the people in the United States. They present an opportunity for us to engage with a dramatic potpourri of life that is found nowhere else.

Rivers also tend to concentrate rare and endangered species, in part because they are such hospitable places, but also because riparian corridors are often the only land that has not been developed. This makes rivers target central for the intense battles over rare or endangered species. When we think of endangered species, the Endangered Species Act (ESA) often comes to mind, but the animals and plants that make it onto the ESA threatened or endangered lists are just a small part of the species in jeopardy worldwide. According to the International Union for Conservation of Nature's "Red List," there are 21,014 species that are threatened, near threatened, or possibly already extinct.[12]

For species to be listed under the ESA, they must undergo a complex application procedure. If they survive that, they are listed as either threatened or, if they are in extreme difficulty, endangered. There are more than 1,900 species listed under the ESA, 1,320 of them in the United States. Of that number, 573 are animal species (the remainder are plant species). Of those 573 animal species on the threatened or endangered lists, about half live in aquatic environments. The largest single group of animals on the list is fish.[13] This explains why so many river-restoration and preservation efforts involve endangered or threatened species. Most of these species come under the jurisdiction of the U.S. Fish and Wildlife Service, but 68 marine species are under NOAA Fisheries. It is not often that a government agency waxes poetic, but the NOAA offers compelling and passionate language to support its ESA work: "The wide variety of species on land and in our oceans has provided inspiration, beauty, solace, food, livelihood, medicines and other products for previous generations. The ESA . . . is to remind us that our children deserve the opportunity to enjoy the same natural world we experience."[14]

The struggle to protect riverine species is taking place all across the nation.[15] Among numerous ESA efforts, several stand out in addition to the "salmon wars" described in earlier chapters (especially chapters 4 and 5). In the Rio Grande a fight to save the silvery minnow has often pitted environmentalists and the ESA against the city of Albuquerque.[16] On the Platte River there are efforts to save the pallid sturgeon, whooping cranes, piping plovers, and least terns.[17] There is a protracted fight over the delta smelt on the Sacramento River.[18] In the Edwards Aquifer in central Texas there are one threatened and seven endangered species that are competing for water with several million people.[19] There are efforts in western watercourses to save the bull trout.[20] There is an ambitious program to restore habitat on the lower Colorado River and protect six threatened or endangered species.[21] And the Gulf sturgeon—one of several types of endangered sturgeon—is imperiled by the Apalachicola Waterway described in chapter 8.[22]

Some of the threats to aquatic wildlife are local, such as a lack of water volume, velocity or temperature problems, upstream sources of pollution, or the destruction of riparian lands. Other threats are more generic. Anthropogenic climate change could dramatically reduce the habitat and range of fish such as salmon and trout.[23] Open-ocean salmon farms present a host of threats to wild salmon.[24] In Alaska it is not uncommon to see bumper stickers that read "Friends don't let friends eat farmed salmon." Farmed salmon is the "hot dog" of seafood; no one would eat it if they knew what is in it—polychlorinated biphenyls, feces, antibiotics, pesticides, and food coloring.[25] Despite the dangers posed by farmed salmon to both wild salmon and people, the George W. Bush administration pushed hard to expand salmon farming in U.S. waters.[26]

Another grave threat to native plants and animals is invasive species. The list is so vast as to resist summary, but it includes such exotic invaders as the vermiculated sailfin catfish, which looks like a pointed brain with wings and whiskers and can walk across land from one watercourse to another while breathing through its anus.[27] Wayne Stroup of the Corps called it "the fire ant of the waterways."[28] Another bizarre invasive fish is the Asian silver carp, which leaps out of the water at the sound of a boat motor, endangering passing motorboaters. Both the sailfin and the Asian carp tend to devour everything in their path. The Asian carp have spread throughout the Mississippi River Basin; if they manage to get past the electric fence in the Chicago Sanitary Ship Canal, they could decimate the Great Lakes fishery.[29] And there is the zebra mussel, which could star in a movie about environmental horror stories.[30] There are so many waterborne invaders that the Corps created an Invasive Species Management Branch.[31]

Another factor that directly affects riverine species is the health of the nation's wetlands—at least the half of them that have not been destroyed. George H. W. Bush announced a "no net loss" wetlands policy, but his son's record is more convoluted. In 2006 his administration announced the first net increase in wetland acreage since wetlands were first inventoried in 1954, totaling 192,000 new acres of wetlands. Some of this new acreage was due to requirements that developers construct wetlands as part of their environmental mitigation. But critics pointed out that the new acreage also includes biologically bare man-made ponds and "borrow pits" dug to build roads.[32]

Wetlands took a potential hit from the U.S. Supreme Court in 2006 in the case *Rapanos v. United States*.[33] The court's decision is muddier than swamp water. Rapanos, a developer, filled in wetlands without a permit from the Corps, which is required by Section 404 of the Clean Water Act for navigable waters. Rapanos argued that because his property was 20 miles from the nearest navigable river, he did not need a permit. Four conservative justices agreed with him, concluding that the Corps' definition of "navigable waters" was much too expansive. The four liberal justices noted that pollutants discharged into intermittent water bodies, even those at a distance from a river or lake, still ended up in those navigable waters. As in so many recent cases, Justice Anthony Kennedy held the trump card. He wrote a separate opinion that recognized Corps jurisdiction if a "significant nexus" could be identified between intermittent wetlands and a navigable watercourse. Because Justice Kennedy sided with the four conservative justices on the question of remanding the case back to a lower court, their decision prevailed. The EPA and the Corps then revised their definitions of wetlands.[34] A bill was introduced in both the 110th and 111th Congresses to overturn the decision but did not pass.[35] It had no chance in the 112th Congress, and instead, House Republicans proposed a bill to substantially weaken the Clean Water Act.[36]

During the Bush years, the *Rapanos* case was only one part of a fierce struggle between the president and proponents of the ESA, especially as it applies to salmon. Perhaps the most controversial move by the Bush administration was to declare that hatchery-raised fish should be counted with wild fish, even though there is nothing wild about them. This would lead to inflated numbers of fish, possibly resulting in the delisting of the twenty-seven threatened or endangered anadromous species. Scientists hired by NOAA Fisheries concluded decisively that hatchery fish should not be counted for several reasons, but the Bush administration ignored their advice.[37] The conflict over this is indicative of the intensity of the "salmon wars" and the politicization of biological science.

While the Bush administration was battling with its critics over the ESA, several river-restoration projects focused on the revival of endangered species and proceeded apace. The projects described in this chapter focus on anadromous fish species and occur on both coasts in the states of Washington, Maine, and California. They are geographically diverse but have a common factor: all are confident that they can overcome the dangers and obstacles described in the previous paragraphs and make these rivers once again teem with salmon and other native species.

GOD BLESS AMERICA

The date was March 13, 2002. The FERC was holding an obligatory public meeting to discuss removal of Condit Dam, a hydropower dam on the White Salmon River, which rises in central Washington on the slopes of Mt. Adams and flows due south to the Columbia. Once teeming with its namesake fish, the upper river has been blocked by Condit Dam since 1913. The dam, now owned by PacifiCorp, completely blocked fish passage to the upper river. The company's FERC license had expired in 1993, and the FERC had made it clear that either a fish ladder had to be installed on the 125-foot dam or the dam had to be removed in order to get another license.[38] PacifiCorp ran the numbers and discovered that fish ladders would cost $30 million, but dam removal would cost only $17 million.[39] An agreement to do just that was signed in 1999, but it was fiercely opposed by some local people—and fiercely supported by others.

At the March 13 meeting, both parties were there in force. At first the air seemed festive, with children carrying placards of colorful fish cutouts. The FERC imposed a strict limit of three minutes for each speaker to ensure that a lot of people could voice their opinions. One of the supporters of dam removal, when it was his turn to speak for three minutes, indicated he had written a song about the White Salmon and would like to sing it during his allotted time. As he began to sing, dam supporters rose out of their seats and began belting out "God Bless America" to drown out the singer. It was an offensive tactic, but it certainly demonstrated the passion that accompanies controversies over rivers and dams.[40]

In 1999, when PacifiCorp worked out a settlement agreement with twenty other parties, it looked like the fate of the dam was sealed. The list of pro-river public interest groups among the intervenors is indicative of the growth of the river movement; there were fifteen of them listed on the agreement. The settlement stated that "an aggressive schedule has been developed to complete the dam removal project within one year."[41] But the settlement turned out to be the beginning, not the end, of the contentious process of removing Condit Dam. In the years immediately after the settlement agreement was signed, the sentiments expressed by dam supporters made it clear that Condit Dam was, for them, merely a symbol of a larger struggle between contrasting belief systems. In letters to local newspapers, dam supporters called removal "wild and crazy" and a "serious hoax."[42] Others used the opportunity to express rage at environmentalists who "squawk about endangered fish. As far as I am concerned, we are the endangered species."[43] Another claimed that "environmentalists are only concerned with the environment when it provides an opportunity to line their pockets."[44]

The two counties bordering on the White Salmon, Klickitat and Skamania, mounted a concerted effort to stop dam removal. One of their principal arguments was that dam removal would release a massive sediment load and thus violate state water-quality standards.[45] They objected vehemently to the removal plan, which they derided as "blow and go," but that is actually a fairly accurate description. The plan of removal was to blast a 15-foot-wide tunnel through the dam and let 'er rip, draining the reservoir in six hours in a giant rush of water, sediment, and debris. The natural flushing action of the river is expected to cleanse the lower channel in three to five years. The remainder of the dam will be cut in chunks and removed by crane.[46] The counties' concerns led to a careful analysis of sediment deposition from the removal process, which found that a temporary reduction in water quality was offset by the long-term environmental advantages of dam removal.[47]

It was in this overheated political environment that PacifiCorp had to make a decision regarding dam removal that really had nothing to do

Condit Dam on the White Salmon River, Washington, before a hole was blasted through its base in 2011.

with politics. Gail Miller and Terry Flores are the one-two punch of the PacifiCorp's hydro relicensing team. Together they delivered a no-nonsense, economically rational argument for removing Condit Dam. Miller explained: "Unlike the other parties, we don't have an ideology on this. We are not pro-dam or anti-dam. We have to protect our customers and our investments, and the best way to do this is to remove Condit Dam."[48] Miller has many years of experience working in project management. She is a businesswoman, but she has spent the last several years of her professional life up to her nose in water politics.

At first, a political resolution of the Condit Dam problem seemed within easy reach. Miller worked diligently with the stakeholders to work out a settlement agreement. Flores made sure everything followed the proper FERC relicensing process. "In 1999, we all signed on the dotted line and everyone went home happy, right?" Gail asked rhetorically. "No! The counties on both sides of the river suddenly decided they wanted to keep the dam."

Klickitat

There was a time when nearly every town in America looked like Goldendale, Washington, in Klickitat County. It is surrounded by vast fields of hay and grain. Without bothering to take a census, I would bet my paycheck that cattle greatly outnumber people in this county. Goldendale is the county seat of Klickitat County, which derives its lyrical name from a local Indian tribe. Main Street is lined with small, locally owned businesses, each one housed in a sturdy brick building with a square-top façade. There is an American Legion post, and the Simcoe Diner promises "high country dining." Many stores cater to farmers and the myriad equipment they need to wrest a living from the ground. The redbrick Methodist church, built in 1948, dominates one end of town. The county courthouse, just a block off Main Street, is an art deco building constructed in 1941 after the old courthouse burned. A recently completed restoration added layers of contrasting colors—pea green, mustard yellow, burnt umber—to accent the concrete moldings and graceful curves of the exterior walls. An obelisk out front honors the original pioneers who came to this area in the 1850s.

But recent years have not been kind to towns such as Goldendale. Indeed, it reminded me of my Dad's hometown of Dana, Indiana. Changes in lifestyle, values, and technology have created a new demographic landscape as people have moved to cities and suburbs. Many storefronts are vacant, their windows papered over with old newsprint or butcher paper—a kind of shroud for small-town businesses. While I was having lunch at the Stone House Café on Main Street, I struck up a conversation with Frank Clements, one of the regulars. He came here in 1962 to be the janitor at the school. He retired in 1982 but decided to stay in Goldendale. He clearly missed the better days, but he was still proud of his town. "Goldendale was named in the 1860s. A big local rancher named Golden donated land here to start this town; he had a daughter named Dale, so they called the town Goldendale. But it ain't what it used to be. We've lost a lot of stores. We used to have a J. C. Penney and a Safeway, but they've all closed. Now everybody goes to The Dalles or Yakima. Now I can't even buy a shirt in this town. This restaurant used to be a feed store."

It seemed to me that a town down on its luck, wishing for better days ahead, would be receptive to change, hoping for a new lease on life. But Goldendale and Klickitat County were the center of opposition to removing Condit Dam. The people here firmly believed that taking out a dam was a step backward, a loss to their community, which had long equated water "development" with prosperity. The most prominent leader of that opposition was Don Struck, a county commissioner when I paid him a visit in 2003 in his office in the county building. A soft-spoken, well-dressed man with a graying buzz cut, he looked more like an urbane professional from Los Angeles than a rural county commissioner. There were pictures of his five daughters all over his office.[49]

He chose his words carefully and began by explaining that he struggled to keep his emotion-laden childhood memories of Northwestern Lake—the reservoir formed by Condit Dam—from affecting his business sense. "I have lived near Northwestern Lake my entire life—nearly fifty years. My grandparents took us there for boating and picnics when I was a kid, and I have to separate that from the purely economic decision of what to do with the dam. As a businessman, I truly believe PacifiCorp should have the option to decide what to do with the dam. It's the manner in which it is slated to be removed that concerns us."

Struck was concerned that sediment released from the dam site would destroy the lower river. He also expressed skepticism that the salmon would actually return, and he was convinced that the loss of hydropower would result in higher energy costs for local residents. These issues had been dealt with extensively in the dam-removal planning process, and great effort had been made to avoid such problems. But I got the sense that Struck, like Frank Clement, whom I had met at the Stone House Café, simply missed the way things were. They loved their community as it had been and did not want it to change. But change is inevitable, especially when the status quo creates significant long-term problems. I suspected that the dam-removal issue was simply the veneer of a much deeper resentment against unyielding outside pressure to change.

Struck became visibly tense as he described his perspective on the environmentalists who support dam removal. "It flabbergasts me that the same environmentalists who I butt heads with on almost any issue—permits for wind-power projects, industrial site development—it seems like there is no end to the torture they like to inflict on us on any issue you can imagine. In this case, the enviros are oddly aligned with big business."[50]

It is true that the "enviros" and PacifiCorp signed an agreement, but only because environmental groups and the federal agencies responsible for enforcing the ESA were successful in convincing PacifiCorp that it was cheaper to remove a dam than to build a fish ladder. Struck saw this as an imposition by a distant federal government, but many local groups participated in a classic grassroots effort to convince the company to let go of Condit Dam. One instigator in particular caught my attention; her name kept popping up in conversations and interviews.

Longevity

I met with Phyllis Clausen, a leader of Friends of the White Salmon, one of the activist groups in the effort to protect and restore the river. She helped lead efforts to stop dams on the upper White Salmon in the late 1970s and to have parts of the upper river designated wild and scenic. She was one of the signatories to the 1999 settlement. We met on the gravel road that runs along a ridge just above Condit Dam. She is the quintessential instigator. Her deeply lined face was framed by curly white hair. She is a petite woman, perhaps 70 years old, but she exuded a finely leavened sense of energy. She talked with a slight tremor, but that did not interfere with her ability to express her position forcefully—something she has been doing for decades. She summarized for me the basic issues regarding the removal of the dam. But her most revealing comment came when I asked her why a broad spectrum of people named her as an effective advocate for dam removal.

> Longevity is part of it; I know what has occurred and why it has occurred. Perhaps it's because the river is so terribly important to me, and people can see that. I also stand up at meetings and speak, and I get out and talk to lots of people. And I hope it's because I try to work with everyone, including the people who are against removing the dam. I don't believe in negative stereotyping of the opponents; that just isn't right.[51]

There, in a succinct paragraph, is the formula for success for an instigator. My students often tell me that they don't get involved in politics
because they can't make a difference, that the whole system is dominated by rich, corrupt special interests. They need to meet Phyllis
Clausen.

In 2005 the original 1999 settlement was amended to give PacifiCorp
more time to obtain all the necessary permits for removal. This meant
even more studies. In the thirteen years following the signing of the
original settlement, there were no less than three major federal studies
of the project; all of them found that removal is the best option. Yet another study was initiated when mercury was discovered in some of the
sediment, but the state of Washington's Department of Ecology found
that dam removal would actually dilute and disperse the mercury and
thus improve the situation.[52] The FERC gave final approval for dam
removal in December 2010.[53] On October 26, 2011, nearly 800 pounds of
high explosives put an end to the debate over Condit Dam. The dramatic
explosion loudly announced the arrival of a new era.[54]

TEN THOUSAND YEARS OF FISHING

I met John Banks at the tribal government building on Indian Island
in central Maine. He is the director of the Department of Natural Resources for the Penobscot Nation, whose reservation consists of a string
of beautifully forested islands in the river of the same name.[55] It is no
coincidence that the river and the people on these islands share a common name; for 10,000 years this tribe has lived here, enjoying the bounty
of the river—a bounty that was lost but will return.[56] Banks was one of
the Penobscot Nation's representatives in a long, complicated series of
negotiations over the future of the river. I was not present at any of those
negotiations, but I can imagine him as a tough, no-nonsense proponent
for his people. He has dark, deep-set eyes, a shock of close-cropped gray
hair, and a rugged countenance.

Maine's rivers have always been viewed as its greatest natural resource.
Native Americans in this region thrived on the fishing and wildlife found
in and around the rivers. The Penobscot River alone supported Atlantic
salmon runs estimated at 50,000 to 70,000 and countless millions of ale-

wives, sturgeon, and shad—a total of ten different sea-run fisheries.[57] The native people here invented birch-bark canoes to travel these rivers easily, devising a design that is still in use today (the Old Town Canoe Company, adjacent to the reservation, still makes canoes based on the Penobscot design). In the early days of European settlement, rivers were used to float giant log rafts, and watercourses were dammed to power sawmills and gristmills. Later, using the rivers meant damming them to create electricity. The rivers also proved useful as a dump for polluted refuse from pulp mills. The Penobscot people and everyone else who lives in this state must now contend with the legacy of these uses. But today the concept of "using" rivers in Maine has broadened considerably. John Banks explained it to me. "We used to just pick rivers apart, but people here are starting to look at rivers differently—as a whole system—for their value as a fishery and for their beauty." Banks knows that perspective well; it has been the philosophy of the Penobscot people since time immemorial. "It's only new to some people," he observed.

Banks showed me two of the dams on the Penobscot slated for removal as a result of a carefully negotiated settlement between the power company PPL, environmental and fishing groups, the state, several federal agencies, and the Penobscot Nation. The settlement agreement, finalized in 2004, provides that the two lowest dams on the river, Veazie and Great Works, will be removed, and another dam, Howland, will be decommissioned and bypassed with fish ladders. A fourth dam on the river, Midland, will remain in place but will be fitted with fish elevators. In exchange for selling three dams, PPL will be allowed to improve six other dams on the river to make up for the lost hydropower, resulting in only a slight drop in generating capacity.[58]

We jumped in Banks's car and headed south. Just downstream from the reservation is Great Works Dam—truly a misnomer. The elements have not been kind to this dam. Slap boards, removable horizontal surfaces that can be placed across the top of a dam to increase its height, covered in white plastic have been placed across part of the dam's lip to raise the water. A concrete abutment in the center of the dam has eroded so badly that it looks more like a sculpted sand castle than weight-bearing concrete. Aging timber cribs support a portion of the dam. It is easy to see why PPL, the power company that owns this dam, could see the benefits of selling it rather than repairing it and adding fish passage.

Howland Dam, Penobscot River, Maine. This dam will be bypassed and decommissioned.

A few miles below Great Works Dam is Veazie Dam, also slated for removal. It is the first dam on the river and stops most anadromous fish from continuing upstream. There is a narrow, aging fish ladder near the center of the dam, and a few salmon manage to climb it. On the day I visited the dam in 2004, a large sign at the top of the ladder announced the salmon count for that day: three.[59] The fish ladder is even less effective in assisting other species of fish over the dam, such as alewives and sturgeon. Banks and I viewed the dam from the vantage point afforded by the cabin of the Veazie Salmon Club, which sits on a bluff overlooking the river just downstream from the dam. He pointed to a sign posted near the cabin that stated, "Closed to all fishing between Veazie Dam and Bangor Water District."

"You can't fish here anymore, even though it's a salmon club," he explained. "Further downstream, on the other side, is the Brewer Salmon Club. You can't fish there either. See that old shade shack across the river? In earlier years there'd be hundreds of fishermen over there. Now there's just not enough fish to support that level of use."

The winters here are harsh, and dams are thrust into the most demanding environment. Rivers flow with great force in the spring and freeze to depth in the winter. The temperature variation is as stunning as the variation in water flows. These harsh conditions dramatically increase the maintenance cost of dams. Ironically, the increased cost assists in efforts to remove dams that no longer create more benefits than costs.

A few miles below Veazie Dam is the site of Bangor Dam, just outside the city of the same name. That dam fell into disrepair about twenty years ago and was removed. Before its removal, spawning Atlantic salmon would come up the Penobscot and thrash wildly below the dam, their swim upstream frustrated by the great wall of concrete. Fishermen used to congregate there by the thousands, snagging the trapped fish. There was a Bangor tradition of sending the first salmon caught to the president of the United States as a symbol of Maine's natural abundance. The Bangor Dam is gone now, but the custom of giving a fish to the president has not been revived. There are simply not enough salmon left to justify it; rather than a symbol of Maine's abundance, sending an Atlantic salmon to the White House these days would be more of a call for help. There are so few salmon left that the Penobscot Nation created a media circus a few years ago by netting just two salmon. John Banks told me that local non-Indians complained about that. "They said Indians should not be allowed to net fish when the stock is so low." Never mind that the Penobscots have been netting fish here since the Stone Age and had nothing to do with the fishes' decline.

The settlement agreement did not automatically take out dams. It merely gave the Penboscot coalition an option to purchase Veazie, Great Works, and Howland dams for $25 million. The coalition formed a nonprofit trust in 2005 and began raising the necessary cash. The trust also had to obtain permits from the FERC, the Corps, and the Maine Department of Environmental Protection and prepare a National Environmental Policy Act (NEPA) document.[60] By 2008 the Penobscot River Restoration Trust had raised the cash through a combination of private donations and some help from Uncle Sam, and it purchased the three dams in 2010.[61] The project received another big boost when it was awarded $6.1 million in 2009 from the American Recovery and Reinvestment Act.

This money will allow the Trust to begin removing Great Works Dam, with demolition scheduled for the summer of 2012.[62]

The Penobscot agreement could serve as a template for other multi-party restoration settlements. It involved a great range of parties, representing very diverse perspectives. But they were able to reach agreement only after long and difficult negotiations. In 2008 the project received a Cooperative Conservation Award from the Interior Department.[63] However, it may be presumptuous to think that such progress will be the norm in most river basins. Most people are much too fond of arguing over water for that. In addition, the people of Maine seem to have a special knack for getting along. There is a mode of considerate, concerned interaction here that is not found in many other states. Perhaps there is still sufficient natural space left here to allow people to appreciate the presence of others. Many Americans live in overcrowded cities and suburbs, with little open space to feed their sense of self. In Maine, encountering another human being, either on the street or in a meeting, is still a respectful exchange in most cases. With that attitude, we could all make greater progress toward resolving our water dilemmas.

A FOOT DEEP WITH FISH

The Presumpscot River, also in Maine, is not as famous as the Kennebec or the Penobscot, but it holds great promise. There was a time when this river was a churning white-water free fall, with a dozen small waterfalls and rapids as it dropped 270 feet in just 25 miles. The Indian name of this river means "many falls." Dam building began early in this area, and the first dam was built on the Presumpscot in 1732. Nine more were to follow, effectively converting the river into a series of stepped reservoirs. Before the dams were built, this river was a rich habitat for fish; an early account says that the "entire surface of the river, for a foot deep, was all fish."[64] As required by colonial and state law, the dams had fish passages, but floods and weather eventually destroyed them, and they were not replaced. By 1900 anadromous fish had virtually disappeared from the river.[65]

The Presumpscot followed a pattern that became common for Maine's rivers. Innumerable dams were built to power sawmills and pulp mills, which in turn polluted the slow-moving water trapped behind the dams. The rivers became linear dead zones. In some areas development did not occur at the river's edge because of the stench and danger posed by the toxic water flows, which included dioxin. This continued until the Clean Water Act and environmental awareness began to change the political equation.[66]

In the case of the Presumpscot, the catalyst for action was a proposal in 1992 to build a huge de-inking plant that would dump millions of gallons of wastewater into the river. Dusti Faucher, who lived near the river, gathered with some neighbors to form what she called a "kitchen table organization," the Friends of the Presumpscot River.[67] Dusti was not wealthy, had no political experience, and was not well connected, but she was a determined instigator. The organization's goal was to stop the de-inking plant. "We raised $20,000 through bean suppers and yard sales to hire a lawyer. We appeared before the Windham town council, but they told us to sit down and shut up." Instead of shutting up, they got the issue placed on a referendum and spent weeks going door-to-door garnering support. They won. This victory was the beginning of a new era in the history of the Presumpscot.[68]

Friends of the Presumpscot, working with half a dozen other groups in a coalition, set its sights on restoring the river as a viable habitat for large numbers of fish. It established an ambitious agenda: remove three dams and install fish passage on three others while improving habitat and water quality. The lowest dam on the river, Smelt Hill Dam, built as Maine's first hydropower dam in 1889, would be its first target. The next dam upriver was Cumberland Mills, which did not produce hydropower and therefore did not require an FERC license. Upriver from there were seven hydro dams, all without fish passage, and a final nonhydro dam. Seven of the ten dams were owned by SAPPI, a South African paper company that had a mill on the river that was the source of much of the pollution in the river. The FERC relicensing process would create an opportunity for the Presumpscot coalition to work for change.[69]

The water quality in the river improved dramatically when SAPPI closed its mill and local towns built more effective water-treatment plants. As the water quality improved, the state of Maine began stocking the various impoundments. A fish lift was installed at Smelt Hill Dam in the 1980s but was demolished by a flood a few years later.[70] The coalition experienced a big victory when Central Maine Power agreed to remove the damaged Smelt Hill Dam in 2002, opening up the lowest 7 miles of the river to anadromous fish.

The Presumpscot flows into Casco Bay. One of the legacies of the effort to restore the Presumpscot and other nearby rivers and estuaries is the Casco Bay Estuary Partnership, with a mission of "preserving the ecological integrity of Casco Bay and ensuring compatible human uses of the Bay's resources."[71] When I visited in 2004, it was directed by Karen Young. Although she was originally from Texas, she had readily adopted that Maine sense of affability and love of the land. To her, "The environment is at the core of what it means to be a Mainer." She spoke highly of all the stakeholders in the process, including those who oppose restoration efforts. She also had an acute understanding of the relationship between the health of the river and the fortunes of the bay. After she described to me, with some understandable pride, the progress the partnership had made on a number of issues, she volunteered to take me to the section of the river that had previously been inundated by the Smelt Hill Dam, which had been removed two years earlier.

The former dam site and lower river were reached by a short but pleasant walk through a wooded area. A land trust called Portland Trails purchased this land along the river and put in the trail, so the restored river and the preserved land create a beautiful little corner of wild Maine at its finest. The lower section of the river runs through a narrow but shallow gorge with several rapids and a small waterfall. The banks at the former dam site have been recontoured, but one of the stone cribs from the dam has been left in midchannel. However, spring floods and ice floes will undoubtedly remove this dam remnant in a few years.

The newly restored section of the Presumpscot and the adjoining land trust created an oasis in the midst of the suburban sprawl of Portland. The former reservoir bed is now a swift-flowing stream coursing between boulders. On the day I visited, the rapids above the former dam site were filled with alewives, literally thousands of them, churning up the water with their energetic swimming. It is amazing how quickly nature reestablishes herself if species are just given a chance to access their natural habitat. The removal of Smelt Hill Dam was just the first step in restoring the Presumpscot River.

As part of the effort to improve habitat and water quality on the river, a collaborative planning process was established. Initially SAPPI, the paper company, participated but withdrew in 2002, apparently more interested in doing battle in court than in reaching a mutually acceptable agreement, as on the Penobscot River. Apparently not everyone gets along well in Maine. SAPPI was forced by the FERC in 2003 to install fish passages on five of its dams as a condition of its license renewal. The FERC also required the company to get permits from the Maine Board of Environmental Protection, as required by Section 401 of the Clean Water Act. Instead of complying, SAPPI sued, arguing that the Clean Water Act did not apply to its five dams because the dams did not "discharge" water into the Presumpscot, a navigable river. That would appear to be a flight of hydrological fantasy, but it took its case all the way to the U.S. Supreme Court, where it lost decisively. It is rare for the Roberts Court to agree on anything, especially environmental matters, but the justices overcame their differences after they looked up the word "discharge" in the dictionary and then ruled unanimously that SAPPI was not, as it claimed, immune from the Clean Water Act.[72]

After its stunning defeat in the Supreme Court, SAPPI became slightly more amenable to discussions. One of the big challenges facing proponents of habitat restoration was Cumberland Mills Dam, which, after the removal of Smelt Hill Dam, was the lowest dam on the river. Because it did not produce hydropower, it did not need an FERC license, so there was no way to force SAPPI to install fish passage. This effectively blocked the entire upper river to fish. River advocates negotiated with SAPPI to remove Cumberland Mills Dam, which led to a preliminary agreement in 2007 to do so. But SAPPI then decided to back out of the negotiations, prompting American Rivers and Friends of the Presumpscot to request that the state of Maine force the company to abide by Maine law and provide fish passage. In June 2009 the Commissioner of Inland Fisheries and Wildlife ordered SAPPI to install fishways at the dam.[73]

River advocates are optimistic that an impressive array of fish will return to the Presumpscot. They predict a dramatic return of Atlantic salmon, shad, alewives, and blueback herring. As successive dams are removed or fitted with fish passage, this river may once again be a "foot deep" with native fish.

THE OTHER SIDE OF THE RIVER

Ken Fletcher, like most Mainers, is hospitable and amiable. One immediately gets the feeling that he likes to interact with all kinds of people. In 2004 I met Ken and his wife Mary Ellen at a small park in Winslow, overlooking the Kennebec River.[74] He is widely recognized as a leader in the effort to stop dam removals in Maine. Ken spent his professional life working in the pulp and paper industry and lived next to the impoundment created by Fort Halifax Dam on the Sebasticook River, a tributary of the Kennebec. He is a quintessential New Englander—hardworking, no-nonsense, loyal to his community. He retired a few years ago and planned to spend most of his time canoeing on the impoundment and taking it easy. But those plans changed when he received a letter asking him to attend a meeting regarding the fate of Fort Halifax Dam. At that meeting he learned that an agreement had been signed that virtually assured that the dam would be removed.

"We realized that dam removal, not fish passage, was their objective. And no one in our community knew about this agreement," he said. "It was a sneaky deal that was pulled, and the community had to deal with it."

The "deal" to which Fletcher referred was a 1998 settlement by the dam's owners, federal agencies, and conservation organizations. In that deal the dam owner committed to either constructing fish passage or surrendering its FERC license.[75] The subsequent owner of the dam, FPL Energy Maine Hydro, then determined that removing the dam was cheaper than building fish passage.[76]

Dam removal is occasionally noncontroversial, but in most situations it signals a change in the status quo and possibly a redistribution in the benefits bestowed by a river. Fletcher and his wife viewed the removal of Fort Halifax Dam as a dramatic loss to them personally and to their community. They met with other people living around the impoundment, formed Save Our Sebasticook (SOS), and began working to save the dam. A year later Fletcher won a seat in the Maine House of Representatives, largely on the basis of the reputation he established working to save the dam.[77] "I had never run for office before. I'd never even paid much attention to politics. But when they started taking out dams, I felt like I couldn't just sit by and let it happen."

Fletcher was very effective in listing the potential drawbacks and costs of removing Fort Halifax Dam.

> We now have a four-season lake: spring is fishing, summer is boating and kayaking, in the fall we hunt waterfowl, and in the winter the lake is the major thoroughfare for snowmobile traffic. There are thousands of people who use it for their snowmobile route. That all will be gone. What will be left is 266 acres of denuded land; we've seen it drained when they did maintenance, so we know what it looks like. There are threatened species in there: yellow lamp mussels, tidewater muckets [a freshwater mussel], and snails. It will take years for it to ever come back and be productive. It will flood in the spring and dry out in summer.

Ken Fletcher's arguments reminded me that all changes in policy incur costs; restoring rivers creates enormous new benefits for society, but there are also losses—that is the nature of political decision-making. Fletcher's preferences and values are no less valid or genuine than those of the persons who want to remove Fort Halifax Dam, and his passion for his point of view serves to remind us that everyone should have a voice, all stakeholders have legitimate claims, and the best restoration projects are those that are designed to minimize the losses, compensate those who incur these losses, and look for win-win combinations. River restoration should always be an exercise in democracy, practiced by a civil society. This is what I mean by a "river republic."

In the case of the Fort Halifax Dam, Ken and Mary Ellen Fletcher fought a valiant fight for their dam and reservoir. But, as Ken predicted, the 1998 settlement set in motion a series of decisions that ultimately led to the removal of the dam. On July 17, 2008, after the FERC issued the necessary orders, heavy equipment pounded a large hole in the dam, and after 100 years of impoundment, the river flowed freely once again.[78]

SHOW ME THE MONEY

Is it worth it to spend billions of dollars restoring rivers and saving fish and riverine wildlife? Aside from the arguments for esthetics, ecological health, and species diversity, that question can be answered with some raw numbers. A 2011 analysis of Department of the Interior lands found that national wildlife refuges generate $3.9 billion in annual economic activity.[79] According to a 2000 study by the Recreational Boating and Fishing Foundation, there are 35 million adult anglers and 15 million youth anglers in the United States. They contributed $116 billion to the national economy and paid $7.3 billion in tax revenues.[80] A national survey in 2004 found that over 80 million Americans had gone fishing.[81] Another survey found that recreational fishing generated $42 billion in 2006.[82] The American Sportfishing Association claims that sportfishing generates $45 billion in retail sales and a $125 billion "impact' on the economy, creating a million jobs.[83] Of course, not everyone fishes just for fun; for many people, fishing is their livelihood. In 1980 the NOAA estimated the value of the domestic commercial catch at $2.2 billion; by 2006 that figure had jumped to $3.9 billion.[84] Fishing supports more jobs than Exxon-Mobil, General Motors, and Ford combined, according to fishing advocates.[85] And the typical fisherman hasn't needed a multibillion-dollar federal bailout or a secret energy deal with the vice president. They just go fishing.

What the nation needs at this critical juncture is a comprehensive study of how much money, how much water, and how many river miles are devoted to fish and wildlife, compared with the extractive uses described in chapters 4 through 8. This comparison should also estimate how much money is generated by those traditional activities in contrast to the economic activity generated by both sport and commercial fishing and by associated wildlife, including an estimate of the damage done to the nation's fisheries by pollutants from these industries.[86] Does it make sense to destroy vast tracts of habitat for the sake of a handful of barges? Should flood-control structures convert species-rich wetlands into fields to grow surplus crops? What is the economic value of a minor increment in hydropower compared with that of a robust run of anadromous fish? What are the health costs of eating fish tainted with pollutants? If we are ever to have a reasonable, economically viable national river policy, we must be able to answer these questions.

The Canary River

Every healthy river is a great metropolis of species, with the attendant crowding, competing for advantage, and status hierarchy and a delicate dance between voracious predators and symbiotic coexistence. Like any vibrant city, the whole is much greater than the sum of its parts. Yet we manage aquatic species as though this complex web can be parceled out to various administrative entities and expected to survive. Heaven forbid if a steelhead trout (a NOAA fish) bumped up against a rainbow trout (a Fish and Wildlife fish). That could lead to a nasty fight between the Commerce Department and the Interior Department. Thank goodness these two fish don't interbreed, or we'd have a major turf war going on.

Would it not make sense to manage interdependent species as a holistic system rather than through piecemeal allocations of authority that create turf battles and disjunctive policy? The advocates of state control over water resources will object to increased federal suzerainty, but rivers and their species do not conform well to straight lines on a map. We should recognize both hydrological and ecological reality rather than perpetuate past models of ineffective balkanized species management. The concept of integrated water-resources management points in this direction, but to be truly effective it must encompass all aquatic species, as well as species that depend on riparian zones.[87] Our goal should be the survival of species, not the survival of every bureaucracy. It is common for animal and plant species to go extinct, much less so for government agencies and programs. Perhaps it should be the other way around.

We often hear the metaphor of a "canary in a coal mine," the idea that certain species give us fair warning of impending doom. I think that rivers serve as the canary for entire ecosystems. When a river dies, along with its host of nonhuman inhabitants, it is time to run like hell, just as one would in a coal mine when the canary pitches off its perch.

It is human nature to have conflicting feelings about wildlife. Some we love, some we poison, and some we eat. The acrimonious debates over habitat and endangered species are indicative of a larger philosophical difference over the proper relationship between humans and other species. To some people, preserving nonhuman species is a moral imperative; who are we to serve as a star chamber for all other forms of life? But to others, it is a question of putting human needs before those of animals and plants. It is not uncommon to hear people in extractive industries declare that they are the endangered species. But if we manage our rivers smartly, there will be enough for all of us, humans and nonhumans alike.

The debate over habitat and species preservation will always be emotional; just witness the current controversy over wolves. A similar level of worked-up antagonism can be found in the salmon wars described earlier, the effort to save four endangered "trash fish" in the Colorado River, and the fight over bull trout in western mountain streams. Public policy regarding riverine species will be shaped by science, special-interest politics, and, at times, the hopelessly inane. A lobbyist for the Washington Association of Wheat Growers demonstrated this point quite clearly when she decried efforts to save wild salmon: "I applaud the people that are trying to save species that are endangered. But it might be good that we don't have dinosaurs now. We've gotten oil from the dinosaurs. If we had preserved the dinosaur, we wouldn't have that oil."[88] As the debate over riverine habitat grinds on, there will undoubtedly be a lot more talk about dinosaurs, endangered species, invasives, sustainable populations, and whether the human race fits into one of those categories.

11

PLAYGROUND ON THE MOVE

River Recreation

I often come to this quiet place,
To breathe the airs that ruffle thy face
—FROM WILLIAM CULLEN BRYANT, "GREEN RIVER"

My earliest memory of fishing begins on my grandmother's front porch in Oakland City, Indiana, when I was about eight or nine years old. She had awakened me early, and I went out to the front-porch swing to eat my oatmeal and psych myself up for the biggest adventure of my life up to that point: Grandpa was taking me fishing. It was going to be just the two of us—my sister was excluded, thank God. The mourning doves were in the maples making that soothing cooing sound when Grandpa came out and sat beside me, mixing peanut butter into his oatmeal, as he had done since he was a kid himself. He had been a coal miner all his life and had retired early because of black lung disease. During the Depression the mines closed, and he fished with a homemade pole and hunted rabbits with a .22 rifle to keep his family from starving. To him, fishing was not a sport, it was part of survival. To me, it was a lark, a grand adventure into the unknown.

Grandpa gathered up the requisite cane poles, bobbers, and hooks and dug a few worms, and off we went in his 1957 Chevy. He parked the car next to a one-lane bridge, and we walked down to the banks of a river. I truly wish I could remember its name, but I cannot. It was muddy and running fast and high. I tried not to show it, but it scared me right to the bottom of my shoes. Grandpa said the river was too high to fish, but I said, bravely, it was still fishable. He said the fish could not see or smell the bait in such murky, quick water, but I said fish weren't that stupid. So he dutifully baited my hook for me and told me to watch the bobber for the telltale sign of a bite. The current was so fast that had I fallen in, I would have drowned before Grandpa could yell out that he couldn't swim, which was true. So he held on to the back of my belt while I tossed my line into the water. The swift water immediately carried my bobber downstream until the line grew taut, and then drove it to bank, not 10 feet from where I was standing. I repeated this ritual several times, Grandpa patiently holding on to me while I awaited the big strike. The fishing trip ended when I got hungry. I didn't even get a nibble. I guess fish really are that stupid.

Back home in Lawrence, Indiana, I fished a few more times after that in Fall Creek. I never caught a keeper, but it didn't matter. I was out there, enjoying the little strip of riparian land that had not been turned into cornfields, hog farms, or suburbs. In central Indiana, rivers provided the only natural respite from the pressures of life. As I grew into my teen years, the "fishing trips" had less to do with catching fish and more to do with the awkward unmentionables of youth, but the sense of place—that combination of green moist bank and light-reflecting water—was at the heart of the experience.

I am not alone in my appreciation of the play among light, land, and water. As we will see in this chapter, people love to recreate on or near water, especially in places that possess natural, untrammeled beauty. Humans are instinctively drawn to nature, which, on "the water planet," is incomplete without that life-giving nectar. That is true even for deserts, where the tiniest rivulet is cause for celebration. Yet as a society we have tended to dismiss recreation and water-related tourism as a lesser "use" of rivers. In the past, it made at least some sense to diminish the beauty of rivers for the sake of crops, barges, hydro, flood control, and sewage disposal, but today we still routinely give priority to extractive river uses even though recreational uses often generate far more money. That is a bit like turning down a hundred-dollar bill because your hands are full of pocket change.

It is impossible to understand the economic impact of recreation and tourism without a barrage of statistics. A litany of such numbers usually produces a case of sudden-onset ADD in the average reader, but the numbers are so stunning that they may in fact hold your interest. I will begin with the larger context of tourism and work toward the statistics for water-specific recreation and natural-area recreation. We will then

Coyote Gulch, Grand Staircase–Escalante National Monument, Utah. The Coyote is a tributary of the Escalante River.

spend some time with river guides in Cataract Canyon on the Colorado and visit a habitat-restoration project in upstate New York.

MOVING-WATER RECREATION

Tourism is big business in the United States. According to the Travel Industry Association of America, tourism is a $1.3-trillion industry that generates $100 billion in annual tax revenues.[1] Commerce Department statistics indicate that tourism employs 8.3 million Americans, and foreigners spend $107 billion in the United States each year.[2] One study referred to travel and tourism as the "largest business on earth."[3] Why do we travel so much? And why do people from other countries visit the United States in droves? Do they travel great distances to see clear-cuts, sprawling suburbs, mine waste, polluted rivers, and de-watered streams? I think not. A poll by the President's Commission on America's

Outdoors found that the single most important criterion for selecting an outdoor travel destination was "natural beauty."[4] In another poll 26.6 million Americans defined themselves as "nature lovers," and another 55 million indicated that they like to engage in the mildly lascivious-sounding "studying nature near water."[5] I think that was what we were doing in high school out by the reservoir.

One of the fastest-growing segments of the tourist industry is eco-tourism, which is described as "responsible travel to natural areas which conserves the environment and sustains the well-being of local people." Two-thirds of all American travelers are interested in ecotourism, and they generate $77 billion a year.[6] That's billions, not millions. In recent years ecotourism has grown at a rate of about 10 to 12 percent annually.[7]

How many of these millions of Americans play in or near the water? The short answer is: most of them. Water-based recreation can be divided into two categories: flat-water recreation and moving-water recreation.

To be sure, flat-water recreation—the kind that takes place on natural lakes and slow-moving rivers, but also on America's thousands of reservoirs—is enormously important. There is considerable demand for this type of recreation. Federal reservoirs have 900 million visits annually and generate $44 billion in recreation spending.[8] Americans own 17.5 million boats.[9] In a recent survey, 65 million Americans said that they participated in motorboating.[10] These data make a compelling argument for the recreational utility of dams and the value of "fishable and swim-mable" water promised by the Clean Water Act.

To meet this demand, there is an enormous supply of flat water. Over 85 percent of the nation's waterways have been developed.[11] There are 1,782 reservoirs behind federal dams.[12] There are 2.6 million small artificial ponds and reservoirs and 28,000 lakes.[13] According to the Environmental Protection Agency (EPA), there are 60 million acres of lakes and reservoirs.[14] In short, there is a big demand for flat water, but plenty of supply, primarily due to our penchant to dam virtually every watercourse.

It is *moving water*—natural, clean, ecologically intact rivers—that is extremely scarce. To a great extent, the reason that there is so little moving water is that most of it has been converted into flat water. Indeed, neither the Corps nor the Bureau of Reclamation has included "moving-water recreation" in its cost-benefit analyses, but they have often inflated estimates of flat-water demand to give their analyses the required positive ratio.[15] In other words, the nation's two biggest water developers do not even conceptualize moving water as a resource, despite the fact that the demand for moving-water recreation is enormous.

According to the 2004 National Survey of Recreation and the Environment, 33 million people go rafting, 15 million like to kayak, and 27 million canoe. But perhaps the greatest use of living rivers is associated with land-based activities that take place within their watersheds. Over 85 million Americans go hiking, 28 million hunt, 72 million visit wilderness areas, and nearly 26 million go backpacking.[16] This is in addition to the tens of millions of Americans who fish and were described in the previous chapter.

These recreationists spend a lot of money. A recent report that focused on the economic benefits of clean water noted that "clean water supports a $50 billion a year water-based recreation industry, at least $300 billion a year in coastal tourism, a $45 billion annual commercial fishing and shellfishing industry, and hundreds of billions of dollars a year in basic manufacturing."[17] According to the 2006 National Survey of Fishing, Hunting, and Wildlife-Associated Recreation, the nearly 88 million Americans who engaged in wildlife-related recreation spent $122 billion.[18] Fishing and hunting alone generate $70 billion.[19] River running, even though it is a highly specialized sport, still makes big money—$70 million in Colorado and $20 million in West Virginia, for example.[20]

Wildland and natural-area visitation also generate huge sums. Hikers spend about $18 billion a year.[21] Muscle-powered recreationists spend $13.2 billion annually on gear and clothing.[22] Bird-watchers are big spenders, dropping a cool $25 billion to get a glimpse of the increasingly rare wild birds of North America.[23] Visitors to state parks in 2006 spent $876 million. The people who engage in "wildlife watching" spent $45.7 billion in 2007.[24] The national parks generate $6.3 billion in labor income and $9 billion in "value added," supporting 244,000 local jobs.[25] Recreational visits to national forests contribute $13 billion to the gross domestic product.[26] Recreation and tourism on Department of the Interior lands generate 388,000 jobs and $4 billion in economic activity.[27] The outdoor industry as a whole in the United States generates about $730 billion and supports 6.5 million jobs.[28] Even in the midst of recession the outdoor industry grew by 4 percent in 2010.[29] The Outdoor Industry Association has 1,200 members.[30] As the "America's Great Outdoors" Commission noted, "Recreation and tourism and related businesses and enterprises have become powerful elements of rural and urban economic development."[31]

Recreational tourism is especially important to the American West.[32] My home state of Utah had nearly 21 million tourist visits in 2010.[33] In 2009, 5.5 million people visited national parks in Utah; can you imagine hiking the Zion Narrows with no water in the Virgin River?[34] Outdoor recreation in Utah generates $5.8 billion annually.[35] In Arizona outdoor recreation supports 82,000 jobs and generates $350 million in annual state tax revenue; in California it produces a whopping $28.1 billion in retail sales and services.[36] In New Mexico the sum is $3.8 billion.[37] In most western states recreational tourism is the largest private industry.[38]

Although most rivers have been compromised, the nation has made quite an effort to provide recreational opportunities and some degree of protection for what is left. The best known is the system created by the Wild and Scenic Rivers Act of 1968, which provides protection for river segments that possess "remarkably scenic, recreational, geologic, fish and wildlife, historic, cultural, or other similar values."[39] There are 12,000 miles on 252 rivers in the system, including the 1,100 river miles protected by the Omnibus Public Lands Management Act of 2009.[40] But there are other forms of river protection, including designation as "outstanding national resource waters" under the Clean Water Act.[41] The Nationwide Rivers Inventory is a list of more than 3,400 free-flowing river segments that deserve special protection.[42] To assist us in enjoying the best parts of America's rivers, we have the North American Water Trails, greenway corridors, and Blue Trails.[43] Congress has designated two national water trails: the Captain John Smith Chesapeake National Historic Trail and the Star Spangled Banner National Historic Trail.[44] And numerous towns and cities are building water parks to attract kayakers and other white-water enthusiasts.[45]

In short, America is having a great time outdoors, and much of that activity is taking place on our rivers or in natural landscapes dependent on healthy rivers. There are 600 professional guiding and outfitting companies in the United States to assist us as we enjoy these great amenities.[46] One of the most exciting and dynamic of these activities is white-water rafting. Let's go for a ride.

Escalante River, Utah. It arises in a national forest, flows through the Grand Staircase–Escalante National Monument and the Glen Canyon National Recreation Area, and into Lake Powell.

AGENTS OF CHANGE

In May 2005 I joined the Colorado Plateau River Guides on its annual training trip through Cataract Canyon.[47] My job was to talk about the politics of water and explain the law of the river—that unique body of interstate compacts, statutory law, case law, and administrative law that controls the Colorado River so strictly that it could justly be called the River of Regulation. My audience would be that unique breed of action entrepreneurs who make a living doing what other people do on vacation.

The professional river runners on the trip were a diverse lot. Some were disfranchised refugees from "civilization." Others were consumed by a love affair with rivers. And a few did it simply because they enjoyed the adrenaline rush during the day and the bibulous socializing in the evening. It was my job to explain to them that they were powerful spokespersons for the new era of the River Republic.

On the first evening of the trip, I began explaining their role in the body politic. They were skeptical, to say the least. "Why should we care about this?" one of them asked. "We can't do anything about it." A young woman who was clearly a leader among the group piped up. "We're just river guides. You should be talking to the big shots who make all the decisions." She did, of course, have a point, but she didn't realize that ecotourism and river recreation are a dominant force in the western economy. River guides are not bit players in a sideshow but rather admired specialists with a captive audience of river-loving travelers. In other words, they were in a position of influence, albeit in a different setting than the six-figure lobbyists plying the halls of Congress.

I talked about how one person can truly change the world. I told them the story of Candy Lightner, the woman who started Mothers Against Drunk Drivers. I told them about Zach Frankel, the ponytailed student described in the preface who changed the nature of the debate over water in Utah by starting the Utah Rivers Council. I mentioned Rosa Parks and the names of other stalwart but "powerless" agents of change. They began to listen. They began to see that they actually could play a role in protecting the rivers they loved—the rivers that give them a living. Such is the nature of a slow, creeping revolution that crawls up the spine of society, one vertebra at a time.

As we descended into Cataract Canyon, the rapids became more entertaining and gradually more intimidating. We were riding in big motorized rafts, favored by commercial river rafters because they can haul large numbers of passengers and huge loads of gear. But they tend to trowel through rapids, which makes them safe and secure at the expense of adventure and thrill. Apparently there are a lot more clients who prefer the former to the latter. I tried to imagine how it would be to run the big springtime rapids of Cataract in my little 14-foot self-bailer, but the experience would be so fundamentally different that I could not bridge the gap. So I sat back and enjoyed the security and comfort of the big rigs, chatting with the guides about everything from white-water technique to the best box wines—both of which have their partisans and their critics.

A trip through Cataract Canyon is a lesson in both geology and the history of river development. We passed the site of a proposed dam that would have inundated the Colorado River up to and including the town of Moab, now a tourist mecca.[48] A National Park Service river ranger with the nickname of T-berry shared with us his encyclopedic knowledge of the history of commercial rafters, explaining the development of J-rigs, G-rigs, and snout-boats. But the best lesson of the trip came from the river corridor itself. As we roared through the "big-drop" rapids, the canyon began to clog with great heaping mounds of sediment left by a retreating Lake Powell. The river cut a trough through this stranded desert soil, which calved into the water with regular frequency. It was a totally artificial landscape—a deeply fissured canyon normally scoured by a mighty river, but now choked with tumescent drifts of misplaced river delta. Downriver, in the Grand Canyon, the river corridor is dying for lack of sediment; here is that missing sediment, pushed upriver hundreds of miles by the plug of Glen Canyon Dam. The beaches of the Grand Canyon are disappearing. The sediment they need exists in great abundance, but it is stalled in Cataract Canyon. The fate of all three of these canyons—Cataract, Glen, the Grand—is tied together by Glen Canyon Dam, a dam built to serve a river compact based on a miscalculation of river flows and a myopic vision of future needs. Cataract Canyon's surfeit of sediment is robbing the Grand Canyon of its future (more about this in the next chapter). What was Floyd Dominy thinking? Surely he and his predecessors did not comprehend the connection between these three canyons—a connection sundered by Glen Canyon Dam that would slowly suffocate all three.

After running the big drops, we enjoyed the rare privilege of running Imperial Canyon Rapids, which had been inundated by Lake Powell for forty years but was exposed by four years of drought and a much-diminished reservoir level. Immediately downriver of this rapid the water began cutting a new channel through the deep beds of sediments, not necessarily its natural course. New rapids would form, only to be eroded in a matter of days. In a sense it was like being a pioneer river runner, not knowing what lay beyond the next bend. Just a few weeks earlier a couple of friends of mine had flipped in a new rapid near the takeout where the river abruptly poured over a break in the wall of crumbling sediment. On our big motor rigs we just watched as the river toyed with us while it moved tons of sediment in search of old haunts.

The National Park Service, which administers the Glen Canyon National Recreation Area, has bulldozed a new takeout ramp for Cataract Canyon river runners on the north side of the river, within sight of the abandoned Hite Marina to the south. Hite used to be crawling with thousands of motorboats; now it was miles from the upper end of the diminished reservoir. Silence had replaced the roar of motors, and a river channel had replaced a broad expanse of sedentary lake. Across from the takeout an unstable 30-foot cliff of congealed sediment pushed the river to the north; the river pushed back, peeling off vulnerable slices of dirt that calved into the river with a splash and continued down-canyon to form yet another nail in the coffin of Lake Powell.

In addition to numerous river guides and a few invited speakers, our river group included some honored guests who had pioneered the business of river running. One of them was Kent Frost. At the time of our trip Kent was 88 years old. On our third evening on the river he explained to me how he got into river running. "I grew up in Monticello, Utah, on the side of a mountain. Unlike most of the farms in the area, we didn't have a spring on our property—couldn't afford it, so we had to haul water. When I was a little boy, it was my job to hitch up the wagon team and take a load of barrels to the nearest spring to be filled. It always seemed like we never had enough water. So I always wanted to be around more water."

As soon as Kent was old enough to start wandering, he did—always in search of more water—which meant that he gravitated toward rivers. In 1937 he heard about a guy, Norm Nevills from Mexican Hat, Utah, who took paying customers for rowboat rides on the San Juan River. Nevills had come up with a wildly unusual idea—that people would actually pay money to be rowed through rough water. It wasn't long before Kent and Norm were working together.

The first boats they used were nicknamed "horse troughs." Long, narrow, and unwieldy, they had a decided tendency to show their bottoms, like so many cancan girls. So Norm came up with a new design, dubbed the San Juan boat, and Kent helped him build several of them and row them. They weighed 700 pounds and were made entirely of wood, but they had a broad flat bottom and a lot of rake (upturned on the ends), so they performed well amidst the churning, rock-laden waters of the San Juan. They could also be rowed easily through the calm waters of Glen Canyon. It wasn't long before Norm and Kent gave them their toughest test and rowed them through the Grand Canyon. This was at a time when taking any kind of craft through the Grand Canyon, especially a homemade wooden rowboat, was still considered a special kind of madness.[49]

After working with Norm Nevills for a couple of years, Kent Frost decided that he wanted to see even more water, so he joined the peacetime Navy. He figured it would be an easy job, without much activity. He could spend his spare time enjoying the water. The year was 1939.

"I figured I'd get to see a lot of water that way. And I sure did. They trained me as a torpedo mechanic. During the war I went to Australia, New Guinea, and the Philippines."

Kent returned from the war in 1945 and started a jeep touring business in an area of southern Utah few people had ever seen—a landscape with names such as the Dolls House, the Needles, and the Maze, a place Craig Childs described as "sheer, unadulterated mystery."[50] With Kent's urging the area was made a national park in 1964; today Canyonlands National Park is visited by nearly half a million people each year, many of them in rafts in Cataract Canyon.[51] Kent Frost was a pioneer, not in "settling" the West or plowing up virgin sod for farms, but in America's newest frontier—recreation and tourism.

The 7,000 people who run Cataract Canyon each year experience not only a wilderness journey (except for the motorized boats) but also a trip

Labyrinth Canyon on the Green River, Canyonlands National Park, Utah.

back in time. Photographs taken on a 1911 expedition can be replicated today. It is one of those rare places where America has not changed much in a century.[52] Half the rapids were drowned by Lake Powell, but as the reservoir recedes as a result of global warming and upstream diversions, these rapids are working their way out from under the blanket of sediment. As a result, the lower half of Cataract Canyon will eventually restore itself.[53] And the upper half is doing just fine, thank you.

Not all river recreation takes place in such dramatic locales. Some of the greatest demand for intact rivers is close to heavily populated areas. In such places even a small project that creates a semblance of natural river is cause for rejoicing.

LUNKERS AND EXTREME INSTANT SHADE

The effort to restore Eighteenmile Creek in western New York has Middle America written all over it. It is the story of everyday people working from the ground up—no high-profile heroes, no movie-star endorsements, no dramatic statements of resolve by national leaders.[54] Instead, the restoration effort was driven by the commitment of a group of citizens who decided to expend a lot of time and energy, take some risks, and endure some criticism to improve their local environment. These are people of modest means but ambitious dreams. Common among them is a hopeful view of how they can improve their community by improving their local river and thus create a better future for the next generation.

This message was driven home when I attended a workshop in 2004 on Eighteenmile Creek organized by the Niagara County Department of Economic Development. The first speaker, Dr. John Syracuse, a Niagara County legislator, held his infant daughter in his arms while he welcomed us to the workshop. He pointed to her and said, "In the future there will be a lot more anglers who will want to utilize Eighteenmile Creek and other rivers in the area. That's why we are doing this." The workshop was held in the 4-H training center at the Cornell Cooperative Extension Center in Lockport. The room in which we met had knotty-pine walls that were embellished with dangling "icicle" Christmas lights. White crepe paper hanging from the ceiling made it appear as though a wedding had just been held there. There was nothing fancy about this setting. Indeed, it spoke loudly of everyday people coming together for public purposes. The people who attended the workshop reflected the wide and diverse support for this project. There were agency people from all levels of government, including the Tuscarora and Seneca Indian Nations, several public interest groups, and local elected officials. The second day of the workshop included a tour of the restored segment of the creek, with dozens of local schoolkids in attendance; a whole new generation of river lovers was created that day.

The effort to restore Eighteenmile Creek went through what might be called serial instigators, each of whom brought the idea to a new step of fruition. One of the earliest supporters was Dick Lang. Lang coached wrestling at the local high school and did so well that the school earned a number of championships and sent one young fellow to the Olympics. In the summer Dick worked as a law-enforcement agent for the New York Department of Environmental Conservation. He took his enforcement job seriously, and when he discovered that a local business was illegally dumping debris into the creek, he vigorously pursued a conviction, resulting in a stout $50,000 fine. However, a deal was worked out so

Restored embankment, Eighteenmile Creek, New York.

that the fine could be paid off in in-kind payments. The only question was, what in-kind work should be performed? At Lang's urging, it was decided that an appropriate payback would be to mitigate some of the damage done to Eighteenmile Creek.[55]

That started the ball rolling, but the county was not all that motivated until Amy Fisk came on the scene as a county planner. Fisk, young and inexperienced, did not yet realize that accomplishing great things was "impossible"; she naïvely thought that you could fight city hall and win. So she simply decided to make it happen and went to work on restoring a segment of the creek with a zealousness that people found hard to resist. It wasn't long before city hall joined her in the effort. But no one in Niagara County had any expertise in restoring badly damaged rivers. Then came the third instigator.

Dave Derrick is a research hydraulic engineer and has worked for the Corps of Engineers for many years at its Vicksburg research center (described in chapter 2). Dave has a set of muttonchop whiskers that would put General Ambrose Burnside to shame, and an aura of dynamism and energy. His specialty is "designing" natural rivers, that is, taking badly damaged watercourses, imagining what they would look like in a natural state, and then re-creating that environment. After spending a couple of days with Dave, I came to the conclusion that he probably talks about river restoration in his sleep; he certainly talks about it all of his waking hours. And it is not idle chatter; he is a one-man brainstorming session, constantly thinking of ways to improve the health and sustainability of rivers. He has more ideas about restoring rivers in five minutes than other people might think up in a lifetime. He exudes enthusiasm and possibilities. The idea that began with a local wrestling coach and was given form by a county planner became a concrete set of instructions under Dave's tutelage.

After Dave came up with the design, Amy Fisk secured $1 million from New York State and a dozen other sources, and it wasn't long before Dave was there in his hip-waders, orchestrating stream-bank stabilization, creating healthy fish habitat, and assuring access for the thousands of fishermen who descend on Eighteenmile Creek when the salmon and steelhead are running. He used twenty-three different restoration techniques, including placing "lunkers"—box structures that create overhanging rock ledges—and pushing trees into a horizontal position over the stream, something Dave calls, with a certain degree of amusement, "extreme instant shade."

If ever there was a stark illustration of not enough river to go around, it is the image of 330 anglers lined up shoulder to shoulder on both banks of Eighteenmile Creek in the half-mile segment just below Burt Dam. Dave Derrick refers to this as "combat fishing." A more abstract term is *resource scarcity*—a condition where demand has outstripped supply to the point of absurdity. The message of this image of fishermen lined up like soldiers in line of battle is clear: there is an acute shortage of healthy, intact river ecosystems in this region. The Eighteenmile restoration project is only about half a mile long; it is essentially an urban pocket park. But this region needs hundreds of such projects. Many of the industries here have closed, the economy is depressed, and people are leaving—Buffalo has lost population. Clearly a new economic paradigm is desperately needed. It will take an enormous effort, but western New York, with a conscientious program of river restoration, could become a fisherman's paradise. But to make that happen, a whole legion of wrestling coaches, local planners, and river-restoration experts will have to step forward and re-create a waterscape that has been damaged by 200 years of abuse and neglect.

Restoring small pockets of the natural landscape in western New York is more than just an ecological challenge; it is a conceptual problem, in two ways. First, local community leaders have to be convinced that what worked in the past is not working now, and a whole new vision is needed if this area is ever again going to see economic stability. Second, the rest of the country has to be convinced that western New York is a great place to fish. An analyst in the Buffalo office of the Corps of Engineers explained the problem this way: "When you think of Florida, you think 'Everglades,' when you think of Utah, you think of national parks, and Virginia has the Blue Ridge and the battlefields, but what do people think of when they picture the Great Lakes Region? They think of rust, toxic waste, Love Canal. That is what we have to overcome."

The restored segment of Eighteenmile Creek is a slender slice of nature amidst a landscape ravaged by an economic system that is a shadow of its former self. It is a natural refuge for thousands of people who want to experience the tranquility of fly fishing or simply enjoy the calm that rivers provide. But it is also a first step into a new economic future. Even a river that has been compromised, such as Eighteenmile, offers a recreational experience that is varied, rewarding, and memorable. When rivers are managed to maximize their attractiveness as a recreational resource, the rewards are plentiful.

River recreation comes in many different varieties; running white water out West and casting a line in an upstate New York stream are just two of the possibilities. For some people, recreating on a river means a weekend getaway or a wild excursion into adventure. For others, it is simply salve for the soul and a chance to explore both the inner and outer worlds of what is possible.

WATER WARRIORS

When my son Weston was four years old, I took him on his first river trip, a four-day venture on the San Juan River. The San Juan courses through some of Utah's finest Red Rock Country, slicing through anticlines and synclines, along cliffs of Navajo sandstone, and past Ancient Puebloan Indian ruins. The kids on the trip outnumbered the adults, although the term *adult* should give way to something like *senior kids* to describe people immersed in the sheer joy of a river. My son and two other children, armed with water guns and unlimited imagination, started calling themselves the "Water Warriors." It had a nice ring to it, but, playing the role of quasi-adult, I asked them what they were warring against.

"Monsters and bad guys," they answered with perfect self-assurance.

A couple of water-development agencies immediately came to my mind, but I decided not to mention them for fear of scaring the hell out of the little juvees. Instead, I asked them why they were fighting this war against monsters and bad guys. Again, they spoke with authority:

"We're protecting the raft so we can come back and go on another river trip."

Obviously they were hooked on rivers.

Our entire nation is hooked on rivers. We love to recreate in them, near them, or among the flora and fauna that are made possible by them. Rivers give stark vertical relief to an otherwise dull landscape; they carve canyons, erode mountains, and snake through narrows. They feed our lakes, keep the wetlands wet, sinuously split our savannas, and give our deserts a sparse taste of blessed dampness. In the midst of all that, we play. There is a saying about meeting one's needs: whatever floats your boat. Think of rivers.

Our nation is in the midst of a difficult economic transition. Our big financial institutions are in havoc, many of our major manufacturing companies are doing the large-scale version of "will work for food," and we appear to be in danger of becoming a wholly owned subsidiary of China. Sometimes it seems that only three things are still made in America: high-fat fast food, suburban sprawl, and a health-care industry designed to treat the maladies caused by the first two. But there is one asset that cannot be outsourced, cannot be transferred offshore, and cannot be produced anywhere else in the world, and that is America's natural beauty. It is no wonder that tourism is the only industry in the United States that has a trade surplus. Our parks, wildlands, forests, mountains, grasslands, and desert ecosystems are an asset of incalculable value. And they are all fed, maintained, and given life by rivers and

12

THE RIVER COMMONS

And I envy thy stream, as it glides along
Through its beautiful banks in a trance of song
—FROM WILLIAM CULLEN BRYANT, "GREEN RIVER"

My first memorable river running experience occurred when I was about 10 years old. I had the good fortune to grow up next to a heavily wooded Boy Scout camp—one of the last remaining patches of natural land in central Indiana. This wooded area was bounded on the northwest by Fall Creek, a meandering little river about twenty paces in width. My fellow boy-adventurers and I spent many summer days testing those waters. On one occasion, with visions of Huck Finn in our heads, three of us decided to build a raft and float all the way to Indianapolis. We had no plan about how to get back home once we got there, but those are mere details to a 10-year-old boy. We lashed together a bunch of downed logs and cut saplings for poles for locomotion. We hoisted a Fruit-of-the-Loom flag. I asked to be named captain, but memory reminds me that I was not granted that exalted rank, and I pulled deck duty along with the other two Hornblowers. With great fanfare we launched from under the bridge on Boy Scout Road.

Fall Creek is dammed above the Boy Scout camp to provide drinking water for Indianapolis, so it was still fairly clear and unpolluted—a rarity in the Midwest in the 1960s. But silt and mud kept it opaque. Our raft did not quite have sufficient buoyancy to support the weight of three piddling boys, and it sank an inch or two below the surface but still kept us afloat. We decided that a wet deck was acceptable and continued downstream. In those days Fall Creek was lined with numerous fishermen, and they looked up from their bobbers to see three kids who were apparently cruising down the river with no visible means of support. We were, it appeared, walking on water. The role delighted us as we realized we were something of a spectacle. We began haranguing our riparian audience with tales of magical water tricks.

The fun did not last long. As the logs in our raft became saturated, they sank further into the muddy water. What had been an ankle-deep perch become a knee-deep submersible. At some point someone mentioned Davy Jones's locker, and we made for shore. We planted our flag in the mud and claimed everything for the queen of Spain. It was a long walk home, but we had tales to tell.

Everyone has a river story—unless you've lived your life locked in a basement (and not one subject to flooding). Rivers are a part of our lives as much as family, home, and place. Through the years I have asked people if they have had some sort of "river experience," and they inevitably break into story. Not all these tales are gleeful reminiscences, but they are usually transformative and powerful. There is a little Huck Finn in all of us. I have many river stories, but perhaps the most telling was provided by the Colorado River in America's biggest canyon and most famous river run—and the dams that almost drowned it.

KITTY CLYDE'S SISTER

On August 30, 1869, John Wesley Powell—one-armed river runner, self-trained scientist and adventurer—must have exhaled one of the greatest sighs of relief of all times. After 100 days of experiencing "an unknown danger, heavier than immediate peril" in the Grand Canyon, he spotted the mouth of the Virgin River. Finally, he had reached safety. His men rowed his boat, the *Kitty Clyde's Sister*, to shore as three startled Mormon fishermen stood there with mouths gaping at the apparitions before them.[1] Powell had just completed one of history's great epic adventures, emerging from the Grand Canyon and completing the first exploration of both the Green River and the Colorado River. Powell was apparently somewhat surprised that he survived his ordeal because he clearly experienced the "life high" that is well known among adventurers who have come a bit too close to the edge of eternity. He effused: "How beautiful the sky, how bright the sunshine, what 'floods of delirious music' pour from the throats of birds, how sweet the fragrance of earth and tree and blossom!"[2] In many ways, rowing a small raft through the Grand Canyon today evokes the same feeling. It remains one of the most rewarding river journeys on the planet, spellbinding for both its audacious beauty and its fearsome waters.

On the fourth day of our three-week Grand Canyon rafting trip we sighted evidence of perhaps the most egregious excess of the big-dam era. What looked like mine shafts were in fact test drillings for yet another dam on the Colorado River, this time deep in the bowels of the Grand Canyon in a section called Marble Canyon.[3] For four days we had rafted through one of the last remaining desert rivers in the American Southwest. We had seen spectacular natural wonders, such as the swirling sandstone alcoves and glistening pools of the Silver Grotto and the great roofed amphitheater Powell named Red Wall Cavern—large enough to engulf a football field. We had filled our water jugs at Vasey's Paradise, where a gushing spring bursts forth from the canyon wall, creating a verdant oasis of native vegetation. We had scrambled up Nautiloid Canyon to see the fossilized impressions of an exquisite squidlike animal that disappeared from the face of the earth in the Paleozoic era.

We had also seen the signature of long-lost peoples who had somehow eked out a living at the bottom of the worlds' largest canyon. We had walked a trail through an ancient village that was flush with people and commotion at the same time at which Charlemagne declared himself holy Roman emperor. We had gazed into Stanton's Cave, where archeologists discovered 4,000-year-old stick figurines.

The Marble Canyon section of the river also offers some of the best white-water rafting in the world, including the "Roaring Twenties," with ten rapids in 10 miles of churning water. These include 24½ Mile Rapid, where veteran river runner Bert Loper, rowing the river for the last time at age 80 in 1949, slumped at his oars, dead of a heart attack—the perfect exit for an old river dog.[4] Marble Canyon also includes House Rock Rapid, where boaters stare into the gaping maw of a frothing mountain of water; misjudge the approach to this rapid or miss a crucial stroke of the oar in the middle of it, and it eats you.

The proposed dam in Marble Canyon was only half of the nightmare planned by the Bureau of Reclamation for the Grand Canyon. A second dam was to be built at Mile 238, called Bridge Canyon. This dam would have inundated the exquisite blue-green waters of Havasu Creek, the narrow slots of Matkatamiba, and the seething waters of Lava Falls—the largest rapid on the river.

All of this—the priceless geologic and archeological treasures, the jaw-dropping scenery, the dramatic river history, and the world-class rapids—would have been drowned if the dam builders had had their way. The dams proposed for the Grand Canyon were planned for the worst of reasons: to generate power and money to support yet another Bureau of Reclamation extravaganza, the Central Arizona Project (CAP). The CAP, an open canal that runs for 300 miles through the Arizona desert, was designed to pump Colorado River water to cotton fields around Phoenix at a time when cotton farmers in the South were going broke because of overproduction. These two "cash register" dams would generate power to pump CAP water over mountain ranges and would earn money from hydropower to subsidize the whole operation so that a handful of farmers—the beneficiaries—would not have to pay for the true cost of their excesses.[5]

Today the idea of damming the Grand Canyon seems obscene on its face, but in 1968 the dams were stopped only because the "environmental extremists" of that era mounted an all-out campaign to convince America, and the U.S. Congress, that one of America's finest Crown Jewels was not a plum to be picked by narrow commercial interests. The Bureau went on to build the CAP, now a source of water for endless subdivisions, but even the most ardent advocates of concrete do not talk publicly today about damming the Grand Canyon.

One of the basic themes of this book is that recreation will become one of the dominant uses of America's rivers. One need look no further than the Colorado River in the Grand Canyon to get a strong taste of what is to come. The river and the canyon reel under the onslaught of industrial tourism; the canyon and its river serve as a portal into the future. Each year over 4 million people stare in disbelief from the canyon rim, and nearly 25,000 crowd into a thin green strip at the bottom of the canyon to run the river. The population density at river campsites far exceeds that of a modern suburb. This is due in part to the Canyon's unique geology and legendary rapids, but the crowding is also abetted by national park policy.

Grand Canyon National Park has always had a permit policy that expressly favored commercial companies over public access. The nation's most vaunted stretch of white water has basically been leased—at well below market prices—to a few well-connected companies with strong political ties to the appropriate bureaucrats and politicians. As a result, 70 percent of the people who raft the Colorado through the Grand Canyon today are paying customers of one of these favored companies.

For many years those who wanted to run the river on their own without paying a commercial group had to face the dreaded "granddaddy of all waiting lists," but that term does not quite do justice to the mind-numbing scale of the wait.[6] When I ran the river a few years ago with a group of friends, we had waited eleven years for a permit. By 2005 the wait was something like twenty years; if a middle-aged person applied for a permit under the old system, that person would need to be a very fit septuagenarian when he or she finally settled in at the oars. When people talked about rowing the Grand Canyon as a once-in-a-lifetime experience, it was a reference to both the intensity of the experience and the scale of the red tape. I know people who applied for permits when their children were young, assuming that those little tots would be grown and capable of handling the oars when the permit finally came in the mail; a permit application was sort of an investment on behalf of the next generation. When our permit application was filed, my son Weston was a little tyke in grade school. By the time our permit made its way through the waiting list, he was a strapping college man who rowed many of the river's biggest rapids.

In 2006, after a seven-year planning effort, the National Park Service changed the permit system to a weighted lottery, meaning that several factors affect one's probability of getting a permit, including past trips. Of course, this applies only to people who are boating the river on their own. If you want to pay a commercial outfitter, you can get on the river tomorrow. The 2006 River Management Plan, like previous plans, gives priority to commercial trips, especially those with motors, allocating 80 percent of summer user days to commercial companies, over half of which use motorized rigs.[7]

Most of the big commercial companies prefer to run giant 30-foot boats made out of military bridge pontoons, with tubes the size of old-growth hemlocks. Carrying as many as thirty-two people and a small mountain of gear, these big rigs are powered by outboard motors. They are not rafts;

they are inflatable tour buses. They do not run rapids but rather plow through them the way a D-9 Cat cuts through sand. For many years these boats relied on powerful two-stroke engines that left a trailing stench of blue smoke. When I first ran the river in 1991, the river corridor stank of burnt oil. The commercials have since switched to cleaner four-stroke engines, but the noise of the motors still breaks the silence with their constant growl. The big motorized rigs push past the other boats on the river, doing in seven days the mileage that requires about sixteen days for a human-powered craft.

They also bring in a large concentration of people; private trips are limited to sixteen people, but the commercial big rigs can carry twice that number. They fit into the natural majesty of the canyon like a nudist at a convent. The Park Service recognized the negative impact of motors in 1972 and began a long, arduous effort to phase them out of the canyon, but Congress, after being lobbied by the big commercial companies, interceded in 1981 and prevented the agency from banning motors.[8] The commercial river companies that run these behemoths do so because they can pack a lot of people into a shorter trip, which increases their profit margin; there is no other reason to use motors. Some commercial companies have recognized that motors and giant pontoon boats compromise the river-running experience for everyone, and they run only human-powered rafts or dories. These companies have put the welfare of the canyon and their fellow boaters before profit. They realize that motors take some of the grand out of the Grand Canyon. River recreation in the Grand Canyon offers an object lesson about the promise and perils of industrial-strength tourism. If it is not done in a careful, sustainable manner, it becomes destructive.

BLACK-SIDE UP

Still, rowing the Grand Canyon is a life-bending experience. A recent survey asked people to identify the "adventure trip of a lifetime," and more people chose rafting the Grand Canyon than any other option.[9] The river lore and the media tend to focus on the rapids, some of the biggest runnable white water in North America. The massive wave train in Hermit, the yawning hole in Crystal, or the crashing maelstrom of Lava Falls evoke a potent mix of exhilaration and dread, overlaid by a thin veneer of studied nonchalance. One's heart rate is further enhanced when a fellow rafter goes "black-side up," river parlance for flipping in a rapid. Indeed, as my son and I learned, rowing a 14-foot raft into a 15-foot wave is mildly entertaining.

On the morning of our tenth day on the river we launched early, perhaps too early, and I was hardly awake when I started rowing out into the current. My son Weston was just as lethargic, but there was no time to gradually make peace with the sun; Tapeats Rapid was coming up fast. Tapeats is rated only a 5 on the Grand Canyon 10-point scale, but it still demands attention. I noticed that Weston's life jacket was not zipped. "Zip up your jacket."

"Why? It's just a 5, it's not a big rapid."

"Son, everything's big in the Grand Canyon."

With desultory motions Weston zipped his jacket. I made sure mine was zipped and then did a haphazard check of the raft to see if everything was secure. I noticed a few loose ends, but there was no time to tidy up before the whitewater. I really could have used another cup of coffee before pushing off.

I rowed out into the current and lined up the boat. There was a long wave train, some of it surly, but that's standard for the Grand Canyon. I took a quick read of the water, made sure my oar-rights were in the locks, and braced myself. Weston slumped on top of the cooler, looking like he was going to take a nap. The first few waves pushed us around, but I kept the raft pointed downstream. I thought, this is becoming routine. Halfway through the rapid a wave just in front of us began building, and the trough in front of it seemed to drop down to the bottom of the river. I should have pushed hard at that point, but instead, I just stared at the wall of water developing in front of us. The raft bit into the huge wave and then started to corkscrew horizontally, at first slowly and then like a bullet going down a rifled barrel. My next sensation was being thrashed by convulsing current, and it dawned on me that I had just flipped my raft. My first thought was "Where is my son?"

I felt my head bump against a hard surface—the floor of the raft—and realized that I had come up under the raft. Again I thought of my son (once a parent, always a parent). Is he under the raft too? Instinctively I clawed my way out from under the raft. I popped out of the water gasping for air and looked around for Weston. At first glance I could not see him, and a sickening wave of panic came over me, like I'd just been poisoned. I screamed his name.

"Dad, get up here!"

I turned 180 degrees and saw Weston kneeling on the upturned raft. Before I could think of what to do, he grabbed me by the life jacket and pulled me onto the overturned raft.

"Oh jeez, I screwed up, I flipped our raft in a number five rapid," I panted.

Weston laughed in a forceful, adrenalin-accented voice. "Don't worry about it, Dad. Everything's big in the Grand Canyon."

Running rapids in the Grand Canyon gets the most attention, but it is just part of the overall experience. The Colorado River is not merely a linear water park. Rather, it is a unique river ecosystem, the golden vein of the Colorado Plateau, an integral part of a complex desert habitat that offers innumerable contrasts. The river has eaten its way down to 1.7-billion-year-old rock but forms new beaches overnight. It can move boulders the size of buildings but still delicately sculpt riverside rock with the finesse of a Michelangelo. It supports myriad forms of life, ranging from desert bighorns to a vast parade of brightly colored and wildly striped lizards. Its archeological record begins 2,000 years before Christ was born, and the Havasupai and Hualapai still live in the canyon.

Spending three weeks on the river in the Grand Canyon gave me new insights into the competing demands of outdoor recreation American style. It is clear that some people want America's natural wonders handed to them on a platter without any effort or discomfort, and our recreation economy must provide for that. We are a nation that craves easy access, whether it is credit, food, or vistas. But in accessing nature in this manner, especially in a place as primordial and pristine as the Grand Canyon, easy, quick access—a motorized in-and-out—robs the recreation experience of some of its significance and meaning. It is like the difference between attending a church service and just watching video outtakes. These competing visions of the "outdoor experience" will play a major role in river recreation in the coming decades as we redesign river policy to fit more closely with our actual needs.

The Grand Canyon belongs to the world, but it is managed by the U.S. government. At times the government has performed great service to the canyon and its river, but there have been stunning lapses in vision, beginning with the very first federal official to visit the canyon. In 1857 a lieutenant of the Army Corps of Engineers, Joseph Ives, became the first known white man to steam up the river and hike into the depths of the canyon. In his official government report he described the canyon as a "profitless locality" and confidently predicted that the canyon would be "forever unvisited and undisturbed."[10] Ives's myopia raises an important question: do our current river policies suffer from a similar lack of imagination about what the future offers for America's rivers? Are we creating sustainable watersheds or sowing the seeds of disaster? Where does the true promise of our rivers lie?

THE KATRINA SYNDROME

The answer to those questions reveals a mixed bag of ideas, proposals, and policies that make up contemporary U.S. water policy. The disjuncture between what we have done to our rivers and what we want from our rivers continues to drive a schizophrenic approach to rivers. First, let's consider the water hubris that led to our current problems and then contrast it with a very different conceptualization. Our flawed current river policies are, to a great extent, a result of what I call the "Katrina syndrome," which is a pronounced tendency to build human-made disasters-in-waiting. Politicians would rather distribute goodies than regulations, so we have focused on government-funded structural approaches to rivers, even when a long-term dependence on that approach creates an enormous potential for catastrophe. This phenomenon is abetted by developers and individuals who detest government regulation but then demand government assistance when disaster strikes. New Orleans, of course, provides a vivid illustration. Rampant development continued even as hurricane-force winds began their fateful course across the Gulf of Mexico. Such events are often described in the media as natural disasters, but in fact they are human engineered. There are many other examples.

The city of Las Vegas is the next catastrophe in the making. It gets 90 percent of its drinking water from Lake Mead on the Colorado River, but the lake is drying up, and all predictions are that climate change and upstream diversions will dramatically reduce the amount of water available to the city.[11] So in utter desperation the city has proposed to suck groundwater out of the central part of Nevada and a portion of western Utah and pump it through an eight-foot-diameter pipeline that is projected to cost somewhere between $3.5 and $15 billion.[12] And when that is no longer enough, there is talk of building a gigantic pipeline from the Missouri River to Colorado's Front Range, which would then allow Denver to reduce its take of Colorado River water so that Vegas could pump more water for more new suburbs. I asked Pat Mulroy, the manager of the Southern Nevada Water Authority, why she thought the federal government, so desperately in need of budgetary liposuction, would foot the bill for this, and her response was that the nation would not dare abandon a major city such as Las Vegas.[13] Like New Orleans, Las Vegas expects the federal government to bail it out with yet another gran mal boondoggle.

The idea of building giant pipelines so that cities can continue non-sustainable growth has become quite popular. The Jordan Valley Water Conservancy District in central Utah plans to build a pipeline that taps into the Bear River—the primary source of water for the Bear River Migratory Bird Refuge.[14] This would allow Utahns to continue over-watering bluegrass lawns in a desert climate.[15] St. George, Utah, not to be outdone in the water-grab department, plans to build a big pipeline that will suck water out of Lake Powell.[16] The problem with that plan is that climate scientists predict that Lake Powell will dry up in our lifetime.[17] Further upstream, a big developer has proposed building a pipeline from the Green River to the Front Range so that Denver and its bedroom communities can continue to sprawl.[18] Of course, the St. George pipeline and the Green River pipeline would reduce flows to all the other cities that depend precariously on Colorado River water—Las Vegas, Los Angeles, San Diego, Tucson, and Phoenix. In other words, the Colorado River is dramatically overallocated—a Katrina in the making.[19]

Another Katrina is the slow buildup of mud and silt in all reservoirs. Even venerable dams such as Hoover and Grand Coulee will one day be useless. What happens to the cities that rely on these dams for water and power? There is, of course, a large degree of water hubris in all this. It assumes that at some point in the future, as cities grow to monstrous proportions, a solution, as yet unfathomable but utterly crucial, will spring forth in timely fashion to rescue them from their folly. Implicit in that assumption is that water can be taken from someone else.

Building giant cities in the desert that have a nonsustainable supply of water is only one example of the Katrina syndrome. Another one is the Ogallala Aquifer. The economy of the Great Plains, from Canada to the Texas Panhandle, is dependent on agriculture that mines groundwater. Everyone agrees that the aquifer is being bled to death, and it is a matter of only about twenty years before groundwater pumping will no longer be feasible.[20] At that time a whole region of the country will be left destitute—and demand a government rescue. Maybe they'll insist on a giant pipeline from the Arctic for their bailout.

Other Katrina-type disasters are waiting to happen because of construction in vulnerable floodplain locations. The Sacramento–San Joaquin Delta, a 1,150-square-mile area surrounded by 1,000 miles of totally inadequate levees, is subsiding (sinking). The area is a "blend of small town communities, busy ship ports, farmlands, industries, highways, historical sites, and marinas."[21] It is also doomed. Much of the state's water supply runs through it. Just one giant downpour that packs sufficient force and volume to breach the levees, and this whole area will once again be wetlands, and millions of Southern Californians will be without drinking water.[22] They won't get any help from Sacramento; it has less flood protection than New Orleans before Katrina.[23] The town of Lewiston, Idaho, is also a sitting duck for a flood, thanks to silt buildup behind the Corps of Engineers' nearby Lower Granite Dam.[24]

Another vast vulnerability concerns water quality. As I pointed out in chapter 8, most Americans are drinking water laced with agricultural chemicals, mercury, antibiotics, and even estrogen. What are the long-term health implications of this? What impact will it have on health-care costs?

This leads to a related point. Water policy cannot be divorced from numerous other issues. The United States appears to be experiencing what I call *interactive nonsustainability*, meaning that scarcity problems in one area affect and enhance scarcity in other areas in a spiraling downward trend. For example, there is a close nexus between water and energy.[25] The obvious link is hydropower and its attendant problems, but coal and nuclear power plants require lots of water, coal-bed methane wells generate tremendous water pollution, and proposed oil-shale development will require enormous amounts of water in the driest areas of the country.[26] Relying on ethanol for energy exacerbates the dead zone in the Gulf of Mexico.[27] The BP oil spill in the Gulf of Mexico and the burst oil pipeline in the Yellowstone River demonstrated how that source of energy poses a threat to rivers and oceans.[28] The flip side to that relationship is that water usage requires energy usage; a recent study found that 13 percent of the nation's energy is used to move, treat, and heat water.[29] In other words, our energy problems exacerbate our water problems, and vice versa.

There are many additional facets to interdependency between water and virtually all other natural resources. Pine beetles are ravaging western forests; this could create dramatic increases in erosion, which means greater sediment loads in waterways, which in turn damage riverine habitat and fill in reservoirs. Dust from construction, dirt roads, and agriculture is causing snowpacks to melt earlier, exacerbating water-supply problems.[30] Sprawl increases both pollution and demand for drinking water, often from the same watercourses, and degrades riparian habitat.[31] Anthropogenic climate change is altering habitats and causing some areas to become drier; snowpacks are growing thin, and glaciers are disappearing—critical sources of water.[32] The age-old response to such problems is to demand government assistance, but the federal budget is also in crisis and cannot afford a whole rash of new catastrophes. It is difficult to escape the conclusion that the nation's hyperconsumptive lifestyle and shortsighted policies have backed us into a corner.

Despite the obvious myopia of current water policy and the long-term disadvantages of a structural approach, there are numerous proposals for new dams, primarily in the West. In recent years new dams have been proposed for the Yakima River in Washington, the Gunnison and the Cache la Poudre rivers in Colorado, a tributary of the Umpqua River in Oregon, the Boise River in Idaho, the Green River in Wyoming, the lower Colorado River and the San Joaquin River, both in California, and tributaries of the much-embattled Columbia and Snake rivers. A huge dam, pushed by Sarah Palin, has been proposed for the Susitna River in Alaska.[33] Some people want to revive the Auburn Dam proposal on California's American River—one of the most egregious dam proposals to come out of the big-dam era. There is even talk of rebuilding Teton Dam; apparently one disaster was not enough.[34] Wyoming is so fond of the idea of new dams that the state government created a new Dam and Reservoir Section to ponder building dams on the Big Horn and Green rivers.[35]

Many new projects are a result of Congress' penchant to use the Corps as a conduit for money sent to home districts and states. The agency's FY 2010 request was $5.125 billion. The Corps has built so many projects that it now costs $2.3 billion a year just to operate and maintain them. And the Corps wants to build more, requesting nearly $1.5 billion for new construction in its proposed budget for 2012.[36]

Calls for more spending on water projects are always accompanied by dire projections of an impending "water crisis." But we do not have a water crisis; we have a water-management crisis. The problem is not a shortage of water, but rather the disjuncture between current policy and the real water needs and priorities of the future. It is clearly time to consider a new approach.

AMENITY RIVERS

The journey to a new water future must begin with the past—with lessons learned. One of the most poignant comes from our greatest river, the Mississippi. George Merrick was a riverboat pilot during what Mark Twain called "the heyday of the steamboating prosperity," when the river "from end to end was flaked with coal-fleets and timber-rafts."[37] Merrick saw the river at its best, then witnessed its rapid destruction. Writing in 1909, many years after his river days, he described what happened with stunning eloquence. He began his book with this eulogy: "The majesty and glory of the Great River have departed."[38] He then explained how that happened:

The primeval forests which spread for hundreds of miles on either side . . . caught and held the melting snows and falling rains of spring within spongy mosses which carpeted the earth; slowly, throughout the summer, were distilled the water from myriad springs, and these, filling brooks and smaller rivers, feeders of the Great River, maintained a mighty volume of water the season through. Upon the disappearance of the forests, the melting snows and early rains have no holding grounds, are carried quickly to the river, which as quickly rises to an abnormal stage in the early part of the season, to be followed by a dearth which later reduces the Mississippi to the dimensions of a second-rate stream, when navigation is impossible for great steamers, and arduous, disheartening and unprofitable for boats of any class.[39]

In other words, the river was destroyed because the riparian lands were destroyed. In the nineteenth century no one understood the connections among land, water, and flora and fauna. They did not comprehend the interdependencies that we now commonly recognize with a word of fairly recent vintage, the *environment*. Merrick's trenchant observations presaged the modern concept of integrated resource management by nearly 100 years. But ironically, Merrick and his adventure-loving compatriots in the steamboat business dealt much of the blow that crippled the Mississippi. He explains that steamboats of his day consumed 25 cords of wood per day. A cord of wood consists of 128 cubic feet. Another riverboat pilot estimated that his seven steam boilers required 70 cords of hardwood every 24 hours.[40] In effect, the steamboats ate their young; they deprived themselves of the sustenance necessary for their own continued existence.

Unfortunately, this is not an isolated example. Rather, it appears that we have a tendency to foul our own nest and destroy what we value the most. I am reminded of an ad in a western newspaper I saw a few years ago: "For Sale: the last unspoiled land on the East Slope [of Colorado]." The seller was inviting someone to come in and spoil the final resting place of nature. Sadly, our species does this repeatedly; we find a habitat, a river, or a "scenic attraction" and overwhelm it with the behaviors and actions that destroy the very quality that attracted us in the first place. Some do this out of love for a particular landscape; others look at the same landscape and think only of how much they can profit from its despoilment. The motives are different, but the results are the same. In Merrick's words, the majesty and glory depart. The great challenge for this generation is to figure out a way to reverse the downward corkscrew of our rivers before we reach a point where there is nothing left to save.

But we can do more than save the last tiny remnant of rivers and streams. With our modern technology and our improved understanding of the natural world, we can reverse some of the damage done by the shortsighted policies of the past. Rivers, unlike people, can be reborn. Rivers, unlike individual species, can reappear after vanishing. This will require a unique set of skills—a whole new field of knowledge regarding how to protect and restore rivers.[41] We have great experience at dismembering rivers, but the fields of restoration biology and deconstruction engineering are fairly nascent. We know how to build a big dam; what is the best way to remove one? We can build levees, but how do we restore floodplains? We can stop fish runs; how do we begin them? The answer is: we are learning—fast. We are also developing great expertise in how to build an amenity economy where restoration plays a central role.[42]

Every river in America is unique, so each one will require a different package of solutions, policies, and plans. But there are some fundamental political, ecological, and economic characteristics of the new era in river policy.

In terms of politics, the River Republic must be open, collaborative, and inclusive. We must democratize the policy-making process and make it open to all stakeholders, with a view to our collective well-being. Joe Six-Pack and Charlene Chablis need to talk. With all of society looking over its shoulder, government will be much more hesitant to give away the store. And each time a narrow special interest attempts to appropriate the river for itself, it will run headlong into a vigilant group of citizens. This newfound inclusiveness must occur across three dimensions. First, the mix of interest groups must represent all stakeholders, with nonprofit public interest groups playing a major role. Second, it must include all levels of government. This is not the top-down river-basin management envisioned in the past, but rather a mix of governments from bottom to top. Third, the process must include those who pay the bills, and those who suffer the consequences of bad river policy. That means that taxpayers must be represented, the people downstream must be at the table, and the people whose livelihoods are compromised in order to protect the livelihoods of others must have a say. It is much harder to rip someone off when he or she is sitting next to you.

The politics of the River Republic will give us rivers that serve all our needs. I call them *amenity rivers* because rivers will be managed to maximize their amenity value to the public as a whole over the long term. There is much evidence that this is already occurring. Some of the examples are described in this book, but there are many others.[43] All over America citizens are forming "watershed groups," "headwaters alliances," and "water roundtables."[44] The river advocacy group American Rivers estimates that there are 2,500 nonprofit river groups.[45] But we need ten times that number; there should be a "Friends of" organization for every stretch of river and stream in America.

At a conceptual level, there are numerous proposals for developing new approaches to water policy, including a "global water commons," "soft-path solutions," an "emerging water paradigm," a "progressive national water policy," "water governments," and "watershed partnerships."[46] The River Republic will require such innovative thinking.

In terms of ecology, we must view rivers as natural systems where everything is linked together in a web of interdependency. Rivers must be managed as a basin, not as a balkanized jumble of bickering jurisdictions that ignore the needs of their neighbors. Our current system, which artificially separates groundwater from surface water, anadromous fish from freshwater fish, river channels from riparian areas, and perennial wetlands from the rivers they feed, is incapable of meeting our needs. We need policy that is based on sound science and hydrological and ecological reality.

In addition, the goal of re-naturing some rivers must take place within political and economic realities. Indeed, on some rivers, dams can actually serve ecosystem needs.[47] Most river-restoration projects will take place within a context that includes dams and other structural developments. This is because most river structures serve humankind well and should remain in place. The debate should never be whether structural development—dams, flood walls, levees, channelization—is "good" or "bad." Rather, each structure should be judged as we should judge people: on an individual level, on the basis of character and contribution to society, and not as part of a group. It is hazardous to guess what percentage of structures do more good than harm, given that we have never had a national assessment that focused on that question, but the goal must be an honest assessment, not a perpetuation of myths, biases, and sacred cows.

The ecology of the River Republic must be mindful that the human species is a big part of the mix, and every river-restoration project must be based on a set of judicious and sensitive choices that reflect the interests of a very broad range of stakeholders. What we need is not a wholesale rejection of the past, but a frank, objective analysis of what best serves society in the long run. That calculus will inevitably include a range of options and a mix of restoration, preservation, and structural devices.

In terms of economics, the River Republic must wean itself from wasteful, environmentally destructive subsidies that destroy rivers and prevent them from meeting the needs of modern society. A systematic economic assessment of our current water projects, as I suggested in chapters 2 and 3, is essential. If we reallocated rivers and water among the various "uses" of rivers in strictly economic terms, the old uses would get much less and the new uses would get much more. But current water policy gives priority to many uses that are no longer economically rational. Continuing with this policy means fewer economic and environmental benefits, less economic stability, and less sustainability. On some rivers there is a very clear choice between what is rational and what is senseless—it's like choosing between Santayana or a room full of birthers. The nation must have the option to choose what works best and not be saddled with the burdens of past mistakes.

The argument for economic rationality does not mean that other nonmarket values and priorities should be ignored. On the contrary, one of the basic premises of this book is that rivers should be managed as a commons—a resource that belongs to society as a whole and not to a narrow group of well-heeled special interests. But in some cases that goal can be met by providing greater opportunities for water marketing and by valuing water according to market prices. This can lead to less waste and greater exchange efficiencies. I am not arguing for the privatization of water, but rather for managing water in ways that mimic market efficiencies while preserving public interests.[48]

Restored and preserved rivers presage the future; they are symbols of a new era just as surely as dams were symbols of progress and the "conquest" of nature in a past era. A compromised river with the potential for restoration is a kinetic form of politics, storing up potential until just the right instigator comes along. As impediments to restoration fall, hopes rise, a perfect exchange of the past with the future.

Amenity rivers are our gooses that lay golden eggs—and will continue to do so indefinitely unless we get greedy and butcher them for one last fat feast of foie gras and confit. Then we will have sacrificed the future well-being of our offspring and our nation simply because we refused to live within our means.

INSTIGATORS REDUX

And who will accomplish this heroic task of remaking the nation's rivers? It will require an army of instigators. It is human nature to want the strength of a Goliath, but we reserve our highest admiration for the Davids. A trenchant example comes from the seemingly impossible fight against the proposed Oahe Irrigation Project in South Dakota, where George Piper organized his fellow small farmers and took on the Bureau of Reclamation:

> Piper and other United Family Farmers were regularly maligned in the state's major newspapers. They faced countless obstacles erected by local, state, and national politicians. They confronted a powerful federal agency.... They were scorned by the business establishment and by every chamber of commerce in South Dakota. During their early years, especially, they stood seemingly alone.[49]

The most remarkable part of that saga was that they won—they stopped the "unstoppable" project. This story and the many other sagas of river restoration and preservation illustrate one of my mother's favorite sayings: "If you aren't rooting for the underdog, you are doing something wrong." Instigators are underdogs on a very long chain.

Instigators from throughout history inspire us because they fulfill a need for heroes and give us confidence that an individual can make a significant contribution to humankind. The instigators who save rivers give substance to Wallace Stegner's observation: "Saved watersheds reflect good all around them."[50] They give us hope that the future will be

Paria River, Vermillion Cliffs National Monument, Arizona.

brighter, and they provide relief from the burden of bad news that makes us stoop under the weight of what Aldo Leopold called "a world of wounds."[51] But first, every instigator must pay his or her dues. Mahatma Gandhi described what instigators go through: "First they ignore you, then they laugh at you, then they fight you, then you win."[52] The "win" in this case will be a river reborn.

Most of the time, this heroic work will be arduous, tedious, and relentless. Sally Bethea, the founder of Friends of the Upper Chatta-hoochee, described what it takes to be a successful instigator: "You just have to keep putting each foot in front of the other. You need the passion and vision, yes, but don't forget to focus on daily perseverance. There will be dark times for sure, but persistence prevails in the end. Our goal is to energize, activate and inspire the community because we can't do it all ourselves." And the first step is always a realization at the individual level that nothing will get done if you simply assume someone else will solve the problem. Chad Pregracke, the remarkable young man who set out to clean all the garbage from America's rivers, put it this way: "If we, the people, don't do it, then who will?"[53]

Being an instigator is fraught with perils, hard work, and unknown hazards. Instigators are typically nonprofessional citizen activists, and they can expect to make blunders; it is truly a learn-as-you-go job. But the bottom line must always be a consciousness that it is all about serving the river commons and the people to whom it belongs. It is about living up to James Madison's notion, expressed in the *Federalist Papers*, that "the public good, the real welfare of the great body of the people, is the supreme object to be pursued."[54] The expression of that lofty goal on an individual level can be found in the epitaph for Jackie Robinson, a man who faced down the insipid wrath of racist America to break the color barrier in major-league baseball. It reads: "A life is not important except in the impact it has on other lives."[55] Those words soar to our highest calling, but we should not underestimate the difficulties. Trying to remake a river and serve an entire community, indeed, all humankind, is daunting but yields the promise of living beyond oneself.[56] Despite the pitfalls, some will stand to the challenge. This book is full of examples of people who have met that challenge.

Being an instigator for rivers is an imposing mission, but we have something of a model in our national parks. When most parks were created, they were fiercely opposed by locals and those who thought that they could enrich themselves from exploiting the land. But now the protection of these last great places is recognized as an act of great foresight—a gift to future generations. The same can be said of the canyons and valleys that are freed by the restoration of rivers. In years to come, when a free-flowing river is as rare as a quiet moment, people will stand on the edge of an unblemished river canyon, marvel at its grandeur, and express thanks for those who sought to protect and restore rivers for an America still unborn.

LOTIC DREAMS

A river is more than just water; it is a linear universe, a constellation of life, matter, and minerals. Yet even that description is incomplete because the history of humankind is mixed into the mud and mood and caprice of rivers. Our cities, trade routes, means of employment, and leisure are as much a part of rivers as the life below the waterline. Marjory

Stoneman Douglas, the woman who saved the Everglades, put it this way: "There's something about a river that's a living thing. It's a stream of history as well as water."[57]

Rivers evoke such powerful emotions that they are the object of everything from lilting poetry and enchanting prose to the occasional maudlin drivel that Twain characterized as "sappy inanities."[58] They provide metaphor and powerful imagery. When President Theodore Roosevelt called all the governors together in 1908 to harangue them about the necessity of conservation, he asked Edward Everett Hale, the chaplain of the Senate, to give the invocation. Hale began his prayer with these words: "The Lord thy God bringeth thee into a good land, a land of brooks of water, of fountains and springs, flowing forth in the valleys and the hills."[59] The Bible admonishes: "But let justice roll on like a river, righteousness like a never-failing stream."[60] For indigenous people, water is "the first living spirit on this earth."[61]

Our landscapes, both natural and human-made, are defined by water. We use only about 1 percent of the earth's water, but that tiny increment makes the difference between life and death. Earth is often referred to as the "water planet," but we could, with equal validity, refer to the human race as the "water species." Biologist Edward Wilson invented the concept of *biophilia*, meaning that we have an innate attraction to nonhuman life.[62] I think that we should consider a similar concept of *hydrophilia*,[63] an innate love of water, a kind of emotive hydro-synchronicity. One can watch a two-year-old in the bathtub, a group of raucous teens in a pool, or a swimmer maundering through a current and get the idea. Water, it seems, is a reflection of being human. Gretel Ehrlich writes that "water can stand for what is unconscious, instinctive, and sexual in us, for the creative swill in which we fish for ideas."[64] That is why rivers innately appeal to what Barry Lopez calls "the interior landscape."[65] There is a good reason that Annie Dillard chose a riparian area for her pilgrimage: "The creeks—Tinker and Carvin's—are an active mystery, fresh every minute. Theirs is the mystery of the continuous creation and all that providence implies."[66]

Throughout human experience rivers have been, as John McPhee characterized them, "the ultimate metaphors of existence."[67] When rivers die, we die, both literally and figuratively. We save rivers because rivers save us—from our foibles, our loneliness, our frantic pace, our boredom. They can provide heart-soothing serenity or heart-pounding excitement. Their languid wandering across a plain or frantic rush through a canyon inspires the whole range of human emotions. Rivers can be a part of our quest for redemption or an avenue of escape. There is always something intriguing downriver.

We are blessed with the gifts bestowed by rivers. The challenge now is to use those gifts gently without destroying them; to make for ourselves both a home and a homeland without diminishing the capacity of our rivers to serve a hundred more generations. We must become partners with our rivers, not exploiters of them.

Let us end with Thoreau. His pensive ramblings provide an incitement to unconventional thought and a future full of possibilities. He ended his essay, *Walking*, with a smiling glance at a destiny of hope:

So we saunter toward the Holy Land, till one day the sun shall shine more brightly than ever he has done, shall perchance shine into our minds and hearts, and light up our whole lives with a great awakening light, as warm and serene and golden as on a bankside in autumn.[68]

NOTES

PREFACE

1. President Obama, much to the consternation of local politicians, tried to cut funding for the CUP in 2010. Judy Fahys, "Central Utah Water Project Funding May Be Plugged," *Salt Lake Tribune*, Oct. 7, 2010.

2. Zach and the Utah Rivers Council are still at it. See "URC Initiates Lawsuit to Protect Diamond Fork & Strawberry Watersheds," *Waterlines: The Newsletter of the Utah Rivers Council* (Apr. 2010): 1.

1. CRUMBLING EDIFICE

1. See U.S. Army Corps of Engineers, *Matilija Dam Environmental Restoration Feasibility Study, Matilija Dam: Ecosystem Restoration Study, Baseline Conditions Draft Report (F3) Milestone* (Aug. 2002); Bureau of Reclamation, *Matilija Dam Removal: Appraisal Report* (Washington, DC: Department of the Interior, 2000).

2. With the removal of Matilija Dam, Casitas will be the only major reservoir in the watershed. According to the 2007 Census of Agriculture, farmers in Ventura County, California, received $554,000 in government payments that year. National Agricultural Statistics Service, U.S. Department of Agriculture.

3. Miguel Cervantes Saavedra, *Don Quixote de la Mancha* (New York: A. L. Burt, n.d.), 617.

4. H. L. Mencken, "Battle Now Over, Mencken Sees; Genesis Triumphant and Ready for New Jousts," *Baltimore Sun*, July 18, 1925.

5. Zeke Barlow, "New Proposal on How to Tear Down Matilija Dam Unveiled," *Ventura County (CA) Star*, Apr. 6, 2011.

6. E-mail from Paul Jenkin to the author, Apr. 25, 2009.

7. David Courtland, "Environmentalists Finally See Progress in Removing Matilija Dam," *Ventura County (CA) Reporter*, Mar. 26, 2009, at www.vcreporter.com/cms/story/detail/?id=6795.

8. U.S. Army Corps of Engineers, *2007 National Inventory of Dams*, at www.nid.usace.army.mil/.

9. National Research Council, Committee on Restoration of Aquatic Ecosystems, *Restoration of Aquatic Ecosystems: Science, Technology, and Public Policy* (Washington, DC: National Academy Press, 1992).

10. Arthur Benke, "A Perspective on America's Vanishing Streams," *Journal of the North American Benthological Society* 9 (1990): 78.

11. U.S. Census Bureau, *Statistical Abstract of the United States: The National Data Book, 1990* (Washington, DC: Government Printing Office, 1991).

12. National Research Council, Committee on Restoration of Aquatic Ecosystems, *Restoration of Aquatic Ecosystems*, 27.

13. "Hydro Facts," *Southwest Hydrology* 8.5 (Sept.–Oct. 2009): 12.

14. See Martin Doyle et al., "Dam Removal in the United States: Emerging Needs for Science and Policy," *Eos* 84.4 (Jan. 28, 2003): 29–33.

15. American Society of Civil Engineers, "Report Card for America's Infrastructure," at www.infrastructurereportcard.org/report-cards; Emily Gordon et al., *Water Works: Rebuilding Infrastructure, Creating Jobs, Greening the Environment*, a report produced for Green for All, at www.greenforall.org/resources/waterworks.

16. Federal Emergency Management Agency, *Dam Safety and Security in the United States: A Progress Report on the National Dam Safety Program, Fiscal Years 2004 and 2005*, FEMA 576 (Sept. 2006).

17. American Society of Civil Engineers, "Report Card," 16.

18. Association of State Dam Safety Officials (ASDSO), press release (Jan. 29, 2009), at www.damsafety.org.

19. Environmental Protection Agency (EPA), *Report on National Water Quality* (Washington, DC: EPA, 1998).

20. Martin Doyle et al., "Dam Removal: Physical, Biological, and Societal Considerations," in Rollin Hotchkiss and Martin Glade, eds., *Building Partnerships: Proceedings of the 2000 Joint Conference on Water Resource Engineering and Water Resource Planning and Management* (Reston, VA: ASCE, 2000), CD-ROM.

21. U.S. Army Corps of Engineers, Interagency Floodplain Management Review Commission, *Sharing the Challenge: Floodplain Management Into the 21st*

Century (1994), 43. After catastrophic losses from 1954 to 1998, the nation actually gained a slight increase in recent years. See EPA, *Highlights of National Trends: EPA's Report on the Environment* (2008), 13.

22. Ralph Waldo Emerson, *The Complete Works of Ralph Waldo Emerson*, vol. 2 (New York: Houghton, Mifflin, 1865), 4.

23. E. S. Bernhardt et al., "Restoration of U.S. Rivers: A National Synthesis," *Science* 308 (Apr. 2005): 636–37. Also see National River Restoration Science Synthesis, at http://nrrss.nbii.gov/.

24. Bernhardt et al., "Restoration of U.S. Rivers," 637. Also see Rocky Barker, "Business Is Booming for Nampa Dam-Blasting Firm," *Idaho Statesman*, Oct. 8, 2011.

25. Margaret Palmer and J. David Allan, "Restoring Rivers," *National Academy of Science Issues in Science and Technology* (Winter 2006): 40–48.

26. See Mark Angelo, "Dam Removals Open Way for Cultural and Habitat Restoration," *National Geographic* (blog), Oct. 22, 2010, at http://newswatch .nationalgeographic.com/2010/10/22/dam-removal-provides-restoration-oppor tunities/; American Rivers, "2011 Dam Removal Resource Guide," at www.ameri canrivers.org/our-work/restoring-rivers/dams/.

27. Edwards Dam on the Kennebec River in Maine was one of the first significant dam removals, a result of the Federal Energy Regulatory Commission's refusal to renew its license. The dam was removed, against the wishes of its owners, in 1999. See Natural Resource Council of Maine, "A Citizen's Guide to Dams, Hydropower, and River Restoration in Maine" (Augusta, ME, 2002). Also see Elisabeth Grossman, *Watershed: The Undamming of America* (New York: Counterpoint, 2002).

28. See, for example, Storm Cunningham, *The Restoration Economy: The Greatest New Growth Frontier* (San Francisco: Berrett-Koehler, 2002); Brian Lavendel, "The Business of Ecological Restoration," *Ecological Restoration* 20.3 (Sept. 2002): 173–78.

29. See, for example, M. A. Palmer et al., "Standards for Ecologically Successful River Restoration," *Journal of Applied Ecology* 42 (2005): 208–17; David Hart et al., "Dam Removal: Challenges and Opportunities for Ecological Research and River Restoration," *BioScience* 52 (Aug. 2002): 669–82; Angela Bednarek, "Undamming Rivers: A Review of the Ecological Impacts of Dam Removal," *Environmental Management* 27.6 (2001): 803–14. Or, visit the annual Berkeley River Restoration Symposium in Berkeley, CA.

30. See, for example, J. V. Ward, "Understanding Natural Patterns and Processes in River Corridors as the Basis for Effective River Restoration," *Regulated Rivers Research and Management* 17 (2001): 311–23; Robert Walter and Dorothy

Meritts, "Natural Streams and the Legacy of Water Powered Mills," *Science* 318 (Jan. 18, 2008): 299–304; Timothy Randle, "Dam Removal and Sediment Management," in William Graf, ed., *Dam Removal Research: Status and Prospects* (Washington, DC: H. John Heinz Center for Science, Economics, and the Environment, 2002).

31. See, for example, William Lowry's excellent analysis, *Dam Politics: Restoring America's Rivers* (Washington, DC: Georgetown University Press, 2003). Other powerful books include Tim Palmer, *Endangered Rivers and the Conservation Movement*, 2d ed. (Lanham, MD: Rowman and Littlefield, 2004); Mary Doyle and Cynthia Drew, eds., *Large-Scale Ecosystem Restoration* (Washington, DC: Island Press, 2008); Grossman, *Watershed*; Patrick McCully, *Silenced Rivers* (London: Zed Books, 1986); Brian Graber, Crystal Yap, and Sara Johnson, *Small Dam Removal: A Review of Potential Economic Benefits* (Madison, WI: Trout Unlimited, Oct. 2001).

32. Elinor Ostrom, *Governing the Commons* (New York: Cambridge University Press, 1990), 7.

33. John Steinbeck, *The Grapes of Wrath* (New York: Viking, 1939), 134.

34. David Noebel, *Rhythm, Riots, and Revolution: An Analysis of the Communist Use of Music, the Communist Master Music Plan* (Tulsa, OK: Christian Crusade Publications, 1966), xx; Amy Joi, "'Million' Pipeline Now Touted as Hydropower Project," *Deseret News*, Aug. 15, 2011; Dave Peterson, "Guest Commentary: Flaming Gorge Pipeline Would Be Disaster," *Denver Post*, Sept. 4, 2011.

35. Not convinced it is a corrupt system? Read Robert Kaiser, *So Damn Much Money* (New York: Knopf, 2009); Kevin Phillips, *Bad Money* (New York: Penguin, 2009); Larry Bartels, *Unequal Democracy* (Berkeley, CA: Russell Sage Foundation, 2008); Elisabeth Drew, *The Corruption of American Politics* (New York: Birch Lane Press, 1999); Larry Sabato, *Dirty Little Secrets* (New York: Crown, 1996); Hedrick Smith, *The Power Game* (New York: Ballantine, 1996); Jonathan Rauch, *Demosclerosis* (New York: Times Books, 1994); William Ashworth, *Under the Influence* (New York: Hawthorn/Dutton, 1981). Or, just go to www.opensecrets.org and watch the money flow.

36. See Russell Martin, *The Story That Stands like a Dam* (New York: Henry Holt, 1989); Jared Farmer, *Glen Canyon Damned* (Tucson: University of Arizona Press, 1999).

37. For an excellent account of how it came to this, see James Lawrence Powell, *Dead Pool* (Berkeley: University of California Press, 2008). Also see Tom Wharton, "Is Lake Powell Slowly Dying?" *Salt Lake Tribune*, Mar. 21, 2011. After years of decline the reservoir received a new lease on life in 2011 with record-breaking precipitation in the Basin due to a powerful El Niño.

38. J. W. Powell, *The Exploration of the Colorado River and Its Canyons* (1895) (New York: Dover, 1961), 227.

39. Katie Lee, *All My Rivers Are Gone* (Boulder, CO: Johnson Books, 1998). Also see Katie Lee, *Sandstone Seduction* (Boulder, CO: Johnson Books, 2004).

40. For increased upstream diversions, see Gary Harmon, "Front Range Projects Are River Threats, Group Says," *Grand Junction (CO) Daily Sentinel*, June 3, 2010; Pete Nickeas, "City Looks at Green River for Water," *Casper (WY) Star-Tribune*, Feb. 10, 2010; Bruce Finley, "Environmentalists Oppose Denver Project to Divert More Water from Western Slope," *Denver Post*, Dec. 9, 2009; Jeff Gearino, "Pipeline Battle Will Be Long and Costly," *Casper Star-Tribune*, Mar. 24, 2010; Bob Berwin, "Trout Unlimited: Upper Colorado 'on the Brink,'" *Aspen (CO) Times*, Nov. 2, 2009.

41. Ingebretsen is a professor of physics and a physician at the University of Utah. His idea of "fill Lake Mead first" is similar to the Bureau's strategy for dealing with the huge flows of 2011.

42. In 2009 some people claimed to have found Ruess's body in the Comb Ridge area of southern Utah, but this claim was later debunked. See David Roberts, *Finding Everett Ruess* (New York: Broadway Books, 2011). Also see Philip Fradkin, *Everett Ruess: His Short Life, Mysterious Death, and Astonishing Afterlife* (Berkeley: University of California Press, 2011).

43. To get a sense of what was lost, see Tad Nichols, *Glen Canyon: Images of a Lost World* (Santa Fe: Museum of New Mexico Press, 1999). To ponder what might be restored, see Annette McGivney and James Kay, *Resurrection* (Seattle, WA: Braided River Books, 2009).

44. See John Keys III, "The Colorado River: Our Heritage and Future," *Spillway* (July–Aug. 2005): 4; Gary Pizor, "Balancing a Complex Set of Interests: Glen Canyon Dam and Adaptive Management," *Colorado River Project River Report* (Winter 2010–2011): 1, 4–8.

45. This motto is on the cover of the agency's 1946 report, *The Colorado River: A Natural Menace Becomes a National Resource* (Washington, DC: Department of the Interior, Mar. 1946).

46. McGee, quoted in Vandana Shiva, *Water Wars: Privatization, Pollution, and Profit* (Cambridge, MA: South End Press, 2002), 54.

47. E. Alexander Powell, "Arizona," *Sunset: The Pacific Monthly* 31 (Oct. 1913): 674.

48. Joseph Stevens, *Hoover Dam: An American Adventure* (Norman: University of Oklahoma Press, 1988), 245.

49. Lane, quoted in Donald Pisani, *Water and American Government* (Berkeley: University of California Press, 2002), 115.

50. William Smythe, *The Conquest of Arid America* (New York: Macmillan, 1899), 329.

51. Bureau of Reclamation, *Lake Powell: Jewel of the Colorado* (Washington, DC: Department of the Interior, 1965), 28.

52. National Irrigation Congress, *Official Proceedings* (Sacramento, CA: National Irrigation Congress, 1907), 119.

2. PLANTERS, SAWYERS, AND SNAGS: THE U.S. ARMY CORPS OF ENGINEERS

1. Wayne Stroupe, in-person interview with author, Vicksburg, MS (Dec. 18, 2003).

2. For an explanation of the origin of the term *pork barrel*, see Chester Collins Maxey, "A Little History of Pork," *National Municipal Review* 8.10 (1919): 691–705.

3. Hay, quoted in Doris Kearns Goodwin, *Team of Rivals* (New York: Simon and Schuster, 2005), 383.

4. Eads, quoted in John M. Barry, *Rising Tide* (New York: Simon and Schuster, 1997), 75.

5. See Jeanne Nienaber Clarke and Daniel C. McCool, *Staking Out the Terrain*, 2d ed. (Albany: SUNY Press, 1996), 17–22.

6. Energy and Water Development Appropriations Bill of 2012, H.R. Rep. 112-118, at 11–12.

7. See Arthur Maass, *Muddy Waters: The Army Engineers and the Nation's Rivers* (Cambridge, MA: Harvard University Press, 1951). For an explanation of the term *iron triangle*, see Daniel McCool, *Command of the Waters* (Berkeley: University of California Press, 1987), chap. 1; Keith Hamm, "Patterns of Influence Among Committees, Agencies, and Interest Groups," *Legislative Studies Quarterly* 8 (Aug. 1983): 379–426.

8. Clarke and McCool, *Staking Out the Terrain*, 31–34.

9. *Weekly Compilation of Presidential Documents* (hereafter, *Weekly Comp. Pres. Doc.*), Jimmy Carter, "The President's Message to the Congress Recommending Deletion of Funds for 19 Projects from the 1978 Fiscal Year Budget" (Feb. 21, 1977).

10. *Weekly Comp. Pres. Doc.*, Jimmy Carter, "Statement by the President Announcing His Decisions on 32 Projects" (Apr. 18, 1977).

11. The quote is from Donald Worster, *Rivers of Empire* (New York: Pantheon, 1985), 326. For a description of the politics, see Daniel P. Beard, "Oral History

Interview" (Washington, DC: Bureau of Reclamation, Oral History Program, 1995), 32–35; Clarke and McCool, *Staking Out the Terrain*, 24–36, 138–43.

12. Clarke and McCool, *Staking Out the Terrain*, 34–36, 144–49.

13. See U.S. Army Corps of Engineers, www.usace.army.mil/history/Pages /Brief/14-water.html.

14. See Martin Reuss, "Reshaping National Water Politics: The Emergence of the Water Resources Development Act of 1986," Office of History, Headquarters, U.S. Army Corps of Engineers, IWR Policy Study 91-PS-1 (Oct. 1993). The WRDAs, usually passed every two years, authorize the construction of projects; it takes a separate bill each year to actually appropriate money to build the projects.

15. T. S. Eliot, "Burnt Norton," in *Collected Poems, 1909–1962* (New York: Harcourt Brace Jovanovich, 1991), 176.

16. Bob Hunter, in-person interview with author, Medford, OR (June 14, 2003); Judy Gove, former board member of the Grants Pass Irrigation District, in-person interview with author, Grants Pass, OR (June 13, 2003).

17. Report by the Comptroller General, Report to the Honorable James H. Weaver, House of Representatives of the United States, *Corps of Engineers Should Reevaluate the Elk Creek Project's Benefits and Costs*, CED 82-53 (Mar. 15, 1982).

18. Anonymous Corps staff person, interview with author.

19. Portland District, U.S. Army Corps of Engineers, "Corps Releases Environmental Assessment Modifying Elk Creek Dam," Press Release 97-111 (Oct. 20, 1997).

20. See Mark Freeman, "Corps Plans Fish Trap at Elk Creek Dam," *Medford Mail Tribune*, Mar. 7, 2003; Matthew Preusch, "Blasting of Elk Creek Dam Will Begin," *Oregonian*, July 15, 2008.

21. National Marine Fisheries Service, *Biological Opinion and Magnuson-Stevens Act Essential Fish Habitat Consultation*, Report OSB2000-0282 (Jan. 23, 2001).

22. WaterWatch and American Rivers, "50 Commercial and Recreational Fishing Groups, Conservation Organizations Blast New Elk Creek Dam Scheme," press release (Sept. 17, 2002).

23. Preusch, "Blasting of Elk Creek Dam Will Begin." Hunter's video of the dam blasting can be found on YouTube.

24. South Florida Water Management District, "Florida Completes 100,000 Acre Land Acquisition for Kissimmee River Restoration," press release (Apr. 11, 2006), at www.floridadep.com/seretary/news/2006/04/0412_01htm.

25. U.S. Army Corps of Engineers, Jacksonville District, Project Execution Branch, Upper East Coast and Kissimmee/Lake Okeechobee Section, *Kissimmee River Restoration* (by Susan Jackson; updated Feb. 24, 2011), at www.saj.usace.army .mil/Divisions/Everglades/Branches/ProjectExe/Sections/UECKLO/KRR.htm.

26. Chuck Wilburn, in-person interview with author, Sebring, FL, and various parts of the Kissimmee River Project (Feb. 10, 2003).

27. Joseph Koebel, in-person interview with author, West Palm Beach, FL (Feb. 13, 2003).

28. Loisa Kerwin, in-person interview with author, Riverwoods Field Laboratory, FL (Feb. 11, 2003).

29. Patrick Luna, in-person interview with author, Kissimmee River job site, FL (Feb. 12, 2003).

30. Now an endangered species.

31. South Florida Water Management District, "Kissimmee Basin," at http://my.sfwmd.gov/portal/xweb%2oprotectingand%2orestoring/kissimmee%2oriver.

32. Water Resources Development Act, Pub. L. 101-640 (Nov. 28, 1990).

33. Water Resources Development Act, Pub. L. 104-303 (Oct. 21, 1996). Water authorization acts are usually passed every two years, but no act was passed in 1994 because of severe disagreements between the protectors of the Corps' old missions and those who wanted to force changes to the agency.

34. This program was officially authorized by Sec. 201 of the 1999 Water Resources Development Act.

35. Nancy Dorn, Assistant Secretary of the Army for Civil Works, quoted in National Research Council, *New Directions in Water Resources* (Washington, DC: National Academy Press, 1999), 66.

36. Quoted in Jeff DeLong, "Projects Turn Back Time on Nation's Rivers," *USA Today*, Oct. 6, 2009.

37. National Research Council, *New Directions in Water Resources*, 8.

38. Michael Grunwald's articles in the *Washington Post* on this topic were "Generals Push Huge Growth for Engineers," Feb. 24, 2000; "Pentagon Vows to Rein in Corps of Engineers," Feb. 25, 2000; "Army Engineers Reforms are Set," Mar. 30, 2000; "Corps of Engineers Reforms Suspended," Apr. 7, 2000; "Senators Seek to Block Corps of Engineers Reforms," May 13, 2000.

39. Domenici retired from the Senate in 2009.

40. Quoted in Grunwald, "Senators Seek to Block Corps of Engineer Reforms."

41. Grunwald also wrote extensively in the *Washington Post* about the rollback of wetlands protection. See "Army Corps Seeks to Relax Wetlands Rules," June 4, 2001; "Army Corps Moves to Ease Wetlands Rules," Aug. 9, 2001; "Interior's Silence on Corps Plan Questioned," Jan. 14, 2002; "White House Relaxes Rules on Protection of Wetlands," Jan. 15, 2002. In 2006, Grunwald penned *Swamp* (New York: Simon and Schuster, 2006) on the Corps' Everglades restoration project.

42. Tancredo's bill was titled the Army Corps of Engineers Reform and Community Relations Improvement Act, H.R. 2353, 107th Cong. (2001).

43. Corps of Engineers Civil Works Independent Investigation and Review Act, S. 2309, 106th Cong. (2000), introduced by Senator Tom Daschle.

44. National Research Council, Panel on Peer Review, *Review Procedures for Water Resources Project Planning* (Washington, DC: National Academies Press, 2002).

45. Corps of Engineers Modernization and Improvement Act of 2002, S. 1987, 107th Cong. (Mar, 5, 2002).

46. *Water Engineering & Management* 149.4 (Apr. 2002): 12–24.

47. "The Budget of the United States, Fiscal Year 2002," Budget Message of President George W. Bush (Feb. 28, 2001).

48. 150.30 Cong. Rec. S2550 (daily ed. Mar. 10, 2004) (statement of Sen. Feingold). It is worth noting that all three of the sponsors lost subsequent elections: Daschle was defeated in 2004, Feingold lost his seat in 2010, and McCain lost his presidential bid in 2008. Apparently, fighting pork barrel is not an effective re-election strategy.

49. S. 2188, 108th Cong. (Apr. 1, 2004).

50. S. 753, 109th Cong. (Apr. 11, 2005).

51. Taxpayers for Common Sense and the National Wildlife Federation, *Crossroads: Congress, the Corps of Engineers and the Future of America's Water Resources*, at www.taxpayer.net/corpswatch/crossroads/index.htm.

52. The GAO completed five studies between 2002 and 2005. The five studies are summarized in *Corps of Engineers: Observations on Planning and Project Management Processes for the Civil Works Program*, GAO-06-529 (Mar. 15, 2006).

53. See Melissa Samet, *A Citizen's Guide to the Corps of Engineers 2009* (Washington, DC and Reston, VA: American Rivers and the National Wildlife Federation, 2009), 104–5.

54. In 2007 the river advocacy group American Rivers compiled a summary of all the studies that had been critical of the Corps—28 of them. See "Flawed Projects, Planning, and Mitigation: A Decade of Analysis Calls for Reforming the Army Corps of Engineers," a Summary of Studies by the National Academy of Sciences, Government Accountability Office, Army Inspector General, and Independent Experts, at www.amrivers.org.

55. Samet, *Citizen's Guide to the Corps of Engineers 2009*.

56. Ronald D. Utt, "The Water Resources Development Act of 2007: A Pork Fest for Wealthy Beach-Front Property Owners," WebMemo #1458 (May 15, 2007), at

http://www.heritage.org/research/reports/2007/05/the-water-resources-develop ment-act-of-2007-a-pork-fest-for-wealthy-beach-front-property-owners.

57. "Veto Message on H.R. 1495: Message from the President of the United States" (Washington, DC: GPO, Nov. 5, 2007).

58. See *Opportunities to Reduce Potential Duplication in Government Programs, Save Tax Dollars, and Enhance Revenue*, "The Corps of Engineers Should Provide Congress with Project-Level Information on Unobligated Balances," GAO-11-318SP (Mar. 2011).

59. Eliot, quoted in Clinton Wallace Gilbert, *Behind the Mirrors* (New York: Putnam, 1922), v.

60. Defense Base Closure and Realignment Commission, at www.brac.gov.

61. Government Accountability Office, Report to Congressional Committees, *Military Base Closures*, GAO-05-138 (Jan. 2005).

3. THE MANLESS LAND:
THE BUREAU OF RECLAMATION

1. Joseph Stevens, *Hoover Dam: An American Adventure* (Norman: University of Oklahoma Press, 1988), 246.

2. *Construction of Hoover Dam*, 13th ed. (Las Vegas, NV: KC Publications, 1976). A historic account prepared in cooperation with the Department of the Interior and the Bureau of Reclamation.

3. Stevens, *Hoover Dam*, 245.

4. Frank Waters, *The Colorado* (New York: Holt, Rinehart and Winston, 1946), 337.

5. See Richard Berkman and W. Kip Viscusi, *Damming the West* (New York: Grossman, 1973); Daniel McCool, *Command of the Waters* (Berkeley: University of California Press, 1987); Donald Pisani, *Water and American Government* (Berkeley: University of California Press, 2002).

6. For a sense of these times, see Edmund Morris, *Theodore Rex* (New York: Random House, 2001). Regarding the population of American Indians, see Charles Mann, *1491* (New York: Vintage, 2006); U.S. Census, "Table 1: U.S. Race and Hispanic Origin, 1790–1990," in "Historical Census Statistics on Population Totals by Race, 1790 to 1990, for the United States, Regions, Divisions, and States."

7. Quoted in Christopher Cerf and Victor Navasky, *The Experts Speak* (New York: Pantheon, 1984), 203.

8. See Richard Hofstadter, *Social Darwinism in American Thought* (Boston: Beacon Press, 1944).

9. From Oliver Wendell Holmes, "The Coming Era," in *The Poetical Works of Oliver Wendell Holmes*, vol. 2 (Boston and New York: Houghton Mifflin, 1892), 226.

10. Bureau of Reclamation, *Brief History of the Bureau of Reclamation, Official History* (Washington, DC: Department of the Interior, July 2000).

11. Quoted in Pisani, *Water and American Government*, 2.

12. Wallace Stegner, *Beyond the Hundredth Meridian* (New York: Penguin, 1953), 2. This is actually Stegner's description of William Gilpin, a zealous advocate of irrigation.

13. Elwood Mead, Commissioner of the Bureau of Reclamation when Hoover Dam was constructed, quoted in Donald Worster, *Rivers of Empire* (New York: Pantheon, 1985), 96.

14. Frederick Newell, the first director of the Reclamation Service, liked to make reference to this slogan. See F. H. Newell, "Engineering Work in Food Supply," paper presented at the annual meeting of the American Society of Agricultural Engineers (Dec. 1916).

15. The Revised Standard Version and the New International Version both say "the desert shall rejoice and blossom like the crocus" (Isaiah 35:1). Even after the Revised Standard Version became popular, the promoters of irrigation continued to use the "rose" reference, apparently preferring that to a crocus.

16. Cong. Rec. 6742 (1902).

17. Cong. Rec. 6766 (1902).

18. See Pisani, *Water and American Government*, xv; and McCool, *Command of the Waters*, 20–24.

19. For a fuller description of the genesis of the Bureau of Reclamation, see McCool, *Command of the Waters*, 20–32, 66–72.

20. Worster, *Rivers of Empire*, 169–70.

21. All of these quotes come from the displays at the visitor center.

22. These phrases and accompanying history are from information panels in the visitor center. Also see Paul Pitzer, *Grand Coulee: Harnessing a Dream* (Pullman: Washington State University Press, 1994), 284–85, 308–9.

23. Timothy Egan, *The Good Rain* (New York: Vintage, 1990), 247.

24. Pitzer, *Grand Coulee*, xii.

25. Bureau of Reclamation (Michael W. Straus, Commissioner), *The Colorado River: A Natural Menace Becomes a National Resource* (Washington, DC: Department of the Interior, Mar. 1946), 25.

26. This last unit was later renamed the Wayne Aspinall Project after the congressman from southwestern Colorado who never saw a water project he didn't like.

27. See Dale Rodebaugh, "Gov. Ritter to Sign A-LP bill in Durango," *Durango (CO) Herald*, June 4, 2010; "Lake Nighthorse Getting Pumped Up Again," *Durango Herald*, Mar. 18, 2011; "Want Water? Take a Number," *Durango Herald*, Aug. 29, 2011. For background, see Daniel McCool, *Native Waters* (Tucson: University of Arizona Press, 2002), 87–99.

28. Marc Reisner, *Cadillac Desert*, rev. ed. (New York: Penguin, 1993), 114.

29. Texas Water Development Board, "The Texas Water Plan, 1968" (Austin, TX, Nov. 1968).

30. William Ashworth, *Nor Any Drop to Drink* (New York: Summit, 1982), 95–97.

31. See John Opie, *Ogallala*, 2d ed. (Lincoln: University of Nebraska Press, 2000), 275–80; Reisner, *Cadillac Desert*, 276–77; Sarah Bates et al., *Searching Out the Headwaters* (Washington, DC: Island Press, 1993), 40.

32. Fred Powledge, *Water* (New York: Farrar Straus Giroux, 1982), 278.

33. Energy and Water Development Appropriations Bill of 2012, H.R. Rep. 112-108, at 62; *Energy and Water Development Appropriations for Fiscal Year 2005: Hearings Before the Subcomm. on Energy and Water Development, Comm. on Appropriations*, pt. 3, 108th Cong. (2004), 736.

34. For background, see Barbara Andrews and Marie Sansone, *Who Runs the Rivers?* (Stanford, CA.: Stanford Environmental Law Society, 1983); and Terry Anderson, *Water Crisis* (Baltimore: Johns Hopkins University Press, 1983).

35. For an excellent critique of Bureau funding, see Richard Wahl, *Markets for Federal Water* (Washington, DC: Resources for the Future, 1989).

36. These critiques are discussed in detail in McCool, *Command of the Waters*, 68–71.

37. See Mark Harvey, *A Symbol of Wilderness: Echo Park and the American Conservation Movement* (Seattle: University of Washington Press, 2000).

38. Bureau of Reclamation, *Lake Powell: Jewel of the Colorado* (Washington, DC: Department of the Interior, 1965), 27.

39. See James Lawrence Powell, *Dead Pool* (Berkeley: University of California Press, 2009).

40. Russell Martin, *A Story That Stands like a Dam* (New York: Henry Holt, 1989), 257.

41. Dominy died in 2010. See Douglas Martin, "F. E. Dominy, Who Harnessed Water in the American West, Is Dead at 100," *New York Times*, Apr. 28, 2010.

42. For a delightful description of Dominy and Brower rafting through the Grand Canyon together, see John McFee, *Encounters with the Archdruid* (New York: Farrar, Straus, and Giroux, 1971), part 3.

43. The Aspinall Projects were named after their chief sponsor, Congressman Wayne Aspinall. All five of them were of dubious benefit. See Steven Sturgeon, *The Politics of Western Water* (Tucson: University of Arizona Press, 2002), 126–27; and Berkman and Viscusi, *Damming the West*, 114–17.

44. Tim Palmer, *Endangered Rivers and the Conservation Movement*, 2d ed. (New York: Rowman and Littlefield, 2004), 117.

45. Examples are the Colorado, the Rio Grande, the Feather, the Salt, and the San Joaquin. But the San Joaquin is getting a new lease on life. See Bettina Boxall, "Water Releases to Bring New Life to Parts of San Joaquin River," *Los Angeles Times*, Oct. 1, 2009.

46. Wendy Espeland, *The Struggle for Water* (Chicago: University of Chicago Press, 1998), 75.

47. Upper Colorado River Commission, *25th Annual Report* (Salt Lake City, UT, Sept. 30, 1973), 25–26.

48. For more on the CUP, see Daniel McCool, ed., *Waters of Zion: The Politics of Water in Utah* (Salt Lake City: University of Utah Press, 1995), chaps, 1, 3, 8, and 9.

49. Isaiah 42:15 (Revised Standard Version).

50. Bruce Babbitt, "A River Runs Against It: America's Evolving View of Dams," *Open Spaces Quarterly* 1.4 (2010): 1–3.

51. Daniel Beard, in-person interview with author, Washington, DC (Nov. 4, 2004).

52. Daniel Beard, Commissioner of Reclamation, "Blueprint for Reform: The Commissioner's Plan for Reinventing Reclamation" (Nov. 1993).

53. See McCool, *Native Waters*.

54. The dam derived its name from a family named Savage who lived on the river at that location; it was not because there was big white water at the dam site.

55. Pacific Northwest Region, Bureau of Reclamation, *Planning Report/Final Environmental Statement, Fish Passage Improvement, Savage Rapids Dam* (Washington, DC: Department of the Interior, Aug. 30, 1995).

56. Dan Shepard, in-person interview with author, Grants Pass, OR (June 13, 2003).

57. Buchal, in his book *The Great Salmon Hoax* (Las Vegas, NV: Iconoclast Publishing, 1997), claims that dams have no impact on anadromous fish.

58. *United States v. Grants Pass Irrigation District*, Civil No. 98-30340HO, Consent Decree (D. Or. Aug. 25, 2001).

59. Bureau of Reclamation, *Draft Environmental Assessment, Fish Passage Improvements, Savage Rapids Dam, Grants Pass Project, Oregon* (Washington, DC: Department of the Interior, Aug. 2005).

60. Associated Press, "River Runs Wilder Now That Dam Is Gone," MSNBC .com (Oct. 20, 2009).

61. See Scott Learn, "After Dam Removals, Oregon's Rogue River Shows Promising Signs for Salmon," *Oregonian*, Oct. 28, 2010.

62. Remarks of Daniel P. Beard, Commissioner, U.S. Bureau of Reclamation, before the International Commission on Large Dams, Durban, South Africa (Nov. 9, 1994).

63. Bureau of Reclamation, *Annual Report* (Washington, DC: Department of the Interior, 1997).

64. Brent Israelsen, "New U.S. Reclamation Boss Promises No Deluge of Changes," *Salt Lake Tribune*, Aug. 13, 2001.

65. Taxpayers for Common Sense, *Waste Basket: A Regular Bulletin on Government Waste* 7.24 (June 13, 2002), at http://www.taxpayer.net/index.php.

66. Bureau of Reclamation, *Water 2025: Preventing Crises and Conflict in the West, 2003* (final report published in 2005), at http://www.worldcat.org/title /water-2025-preventing-crises-and-conflict-in-the-west/oclc/78243349& referer=brief_results.

67. Bureau of Reclamation, *Reclamation: Managing Water in the West, Managing for Excellence, an Action Plan for the 21st Century* (Washington, DC: Department of the Interior, Feb. 2006).

68. On the President's Fiscal Year 2012 Budget Request: Before the Appropriations Comm., Subcomm. on Energy and Water Development, U.S. Senate (statement of Michael Connor, Commissioner, Bureau of Reclamation, Department of the Interior, Apr. 13, 2011). The Bureau's budget declined for 2012, but the agency received a whopping $950 million in one-time funds from the 2009 Recovery Act. See "Remarks by Michael Connor" at the Four States Irrigation Council's 58th Annual Meeting, Fort Collins, CO (Jan. 13, 2011). For a summary of the Bureau's restoration activities, see Bureau of Reclamation, "Bureau of Reclamation River Restoration Programs: A Summary of 16 Programs and Shared Institutional Challenges," prepared for "River Restoration: Exploring Institutional Challenges and Opportunities," sponsored by the Utton Transboundary Resources Center, University of New Mexico School of Law, Sept. 14–15, 2011, Albuquerque, NM.

69. Stegner, *Beyond the Hundredth Meridian*, 20; Address by John Wesley Powell, *Official Report of the International Irrigation Congress* (1893), 112.

70. For an incisive political critique of the Bureau, see George Miller, "In a Drying West, Dams Are No Longer the Answer," *San Francisco Chronicle*, Jan. 8, 2009. Also see Marc Reisner, *Cadillac Desert*; Espeland, *Struggle for Water*; Wahl, *Markets for Federal Water*; Andrews and Sansone, *Who Runs the Rivers?*; Robert

Gottlieb, *A Life of Its Own: The Politics and Power of Water* (New York: Harcourt Brace, 1998); Berkman and Viscusi, *Damming the West*.

71. Jeanne Nienaber Clarke and Daniel McCool, *Staking Out the Terrain*, 2d ed. (Albany: SUNY Press, 1996), 31.

72. For reference, see Dan McCool, "Warning: Water Policy Faces an Age of Limits," *High Country News* (Paonia, CO) (Apr. 22, 2010); "Rethinking the Future of the Colorado River" (draft interim report for the Colorado River Governance Initiative, Dec. 2010).

73. After our discussion I realized that we both loved rivers, albeit in different ways. Former Commissioner Keys died tragically in a plane crash in 2008.

4. HANDOUT HORTICULTURE: FARMING AND THE FEDS

1. Quoted in Francis Newton Thorpe, *The History of North America*, vol. 15 (Philadelphia: George Barrie's, 1906), 510.

2. U.S. Department of Agriculture, National Agricultural Statistics Service, *2007 Census of Agriculture*, http://www.agcensus.usda.gov/Publications/2007/Full _Report/index.asp.

3. U.S. Geological Survey, "Estimated Use of Water in the U.S. in 2000," Survey Fact Sheet 2005-3051 (Washington, DC: U.S. Department of the Interior, 2005).

4. Dana was also home to World War II correspondent Ernie Pyle. See "Pyle's Legacy Too Important to Close Dana Historic Site," *Indianapolis Star*, June 13, 2010.

5. From 1929 to 1932, U.S. farm income fell by 52 percent. George Pyle, *Raising Less Corn and More Hell* (New York: Public Affairs, 2005), 26.

6. See Chris Edwards, "Agricultural Subsidies," Cato Institute (June 12, 2007).

7. "Concentration in Agriculture," in U.S. Department of Agriculture, National Agricultural Statistics Service, *2007 Census of Agriculture*.

8. Robert Atkinson, "Reversing Rural America's Economic Decline," *Policy Report* (Feb. 2004): 9.

9. Paul Faeth, *Growing Green: Enhancing the Economic and Environmental Performance of U.S. Agriculture* (Washington, DC: World Resources Institute, 1995).

10. Environmental Working Group, "EWG Farm Subsidy Database," at http:// farm.ewg.org/region.php?fips=00000&regname=UnitedStatesFarmSubsidySum

mary. Also see Taxpayers for Common Sense, "Cultivating Cash," at http://tax payer.net/sarticle.php?proj_id=840&t=Cultivating.

11. Letter to the Honorable Paul Ryan, Chairman, House Committee on the Budget, from Congressman Frank Lucas, Chairman, and Collin Peterson, Ranking Minority Member, Committee on Agriculture, U.S. House of Representatives, Mar. 15, 2011.

12. GAO, *Farm Program Payments: USDA Needs to Strengthen Regulations and Oversight to Better Ensure Recipients Do Not Circumvent Payment Limitations*, GAO-04-407 (Washington, DC, Apr. 30, 2004).

13. Dan Morgan, "$1.3 Billion Crop Subsidies Shelled Out to Nonfarmers," *Washington Post*, July 2, 2006.

14. GAO, *Federal Farm Programs: USDA Needs to Strengthen Controls to Prevent Improper Payments to Estates and Deceased Individuals*, GAO-07-818 (Washington, DC, July 9, 2007).

15. Pyle, *Raising Less Corn*, 35–36.

16. *U.S. Budget for Fiscal Year 2004*, at www.whitehouse.gov/omb/budget /fy2004.

17. Brian Riedl, "Another Year at the Federal Trough: Farm Subsidies for the Rich, Famous, and Elected Jumped Again in 2002," The Heritage Foundation, Policy Research and Analysis, Backgrounder #1763 (May 24, 2004).

18. GAO, *Reducing Some Farm Program Payments Could Result in Substantial Savings: Opportunities to Reduce Potential Duplication in Government Programs, Save Tax Dollars, and Enhance Revenue*, sec. II: "Other Cost Savings and Revenue Enhancements," GAO-11-318SP (Washington, DC, Mar. 2011).

19. Riedl, "Another Year at the Federal Trough," 7.

20. David Weigel, "Tea Party Candidate Received Farm Subsidies, but He's Ready to Eliminate Them," *Washington Post*, Apr. 8, 2010.

21. Study cited in Taxpayers for Common Sense, "A Litmus Test on Spending," *Waste Basket* 10.6 (Feb. 18, 2004).

22. Quoted in Robert Pear, "Bush Said to Seek Sharp Cuts in Subsidy Payments to Farmers," *New York Times*, Feb. 6, 2005.

23. See three reports by the Environmental Working Group: *Soaking Uncle Sam: Why Westlands Water District's New Contract Is All Wet* (Sept. 14, 2005); *Double Dippers: How Big Ag Taps Into Taxpayers' Pockets—Twice* (Aug. 3, 2005); *Power Drain: Big Ag's $100 Million Energy Subsidy* (May 29, 2007).

24. GAO, *Views on Proposed Reclamation Reform Legislation*, GAO/RCED-91-90 (Washington, DC, Sept. 12, 1991); GAO, *Water Subsidies: Basic Changes Needed to Avoid Abuse of the 960-Acre Limit*, GAO-RCED-90-6 (Washington, DC, Oct. 12, 1989).

25. Cited in Dave Aftandilian, "Farm Bill 2002: Corporate Welfare or Farmer's Friend?" *Conscious Choice* (July 2002), at http://www.consciouschoice.com/note /note1507.html.

26. Dan Morgan et al., "Powerful Interests Ally to Restructure Agriculture Subsidies," *Washington Post*, Dec. 22, 2006.

27. GAO, *Federal Farm Programs: USDA Needs to Strengthen Controls to Prevent Payments to Individuals Who Exceed Income Eligibility Limits*, GAO-09-67 (Washington, DC, Oct. 2008).

28. Rob Inglis, "More Skepticism of Obama's Farm-Subsidy Reforms," *New Republic*, Mar. 3, 2009, at http://www.tnr.com/blog/the-vine/more-skepticism -obamas-farm-subsidy-reforms.

29. Environmental Working Group, Farm Subsidy Database, "Full Disclosure: Who Really Benefits from Federal Farm Subsidies" (Washington, DC, 2007).

30. Oregon State University, Extension Service, Rural Studies, Total direct and indirect economic activity generates a total of just under 10 percent. See *Oregon Agriculture and the Economy*, Special Report 1080 (Corvallis, OR, Feb. 2008).

31. Governor's Office of Planning and Budget, *Economic Report to the Governor, 2010*, State of Utah, Gary Herbert, Governor, 78. Utah still managed to collect $54 million in federal farm subsidies in 2002.

32. Frank Ackerman and Elizabeth Stanton, *The Last Drop: Climate Change and the Southwest Water Crisis* (Stockholm: Stockholm Environment Institute, Feb. 2011), 6–7; Shaun McKinnon, "Water-Demanding Farms Looked At as Resources Vanish," *Arizona Republic*, Oct. 25, 2009.

33. An acre-foot is the amount of water it takes to flood an acre to a depth of one foot, about 326,000 gallons. These figures are from the 1998 NASS survey, U.S. Department of Agriculture. A new survey is currently being conducted.

34. Environmental Working Group, "California Water Subsidies," at http:// archive.ewg.org/reports/Watersubsidies/.

35. Philip Garone, *The Fall and Rise of the Wetlands of California's Great Central Valley* (Berkeley: University of California Press, 2011).

36. Serge Birk, in-person interview with author, Dairyville, CA (June 9, 2003). Serge passed away in November 2010.

37. Tehama Colusa Canal Authority, "Tehama Colusa Canal Authority Commends President Obama and His Administration for Their Commitment to the Economy and the Environment," press release (Apr. 15, 2009).

38. See *Field Hearing—Creating Jobs by Overcoming Man-Made Drought: Time for Congress to Listen and Act: Before the Natural Resources Comm., Subcomm. on Water and Power, U.S. House of Representatives* (statement of Michael Connor, Commissioner, Bureau of Reclamation, Apr. 11, 2011).

39. A NOAA (National Oceanic and Atmospheric Administration) Fisheries biological opinion cut an additional 5–7 percent of water from agriculture, to be used for increased flows for endangered species. Julie Cart, "Federal Directive to Cut California Water Deliveries," *Los Angeles Times*, June 5, 2009.

40. California Water Conservation Act, signed by Governor Arnold Schwarzenegger on November 10, 2009. See Peter Klebnikov, "Calming the Water Wars," *Solutions* 42.1 (Winter 2011): 6–9. Also see Peter Gleick, "Sustaining Agriculture in an Uncertain Future: The Role of Water Efficiency," *Southwest Hydrology* (Nov.–Dec. 2009): 18–19, 30.

41. Trinity River Division Act of August 12, 1955, chap. 872, Pub. L. No. 84-386, 69 Stat. 719 (1955).

42. U.S. Department of the Interior, Record of Decision, *Trinity River Mainstem Fishery Restoration: Final Environmental Impact Statement/Environmental Impact Report* (Washington, DC, Dec. 2000).

43. Bureau of Reclamation, Mid-Pacific Region, Sacramento, CA, "Reclamation Announces Final Schedule for High Flow Releases Into Trinity River as Part of Restoration," press release (Apr. 19, 2011).

44. Arnold Whitridge, in-person interview with author, Trinity County (June 10, 2003).

45. The truth is somewhere in between those two extremes. For a succinct description of the District and its problems, see Matt Jenkins, "Breakdown," *High Country News* (Paonia, CO), Jan. 18, 2010.

46. Mr. Stokely retired in 2008.

47. Tom Stokely, in-person interview with author, Hayfork, CA (June 10, 2003).

48. Mike Orcutt, in-person interview with author, Hoopa, CA (June 11, 2003). The name of the tribe is spelled "Hupa," but its namesake valley and reservation are spelled "Hoopa."

49. Sally Morris, "$4.5 Million for the Trinity River," *Trinity Journal* (Weaverville, CA), Apr. 22, 2009; *Friends of the Trinity River Newsletter* (Apr. 10, 2007).

50. *Friends of the Trinity River Newsletter* (Sept. 2010).

51. The latest count was a run of about 30,000 fish in the Trinity (Hoopa Valley Tribal Fisheries, Mar. 7, 2011); at www.hoopafisheries.org/3901.html.

52. Environmental Working Group, 2007 Direct Payments, Ken Cook and Chris Campbell, "Amidst Record 2007 Crop Prices and Farm Income Washington Delivers $5 Billion in Subsidies," press release, at www.farm.ewg.org/farm/dp _text.php.

53. Ibid.

54. Renee Johnson, *What Is the "Farm Bill"?* CRS Report for Congress, Congressional Research Service, Order Code RS22131 (Sept. 23, 2008).

55. *Weekly Compilation of Presidential Documents*, "President Bush's Farm Bill Veto Message to House of Representatives" (May 21, 2008). See "Pork Gridlock," *Washington Post*, Apr. 24, 2008.

56. Finlay Lewis, "Hard Road Ahead for Plan to Cut Big-Farm Subsidies," *CQ Politics* (Apr. 4, 2009): 766–67.

57. Taxpayers for Common Sense, "FY Budget Cuts Agriculture Waste" (Feb. 2, 2010), at http://www.taxpayer.net/projects.php?action=view&category=&type=Project&proj_id=3161.

58. See, for example, "Green Scissors, 2010: More Than $200 Billion in Cuts to Wasteful and Environmentally Harmful Spending," produced by Friends of the Earth, Taxpayers for Common Sense, Environment America, and Public Citizen.

59. Taxpayers for Common Sense, "TCS Budget Cut List for the 112th Congress: Agriculture" (Feb. 10, 2011), at http://taxpayer.net/sarticle.php?proj_id=4354&t=TCS%20Budget%20Cut%20List; letter to Honorable Paul Ryan from Congressmen Frank Lucas and Collin Peterson, U.S. House of Representatives, Mar. 15, 2011; Claire Courchane, "Cutting Subsidies for Farms Feasible," *Washington Times*, July 14, 2011. Also see: Felicity Barringer, "Empty Fields Fill Urban Basins and Farmers' Pockets," *New York Times*, Oct. 23, 2011; Barry Goodwin, Vincent Smith, and Daniel Sumner, "American Boondoggle: Fixing the 2012 Farm Bill" (Washington, DC: American Enterprise Institute, July 12, 2011), at http://www.aei.org/article/american-boondoggle-fixing-the-2012-farm-bill-1/.

60. John Leshy makes a similar point about using Colorado River Basin waters to grow beef; John Leshy, "Notes on a Progressive National Water Policy," *Harvard Law and Policy Review* 3 (2009): 154. Also see Libby Quaid, "Feds Aren't Subsidizing Recommended Foods," Associated Press, Aug. 11, 2005; Tom Philpott, "I'm Hatin' It: How the Feds Make Bad-for-You Food Cheaper than Healthful Fare," *Grist* (Feb. 22, 2006); Tom Philpott, "Why Are We Propping up Corn Production, Again?" *Grist* (Mar. 25, 2010). For a debate on the health effects of industrial foods, see Michael Pollan, *Omnivore's Dilemma* (New York: Penguin, 2006); Blake Hurst, "The Omnivore's Delusions: Against the Agri-intellectuals," *American* (July 30, 2009); Tom Philpott, "An 'Agri-intellectual' Talks Back," *Grist* (Aug. 14, 2009).

61. The cost estimate is from the Centers for Disease Control, cited in *Newsweek*, Mar. 22, 2010, 43. Also see C. L. Ogden et al., "Prevalence and Trends in Overweight Among U.S. Children and Adolescents, 1999–2000," *Journal of the American Medical Association* 288.14 (Oct. 9, 2002): 1728–32; C. L. Ogden et al., "Prevalence of Overweight and Obesity in the United States, 1999–2004," *Journal of the American Medical Association* 295.13 (Apr. 5, 2006): 1549–55; White House Task Force on Childhood Obesity, *Report to the President* (2010), at http://www.letsmove.gov.

5. FALLING WATERS:
HYDROPOWER AND RENEWABLE ENERGY

1. National Hydropower Association, "Hydro Facts," at www.hydro.org. One kilowatt equals 1,000 watts, 1 megawatt equals 1 million watts, and 1 gigawatt equals 1 billion watts.

2. Ibid.

3. See, for example, Utah Division of Natural Resources, *Managing Sediment in Utah's Reservoirs*, Utah State Water Plan, 2010.

4. http://www.glencanyon.org/about/faq. Also see National Research Council, *Downstream: Adaptive Management of Glen Canyon Dam and the Colorado River Ecosystem* (Washington, DC: National Academy Press, 1999); Patty Henetz, "Glen Canyon Flush Shows Dam Remains a Sand Trap," *Salt Lake Tribune*, Feb. 3, 2010; Upper Colorado Region, "Reclamation Releases Environmental Assessment for Public Review and Comment on a Protocol for Experimental High-flow Releases from Glen Canyon Dam," press release (Jan. 18, 2011).

5. The target date for the Record of Decision (the ROD) on the removal agreements is expected in March 2012. See "Klamath Hydroelectric Settlement Agreement" (Feb. 18, 2010), at http://67.199.95.80/Klamath/Klamath%20Hydro electric%20Settlment%Agreement2-18-10.pdf ; "Klamath Basin Restoration Agreement for the Sustainability of Public and Trust Resources and Affected Communities" (Feb. 18, 2010), at http://klamathrestoration.gov/sites/klamathrestoration .gov/files/Klamath-Agreements/Klamath-Basin-Restoration-Agreement-2-18 -10signed.pdf.

6. For an excellent analysis of the problems in the Klamath Basin, see Holly Doremus and A. Dan Tarlock, *Water War in the Klamath Basin* (Washington, DC: Island Press, 2008).

7. The Electric Consumers Protection Act of 1986, Pub. L. 99-495, 100 Stat. 1243, Sec. 3.

8. Tim Palmer, *Endangered Rivers and the Conservation Movement*, 2d ed. (Lanham, MD: Rowman and Littlefield, 2004), 267. Also see William Lowry, *Dam Politics: Restoring America's Rivers* (Washington, DC: Georgetown University Press, 2003), 51–54.

9. Federal Energy Regulatory Commission, www.ferc.gov/industries/hydro power/gen-info/licensing/issued-licenses.asp.

10. The other two power administrations are the Southeastern, which operates in the Deep South, and the Southwestern, which is primarily in Texas. A fifth power administration in Alaska was sold to private investors and municipalities

a few years ago. For an interesting perspective on the BPA, see its report, *What Led to the Current BPA Financial Crisis*, a BPA Report to the Region (Portland, OR, Apr. 2003).

11. Northwest Power and Conservation Council, *Briefing Book*, Council Document 2005-1 (2005), 7. The high estimate for historic flows is discussed in Steven Hawley, *Recovering a Lost River* (Boston: Beacon Press, 2011), 40–50.

12. Nearly all the downstream barge shipments are food commodities for export. Jim Shaw, "The Columbia Snake River System," *Pacific Maritime* (June 2006): 1–3; National Research Council, Committee on Water Resources Management, *Managing the Columbia River: Instream Flows, Water Withdrawals, and Salmon Survival* (Washington, DC: National Academies Press, 2004), 155–60.

13. *Idaho Dept. of Fish and Game v. NMFS*, 850 F. Supp. 886 (D. Or. 1994); Federal Columbia River Power System Adaptive Management Implementation Plan, *Lower Snake River Fish Passage Improvement Study: Dam Breaching Update*, U.S. Army Corps of Engineers, Walla Walla District (Mar. 2010).

14. Greg Graham, in-person interview with author, Walla Walla, WA (Aug. 9, 2006).

15. See Lisa Mighetto and Wesley Ebel, *Saving the Salmon: A History of the U.S. Army Corps of Engineers' Efforts to Protect Anadromous Fish on the Columbia and Snake Rivers* (Seattle: Historical Research Associates, Sept. 6, 1994).

16. U.S. Army Corps of Engineers, Walla Walla District, *Lower Snake River Juvenile Salmon Migration Feasibility Report*, interim report (Dec. 1996).

17. Steven Hawley notes that an even earlier incarnation of the breaching idea came from Reed Burkholder, a piano teacher turned instigator. See Steven Hawley, *Recovering a Lost River* (Boston: Beacon Press, 2011), 73.

18. Ibid., 185.

19. U.S. Army Corps of Engineers, *Summary, Final Lower Snake River Juvenile Salmon Migration Feasibility Report/Environmental Impact Statement* (Feb. 2002), 22.

20. Steve Pettit, in-person interview with author, at Pettit's home, east of Lewiston, ID (Aug. 9, 2006).

21. Anonymous Corps of Engineers official, interview with author.

22. Institute for Fisheries Resources, *The Cost of Doing Nothing*, Report No. 1 of 3 (San Francisco, Oct. 1996).

23. Philip Lansing, *Restoring the Lower Snake River: Saving Snake River Salmon and Saving Money*, a report for the Oregon Natural Resources Council Fund (Portland, OR, 1998). Also see Earthjustice, Oakland, CA, "Victories: Judge

Finds Dam Operations Illegal" (May 31, 2005), at www.earthjustice.org/our_work /victory/judge_finds_dam.

24. Taxpayers for Common Sense, Republicans for Environmental Protection, Pacific Coast Federation of Salmon Fisheries Associations, Northwest Sportfishing Council, NW Energy Coalition, Save our Wild Salmon, and American Rivers, *Revenue Stream: An Economic Analysis of the Costs and Benefits of Removing the Four Dams on the Lower Snake River* (n.d.).

25. Independent Economic Analysis Board of the Northwest Power and Conservation Council, *Review of the SOS Revenue Stream Report*, Document IEAB 2007-1 (Portland, OR, Feb. 25, 2007).

26. Pacific Northwest Waterways Association, "Dam Breaching and Drawdowns Are NOT the Answer," at www.pnwa.net.

27. U.S. Army Corps of Engineers, Walla Walla District, "Fish Myths and Facts" (n.d).

28. Scott Learn, "Habitat Restoration Soars on Columbia River, but Fish Benefits Are Murky," *Oregonian*, May 8, 2011.

29. Samantha Mace, in-person interview with author, Spokane, WA (Aug. 7, 2006).

30. U.S. Army Corps of Engineers, *Summary*, 23.

31. Christopher Pernin et al., *Generating Electric Power in the Pacific Northwest*, RAND Science and Technology MR-1604 (Arlington, VA, 2002).

32. Institute for Water Resources, Navigation Data Center, U.S. Army Corps of Engineers, Waterborne Commerce Statistics Center, *Final Waterborne Commerce Statistics, for Calendar Year 2004. Waterborne Commerce National Totals and Selected Inland Waterways for Multiple Years* (n.p., n.d.), 16.

33. Arvid Lyons, in-person interview with author, Lewiston, ID (Aug. 9, 2006).

34. U.S. Army Corps of Engineers, *Summary*, 39.

35. Don Reading, *The Potential Economic Impact of Restored Salmon and Steelhead Fishing in Idaho* (Boise, ID: Ben Johnson Associates, Feb. 2005), 2.

36. Witt Anderson, in-person interview with author, Portland, OR (Aug. 11, 2006).

37. Rod Condo, in-person interview with author, Portland, OR (Aug. 10, 2006).

38. Idaho Department of Fish and Game, *Idaho's Anadromous Fish Stocks: Their Status and Recovery Options*, Report to the Director, IDFG 98-13 (Boise, ID, May 1, 1988), 6.

39. NOAA Fisheries, *Programmatic Biological Opinion and Magnuson-Stevens Act Essential Fish Habitat Consultation for Standard Local Operating Procedures for Endangered Species (SLOPES) for Certain Activities Requiring Department of*

Army Permits in Oregon and the North Shore of the Columbia River (SLOPES I) (June 14, 2002, NOAA Fisheries No.: 2002/00976), 9–144.

40. Liz Hamilton, in-person interview with author, Portland, OR (Aug. 10, 2006).

41. Gina Baltrusch, in-person interview with author, Walla Walla, WA (Aug. 8, 2006).

42. *National Wildlife Federation v. National Marine Fisheries Service*, 254 F. Supp. 2d 1196 (D. Or. 2003).

43. Joel Roberts, "Bush Fishes for Votes Out West," CBS News, Aug. 22, 2003.

44. *National Wildlife Federation v. National Marine Fisheries Service*, 2005 WL 1278878 (D. Or., 2005).

45. *National Wildlife Federation v. National Marine Fisheries Service*, CV 05-23-RE, Opinion and Order of Remand (D. Or., Oct. 7, 2005).

46. Pacific Northwest Waterways Association, "Dam Breaching and Drawdowns Are NOT the Answer." Also see Joint Columbia River Management Staff, Oregon Department of Fish and Wildlife, Washington Department of Fish and Wildlife, *2009 Joint Staff Report: Stock Status and Fisheries for Spring Chinook, Summer Chinook, Sockeye, Steelhead, and Other Species, and Miscellaneous Regulations* (Jan. 26, 2009).

47. Fish Passage Center, *Annual Report for 2005* (Portland, OR), 37. Also see Independent Scientific Group, *Return to the River: Restoration of Salmonid Fisheries in the Columbia River Ecosystem, Development and Synthesis of Science Underlying the Columbia River Basin Fish and Wildlife Program of the Northwest Power Planning Council* (Sept. 10, 1996).

48. See Blaine Harden, "Court Rejects Senator's Bid to Eliminate Fish Agency," *Washington Post*, Jan. 25, 2007.

49. As of 2011, DeHart was still the manager of the Fish Passage Center.

50. See Fish Passage Center, "Columbia River Mouth Fish Returns," at http://www.fpc.org/adultsalmon/summary_actualreturnbyspecies.html, And "Columbia River Mouth Fish Returns 2010 Actual and 2011 Forecasts," at http://www.fpc.org/adultsalmon/preseason11forecastsv2.html.

51. See Federal Caucus, "Adult Salmon and Steelhead Returns at Lower Granite Dam" (Dec. 2010), and "Adult Salmon and Steelhead Returns at Bonneville Dam" (Dec. 2010). The Federal Caucus is a group of ten federal agencies working to protect Endangered Species Act–listed fish in the Columbia Basin. www.salmonrecovery.gov/homepage.aspx.

52. Idaho Department of Fish and Game, *Idaho's Anadromous Fish Stocks*, iii.

53. Steve Pettit, in-person interview with author, at Pettit's home, east of Lewiston, ID (Aug. 9, 2006).

54. Jeremy Five Crows, in-person interview with author, Portland, OR (Aug. 11, 2006).

55. Columbia River Inter-Tribal Fish Commission, "Spirit of the Salmon: The Columbia River Anadromous Fish Restoration Plan of the Nez Perce, Umatilla, Warm Springs and Yakama Tribes," at www.critfc.org/TEXT/TRP.HTM.

56. William Yardley, "Deal Gives Money to Tribes to Drop Role in Fish Lawsuits," *New York Times*, Apr. 8, 2008; "Dams Remain in $900 Million Columbia River Salmon Deal," *Environment News Service*, Apr. 7, 2008; Hal Bernton, "4 Tribes Agree to Settle on Restoring Salmon Runs," *Seattle Times*, Apr. 5, 2008.

57. Mike Lee and Kim Bradford, "Thousands Rally to Save Snake Dams," *Tri-City Herald* (Kenniwick, WA), Feb. 19, 1999, at www.tri-cityherald.com/news.

58. U.S. Army Corps of Engineers, "Draft Lower Snake River Juvenile Salmon Mitigation Feasibility Report/Environmental Impact Statement, with Federal Caucus Conservation of Columbia Basin Fish 'All-H Paper,'" public comment session, Lewis-Clark Convention Center, Clarkston, WA (Feb. 10, 2000), 9, 50.

59. NOAA Fisheries, "NOAA's Fisheries Service Issues Far-Reaching Plans for Protecting Northwest Salmon," News from NOAA, press release (May 5, 2008).

60. NOAA Fisheries, *Issue Summaries of the FCRPS Biological Opinion* (May 5, 2008).

61. Fourth Supplemental Complaint for Declaratory and Injunctive Relief, *National Wildlife Federation v. National Marine Fisheries Service*, Civ. No. 01-0640-RE. Dist. Ct. OR, (July 2008). Also see Rocky Barker, "Idaho Lt. Gov. Little Urges Obama Administration to Back Salmon Collaboration," *Idaho Statesman*, May 26, 2009.

62. Letter from Judge James Redden to "Counsel of Record, *National Wildlife Federation v. NFMS*, CV 01–640 RE," re "2008 Biological Opinion" (May 18, 2009).

63. "Our View: Salmon Judge's Message Is 'Solve This Yourselves,'" *Idaho Statesman*, Mar. 11, 2009; Rocky Barker, "Sen. Crapo Says All Options Including Breaching Must Be on Table for Regional Salmon Collaboration," *Idaho Statesman*, May 29, 2009.

64. NOAA, "FCRPS Adaptive Management Implementation Plan," 2008–2018 Federal Columbia River Power System Biological Opinion (Sept. 11, 2009), 37.

65. NOAA Fisheries, Northwest Regional Office, *2010 Supplemental Consultation on Remand for Operation of the Federal Columbia River Power System (FCRPS) Biological Opinion* (May 19, 2010).

66. U.S. Bureau of Reclamation, "Columbia and Snake River Salmon Recovery Project," press release (Feb. 14, 2011).

67. *National Wildlife Federation v. National Marine Fisheries Service*, U S. Dist. Ct. OR "Opinion and Order," CV 01-00640-RE (Aug. 2, 2011). This was Judge

Redden's last decision on the BiOp. On his request to be removed from the case, see Jeff Barnard, "Judge Redden Taking Himself Off Salmon Case," *The Columbian*, Nov. 23, 2011; Op-ed, "A Judge Has Stepped Up for Idaho's Fish. Now It's Our Turn," *The Idaho Statesman*, Dec. 4, 2011.

68. Earthjustice, "Columbia and Snake River Power Plan Challenged" (July 7, 2010), at http://earthjustice.org/news/press/2010/Columbia-and-snake-river-power -plan-challenged.

69. *National Wildlife Federation v. National Marine Fisheries Service*, "Reply in Support of Federal Defendants' Supplemental Cross Motion for Summary Judgment and Combined Opposition to Plaintiffs' Supplemental Motions for Summary Judgment," 19, Civil No. 01-CV-640-RE, Dist. Ct. OR (Feb. 11, 2011).

70. David James Duncan, *My Story as Told by Water* (San Francisco: Sierra Club Books, 2001), 184.

71. Timothy Egan, *The Good Rain* (New York: Vintage Books, 1990), 182.

72. Doug Morrill, in-person interview with author, Lower Elwha Klallam tribal headquarters (June 18, 2004).

73. Pub. L. 102-495 (Oct. 24, 1992).

74. Robert Elofson, in-person interview with author, Lower Elwha Klallam tribal headquarters (June 21, 2004).

75. Russ Veenema, in-person interview with author, Port Angeles, WA (June 21, 2004).

76. Heather Weiner, in-person interview with author, Seattle, WA (June 23, 2004).

77. Bryan Johnson, "Power Shut Off at Elwha, Glines Canyon Dams for Good," *Komo News*, June 1, 2011; Kim Murphy, "Dam Removal Begins, and Soon the Fish Will Flow," *Los Angeles Times*, Sept. 17, 2011.

78. Lynda Mapes, "Contract to Remove Elwha Dams Goes to Montana Firm," *Seattle Times*, Aug. 26, 2010. Also see Juliet Eilperin, "Elwha Dam Removal Illustrates Growing Movement," *Washington Post*, Sept. 16, 2011; Ker Than, "Largest Dam Removal to Restore Salmon Runs," *National Geographic Daily News*, Aug. 31, 2011.

79. Greg Delwiche, "Go Beyond Dams to Save Salmon," *High Country News* (Paonia, CO), July 2, 2009.

80. Kate Galbraith, "As Wind Power Grows, a Push to Tear Down Dams," *New York Times*, June 12, 2009; "BPA: Wind Farm System Sets Output Milestone," *Oregon Business News*, Aug. 12, 2009.

81. Bob Moen, "Wyoming Back in Top 10 for Wind Power Production," *Casper (WY) Star-Tribune*, July 15, 2011.

82. "Wyoming Among Top States Adding Wind Power," *Casper-Star Tribune*, Oct. 25, 2009.

83. Linda Halstead-Acharya, "Study Bumps Montana's Ranking to No. 2 for Wind Potential," *Billings Gazette*, Aug. 25, 2009.

84. Bonneville Power Administration, *2008 BPA Facts*; Clint Wilder, "Winds of Change Blowing Through the Heartland and Beyond," *Clean Edge* (Oct. 26, 2011), at www.cleanedge.com/views; "Wyoming Among Top States Adding Wind Power."

85. U.S. Department of Energy, "Building a New Energy Future with Wind Power" (May 2011), at http://www1.eere.energy.gov/wind/pdfs/51240.pdf. Also see Synapse Energy Economics, *A Responsible Electricity Future* (Cambridge, MA, June 11, 2004).

86. Worldwatch Institute, Center for American Progress, *American Energy: The Renewable Path to Energy Security* (Washington, DC, September 2006), 20.

87. Ibid., 27.

88. U.S. Department of the Interior and U.S. Department of Agriculture, *New Energy Frontier: Balancing Energy Development on Federal Lands*, a Joint Report to Congress on Siting Energy Development Projects on Federal Lands (May 2011), 13.

89. Ibid., 14.

90. Xi Lu et al., "Global Potential for Wind-Generated Electricity," *Proceedings of the National Academy of Sciences* 106.27 (July 7, 2009): 10933–38.

91. Ehren Goosens, "Wind Power's Best Projects Rival Costs of New Coal-Fired Plants, BNEF Says," *Bloomberg* (April 5, 2011), http://www.bloomberg.com /news/2011-04-05/wind-power-s-best-projects-rival-costs-of-new-coal-fired -plants-bnef-says.html.

92. "Energy Foundation Study Finds U.S. Rooftops Could Support Vast Market for Solar Power," *Clean Edge News* (March 1, 2005), http://www.cleanedge .com/taxonomy/term/2/all. Also see Kees van derLeun, "Solar PV Rapidly Becoming the Cheapest Option to Generate Electricity," *Grist* (Oct. 11, 2011), at http://www.grist.org/solar-power/2011-10-11-solar-pv-rapidly-becoming-cheapest -option-generate-electricity.

93. Western Governors' Association, "Goal for Western US Is 8GW of Solar Electricity by 2015, Says Western Governors' Association Solar Task Force," press release (June 2006).

94. "Skyrocketing Solar Projects Create Demand for Desert Land," *Arizona Republic*, June 4, 2009.

95. U.S. Department of the Interior and U.S. Department of Agriculture, *New Energy Frontier*, 17.

96. Ibid., 21; Andew Webster, "Advanced Map Data Shows Enhanced Potential for Geothermal in U.S.," *Good Clean Tech* (Oct. 26, 2011), at http://goodcleantech .pcmag.com/science/289666-advanced-map-data-shows-enhanced-po . . .

97. U.S. Department of the Interior and U.S. Department of Agriculture, *New Energy Frontier*, 22.

98. John Farrell and David Morris, *Energy Self-Reliant States*, 2d expanded ed. (Minneapolis: New Rules Project, updated May 2010). Also see Mark Jacobson and Mark Delucchi, "A Plan to Power 100 Percent of the Planet with Renewables," *Scientific American* (Oct. 26, 2009): 169–74.

99. The data are from the Northwest Power and Conservation Council, quoted in Steven Weiss, *Bright Future*, a report by the Northwest Energy Coalition (March 2009), 19.

100. Hannah Choi Granade, *Unlocking Energy Efficiency in the U.S. Economy*, McKinsey Global Energy and Materials (July 2009).

101. Matthew Wald, "U.S. Alliance to Update the Light Bulb," *New York Times*, Mar. 14, 2007.

102. Alliance to Save Energy, "Appliance Standards," at www.ase.org/section /_audience/policymakers/applstds.

103. Power companies encourage marginal efficiencies when demands exceed in-house supply and they have to buy energy from other sources. But their goal is to match energy demand with existing generation capacity; if demand falls far below existing capacity, the company is left with idle power plants that do not make money.

104. Joel Makower, Ron Pernick, and Clint Wilder, "Clean Energy Trends, 2004," *Clean Edge*. Also see "New Energy for America, The Apollo Jobs Report: Good Jobs and Energy Independence," a Report to the Apollo Alliance (January 2004), at www.apolloalliance.org.

105. Edmund Burke, letter to Sir Hercules Langrishe (1792), in *The Works of the Right Honourable Edmund Burke*, vol. 6 (London: Thomas McLean, 1823).

106. Carl Sandburg, in "Four Preludes on Playthings of the Wind," from "Smoke and Steel," in *The Complete Poems of Carl Sandburg* (Boston: Houghton Mifflin, 2003), 183–84.

6. RIVERS INTO WATERWAYS: BARGING, LOCKS, AND DAMS

1. Charles Beard and Mary Beard, *History of the United States* (New York: Macmillan, 1928), 234.

2. Thomas Jefferson, *Notes on the State of Virginia* (Richmond, VA: J. W. Randolph, 1853), 131.

3. *Preliminary Report of the Inland Waterways Commission*, 60th Cong., 1st sess., Doc. 325 (Feb. 26, 1908), 204.

4. Quoted in: Ronald White Jr., *The Eloquent President* (New York: Random House, 2005), 207.

5. Jack Kerouac, *On the Road* (New York: Viking, 2007), 141.

6. Mark Twain, *Life on the Mississippi* (New York: Collier and Son, 1874), 2.

7. John Anfinson, *The River We Have Wrought* (Minneapolis: University of Minnesota Press, 2003), 110–13.

8. Another 14,000 miles of waterways follow the East Coast and the Gulf Coast. U.S. Army Corps of Engineers, Navigation Data Center, "Did You Know? Facts About Locks" (Alexandria, VA, 2003).

9. Congressional Budget Office, *Paying for Highways, Airways, and Waterways: How Can Users Be Charged?* (Washington, DC: Congressional Budget Office, May 1992), 54.

10. Inland Waterways Users Board, *16th Annual Report to the Secretary of the Army and the U.S. Congress* (May 2002); U.S. Army Corps of Engineers, Navigation Data Center, "The U.S. Waterway System—Transportation Facts" (Dec. 2008).

11. U.S. Army Corps of Engineers, Institute for Water Resources, Navigation Data Center, Waterborne Commerce Statistics Center, "Waterborne Commerce National Totals and Selected Inland Waterways for Multiple Years," *Final Waterborne Commerce Statistics for Calendar Year 2004*.

12. Ibid., 23, 28, 36.

13. U.S. Army Corps of Engineers, Navigation Data Center, "Did You Know? Facts About Port Facilities and Inland Waterways" (Alexandria, VA, n.d.); Ken Casavant, *Inland Waterborne Transportation—An Industry Under Siege* (U.S. Department of Agriculture, Agricultural Marketing Service, Nov. 2000); Taxpayers for Common Sense, "River of Subsidy Washes Tax Dollars down the Mississippi," *Waste Basket* 2.36 (Oct. 27, 1997): 1–3.

14. U.S. Army Corps of Engineers, Navigation Data Center, "Did You Know? Facts About Port Facilities and Inland Waterways."

15. See T. R. Reid, *Congressional Odyssey* (San Francisco: Freeman and Co., 1980). The fuel tax was authorized by the 1978 Inland Waterways Revenue Act and the 1986 Water Resources Act.

16. Congressional Budget Office, *Paying for Highways, Airways, and Waterways*, 53.

17. U.S. Army Corps of Engineers, Navigation Data Center, "U.S. Waterway System," 3. The Corps spends about $2.5 billion annually for the operation and

maintenance of its projects; its budget typically does not separate out the portion of that money that is used to maintain inland waterways.

18. This effort was led by the National Waterways Alliance, a coalition of businesses that utilize federally maintained waterways, and the American Waterways Operators.

19. U.S. Army Corps of Engineers, Navigation Data Center, "U.S. Waterway System"; U.S. Army Corps of Engineers, "Lock Characteristics Operational Statistics—1999," at www.iwr.usace.army.mil/ndc/lockchar/pdf/koper99.pdf; Michael Grunwald, "Corps' Taming of Waterways Doesn't Pay Off," *Washington Post*, Jan. 9, 2000.

20. *Mississippi v. United States*, 498 U. S. 16 (1990).

21. U.S. Army Corps of Engineers, "Waterborne Commerce National Totals," 31. Also see U.S. Army Corps of Engineers, Northwestern Division, *Summary, Missouri River Final Environmental Impact Statement* (Mar. 2004), 14.

22. U.S. Army Corps of Engineers, Navigation and Water Resources Application Division, Institute for Water Resources, *Projected and Actual Traffic on Inland Waterways* (Alexandria, VA, Aug. 2000).

23. Taxpayers for Common Sense and the National Wildlife Federation, "Crossroads: Congress, The Corps of Engineers and the Future of America's Water Resources" (2003), at www.taxpayer.net/corpswatch/crossroads/index.htm.

24. American Rivers, "The Truth About Rivers, Barges, and Navigation," at www.amrivers.org/corpsereformtoolkit/riverbarges.htm.

25. The last Tallstacks Festival was held in 2006. The event scheduled for 2010 was scrapped because of the poor economy.

26. George Byron Merrick, *Old Times on the Upper Mississippi* (1909) (Minneapolis: University of Minnesota Press, 2001), 79.

27. Cited in National Research Council, Committee to Review the Upper Mississippi–Illinois Waterway Navigation System Feasibility Study, *Inland Navigation System Planning: The Upper Mississippi River–Illinois Waterway* (Washington, DC: National Academy Press, 2001), 10.

28. Anfinson, *River We Have Wrought*, 158–60.

29. Janice Petterchak, *Taming the Upper Mississippi: My Turn at Watch, 1935–1999* (Rochester, IL: Legacy Press, 2000), 56–57.

30. Water Resources Development Act of 1986, Pub. L. 99-662 (1986).

31. U.S. Army Corps of Engineers, "Upper Mississippi River–Illinois Waterway System Navigation Study Feasibility Report" (draft) (Rock Island, IL, 2000).

32. Dr. Donald Sweeney, Associate Director, Center for Transportation Studies, University of Missouri–St. Louis, telephone interview with author (Apr. 15, 2009).

33. See Affidavit of Donald C. Sweeney, submitted to the U.S. Office of Special Counsel (Feb. 1, 2000), 26–27.

34. Ibid., 2.

35. Briefing of MG Fuhrman and staff; Guidance Received (Sept. 23, 1998), included with Sweeney Affidavit.

36. Sweeney Affidavit, 30–33.

37. E-mail from Ronald Conner, HQ02 (Sept. 3, 1998), included with Sweeney Affidavit.

38. The *Sultana* was a riverboat on the Mississippi River that exploded and sank in 1865 with the loss of perhaps 1,500 people, mostly Union soldiers. It is still the worst disaster to ever occur on America's rivers, and resulted in public outrage and an official investigation. See Shelby Foote, *The Civil War: A Narrative*, vol. 3 (New York: Random House, 1974), 1027.

39. Statement of Elaine Kaplan, Special Counsel, U.S. Office of Special Counsel, press conference (Dec. 6, 2000).

40. Department of the Army, Office of the Inspector General, *U.S. Army Inspector General Agency Report of Investigation*, Case SAIG-IN (201b) (00–019) (2000), 4–6.

41. National Research Council, *Inland Navigation System Planning*, 3.

42. Ibid., 32–33.

43. Ibid., 46–47.

44. Ibid., 37–40.

45. Ibid.

46. Ibid., 41–42.

47. Ibid., 4.

48. Ibid., 52.

49. Ibid., 16.

50. Letter from Richard Nelson, Supervisor, U.S. Fish and Wildlife Service, to Colonel William Bayles, District Engineer, Rock Island District, U.S. Army Corps of Engineers (Aug. 31, 2000).

51. National Research Council, *Inland Navigation System Planning*, 7.

52. Ibid., 60.

53. Ibid., 54.

54. Ibid., 17.

55. Taxpayers for Common Sense, *River of Subsidy: How Taxpayer Investments Are Wasted in the Mississippi River Basin* (Washington, DC, 1998).

56. *UMR-IWW System Navigation Study Newsletter* 9.2 (Sept. 2003).

57. U.S. Army Corps of Engineers, *Interim Report for the Restructured Upper Mississippi–Illinois Waterway System Navigation Feasibility Study*, Rock Island District (2002).

58. Sweeney telephone interview.

59. National Research Council, Committee to Review the Corps of Engineers Restructured Upper Mississippi River–Illinois Waterway Feasibility Study, *Review of the U.S. Army Corps of Engineers Restructured Upper Mississippi River–Illinois Waterway Feasibility Study* (2004) and *Review of the U.S. Army Corps of Engineers Restructured Upper Mississippi River–Illinois Waterway Feasibility Study, Second Report* (2004); National Research Council, *Water Resources Planning for the Upper Mississippi River and Illinois Waterway* (2005).

60. Half the commercial shipments on the upper Mississippi/Illinois system consist of corn, soybeans, and wheat—all heavily subsidized commodity crops. If dramatic cuts in subsidies were to occur, they would be followed by dramatic reductions in the production of these crops and thus the need for barge shipments.

61. Nicollet Island Coalition, *Big Price—Little Benefit: Proposed Locks on the Upper Mississippi and Illinois Rivers Are Not Economically Viable* (Feb. 2010).

62. Many thanks to Mary O'Rourke.

63. Dan Owen, *Inland River Record, 1996* (St. Louis: The Waterways Journal, 1996). By far the most numerous boats in the *Record* are pushboats designed to move barges.

64. A wicket dam can be retracted during periods of high water. These dams are capable of raising the water level by only a few feet.

65. Walter Havighurst, *Voices on the River: The Story of the Mississippi Waterways* (1964) (Minneapolis: University of Minnesota Press, 2003), 260–61. This is why there are two numbered dams on the Ohio, 52 and 53, but no others; all but those two have been replaced with dams that have names rather than numbers. The forty-six dams were not numbered in sequence because of the expectation that more dams would be inserted between existing dams; thus the numbers run to 53.

66. Ibid., 277. Cairo narrowly avoided complete destruction by the 2011 flood when the Corps opened up the levees to release the excess water on the other side of the river. See George Sorvalis, "Grave Decisions," *Water Protection Network* (May 3, 2011), at www.waterprotectionnetwork.org.

67. Beltrami later claimed to have discovered the source of the Mississippi. His claim was never verified, but he did get a county in Minnesota named after him. See Havighurst, *Voices on the River*, 112.

68. Twain, *Life on the Mississippi*, 472.

69. The Mark Twain Complex has five units scattered along the river between Illinois on the east bank and Missouri and Iowa on the west bank. Farther upstream is the Upper Mississippi River Fish and Wildlife Refuge, which has the greatest visitation of any unit in the national wildlife system, according to the National Research Council, *Review of the U.S. Army Corps of Engineers Restructured Upper Mississippi River–Illinois Waterway Feasibility Study*, 17.

70. Petterchak, *Taming the Upper Mississippi*, 15.

71. This story was told to me by members of the crew from Red Wing.

72. Upper Midwest Environmental Sciences Center, U.S. Geological Survey, "About the Upper Mississippi River System," at www.umesc.usgs.gov/umese_about /about_umrs.html.

73. Ron Way, "River Restoration: Should We Bring Back Mississippi's Roaring White-Water Rapids?" *Minneapolis Post*, May 26, 2009.

74. See Clarence Jonk, *River Journey* (St. Paul, MN: Borealis Books, 2003). Jonk built a raft in the 1930s and attempted to float to New Orleans. He made it only 150 miles, but he squeezed a good book out of that adventure. Captain Halter told me that a group of Swiss college students floated the entire length of the Mississippi on a raft in the early 1980s; since then regulations have changed.

75. "Voices of Democracy." The U. S. Oratory Project, at: http://voicesof democracy.umd.edu/theodore-roosevelt-conservation-as-a-national-duty-speech -text/.

76. U.S. Army Corps of Engineers, Institute for Water Resources, *Waterborne Commerce of the U.S. Calendar Year 2009, Part 5—Summaries*, table 3-18.

77. Ibid., table 3-4.

78. American Waterways Operators, "Facts About the American Tugboat, Towboat and Barge Industry" (n.d.), at www.americanwaterways.com.

79. U.S. Army Corps of Engineers, Northwestern Division, *Summary, Missouri River*, 14, 18.

80. U.S. Army Corps of Engineers, Navigation Data Center, "U.S. Waterway System," 3.

81. Center for Ports and Waterways, Texas Transportation Institute, *A Modal Comparison of Domestic Freight Transportation Effects on the General Public*, prepared for the U.S. Department of Transportation and the National Waterways Foundation (Nov. 2007), 3.

82. The railroads face the same unfair subsidy for trucking. See Oxford Analytica, "U.S. Railroads Profits Tied to New Investment," *Forbes.com* (June 14, 2006), at http://www.forbes.com/2006/06/14/rail-profits-invest-cx_np_0615 oxford.html.

83. Railroads still carry 41 percent of the nation's freight, compared with 14 percent on barges. See Association of American Railroads, Policy and Economics Department, "Class I Railroad Statistics" (Jan. 15, 2008); Association of American Railroads, "A Short History of U.S. Freight Railroads" (May 2010).

84. Barge companies prefer to call user fees by the less accurate but more politically charged name of "user taxes." American Waterways Operators, *2007 Annual Report* (Arlington, VA, 2007). See 2009 U.S. Budget, Corps of Engineers, Civil Works.

85. Staff Reports, "Inland Waterways Trust Fund," *National Waterways Conference Newsletter* (July 2009): 9.

86. American Waterways Operators, "Obama Administration FY 2010 Budget Continues to Push Lock Tax," *AWO Letter* 66.8 (May 22, 2009): 1.

87. Taxpayers for Common Sense, "Taxpayers Oppose Barge Company Bailout" (May 19, 2010).

88. U.S. Army Corps of Engineers, Institute for Water Resources, *Waterborne Commerce of the U.S. Calendar Year 2006*, table 1-3.

89. Patrick O'Driscoll and Tom Kenworthy, "Western Drought Shrinking Big Muddy," *USA Today*, Apr. 29, 2005; Ryan Gentzler, "Missouri/Kansas Barge Traffic Overview," KC Smart Port (2009), at www.kcsmartport.com/sec_news/trans _outlook/BargeIndustryReport.htm.

90. "Fewer Boats, Barges Use Upper Mississippi Locks," *StarTribune.com*, Jan. 6, 2009, at http://www.startribune.news; Friends of the Mississippi River, "Barge Traffic Declines on Upper Mississippi" (2009), at www.fmr.org/news/current/barge _traffic_declines-2009-01.

91. Inland Waterways Users Board, www.waterwaysusers.us?IWTF_Status _64.pdf. In 2008 the Inland Waterways Users Board declared that the trust fund is "in danger of running out of funds." Inland Waterways Users Board, *22nd Annual Report to the Secretary of the Army and the United States Congress* (May 2008), 3.

7. BLACK WATER RISING:
THE MYTH OF FLOOD CONTROL

1. Toni Morrison, "The Site of Memory," in *Inventing the Truth*, edited by William Zinsser (Boston: Houghton Mifflin, 1987), 113.

2. National Flood Insurance Program, www.floodsmart.gov.

3. This story is told with graphic perfection in John Barry, *Rising Tide* (New York: Simon and Schuster, 1997).

4. Ibid., 201–2.

5. Ibid., 285–86; Stephen Ambrose and Douglas Brinkley, *The Mississippi and the Making of a Nation* (Washington, DC: National Geographic Press, 2002), 125; Upper Mississippi River Division, "1927 Mississippi River Flood," at http://mst .edu/-rogersda/um/courses/ge342/1927%20Mississippi.

6. National Research Council, Committee on Restoration of Aquatic Ecosystems, *Restoration of Aquatic Ecosystems: Science, Technology, and Public Policy* (Washington, DC: National Academy Press, 1992), 27.

7. Interagency Levee Policy Review Committee, *The National Levee Challenge: Levees and the FEMA Flood Map Modernization Initiative*, report of the Interagency Levee Policy Review Committee (Washington, DC, Sept. 2006), E-1. All the federal levees, and some nonfederal levees, are cataloged in U.S. Army Corps of Engineers, *National Levee Data Base*, at //nld.usace.army.mil/nld/production /images/NLD%PublicMedia10202011.pdf.

8. James Kahan et al., *From Flood Control to Integrated Water Resource Management*, Rand/Gulf States Policy Institute, Occasional Paper (2006), 18.

9. Federal Emergency Management Agency, Federal Insurance and Mitigation Administration, "National Flood Insurance Program: Program Description" (Aug. 2002). The fund was devastated by claims from the Katrina disaster.

10. See Gloria Bucco, "Flood of the Century: Remembering the Great Midwest Flood of 1993," at www.dnr.ne.gov/floodplain/PDF_files/floodupdatestory_Rev3 .pdf; Interagency Floodplain Management Review Committee, *A Blueprint for Change: Sharing the Challenge: Floodplain Management into the 21st Century*, Report of the Interagency Floodplain Management Review Committee to the Administration Floodplain Management Task Force (Washington, DC, June 1994), ix, 18; Interagency Floodplain Management Review Committee, *A Blueprint for Change, Part V: Science for Floodplain Management into the 21st Century*, Preliminary Report of the Scientific Assessment and Strategy Team Report of the Interagency Floodplain Management Review Committee to the Administration Floodplain Management Task Force (Washington, DC, June 1994), 49. These reports became known as the "Galloway Report."

11. Douglas Clement, "Dam It All," *Fedgazette* (Sept. 2001), at http://www .minneapolisfed.org/publications_papers/pub_display.cfm?id=2104.

12. Interagency Floodplain Management Review Committee, *Blueprint for Change: Sharing the Challenge*, v.

13. Ibid., viii.

14. Professor White is often referred to as "the father of floodplain management." His first study of flood management was published in 1945. His treatise is

Strategies of American Water Management (Ann Arbor: University of Michigan Press, 1969).

15. American Rivers prefers the term *green infrastructure approach* to flood management. See American Rivers, "Natural Defenses: Safeguarding Communities from Floods" (2010).

16. Dr. Gerald Galloway, telephone interview with author (Aug. 6, 2009). The Congress in 2010 voted to place a temporary moratorium on earmarks. See Taxpayers for Common Sense, "Earmarks and Earmarking: Frequently Asked Questions, Updated for the 112th Congress" (March 2011).

17. Jim Johnson, in-person interview with author, U.S. Army Corps of Engineers, Washington, DC (Sept. 16, 2002).

18. "The Water Marks Interview," with Dr. Bob Thomas, Environmental Communications Department, Loyola University in New Orleans, *WaterMarks*, no. 21 (Sept. 2002): back cover.

19. This was the Mississippi River Gulf Outlet, completed in 1968 at a cost of $92 million.

20. Byron Fortier, U.S. Fish and Wildlife Service, e-mail correspondence with author (Apr. 22, 2009).

21. Ibid.

22. "The Coastal Crisis and Louisiana's Response," *WaterMarks*, no. 21 (Sept. 2002): 5.

23. Coastal Protection and Restoration Authority of Louisiana, *Fiscal Year 2010 Annual Plan* (Apr. 27, 2009), 1.

24. U.S. Army Corps of Engineers, New Orleans District, Team New Orleans, "Louisiana Coastal Area Ecosystem Restoration Study (LCA)," at www.mvn .usace.army.mil/environmental/lca.asp.

25. Sid Perkins, "Losing Louisiana," *Science News* 176.2 (July 18, 2009): 15.

26. Louisiana Coastal Wetlands Conservation and Restoration Task Force and Wetlands Conservation and Restoration Authority, *Coast2050: Toward a Sustainable Coastal Louisiana* (Baton Rouge, LA, 1998).

27. Donald F. Boesch et al. "Scientific Assessment of Coastal Wetland Loss, Restoration and Management in Louisiana," *Journal of Coastal Research*, special issue no. 20 (May 1994): 18.

28. "Looking to the Future," *WaterMarks*, no. 21 (Sept. 2002): 10.

29. See Campbell Robertson, "Ruling on Katrina Flooding Favors Landowners," *New York Times*, Nov. 18, 2009; *In re Katrina Canal Breaches Consolidated Litigation*, Civil Action No. 05-4182 (E.D. La., Nov. 18, 2009).

30. See New Orleans District, U.S. Army Corps of Engineers, *Mississippi River-Gulf Outlet (MRGO) Ecosystem Restoration Plan*, Draft Feasibility Report

(Dec. 2010); "Closing Louisiana's Hurricane Highway," *American Rivers* (Summer/Fall 2006): 10.

31. This cultural milieu is described with great affection by Mike Tidwell in *Bayou Farewell* (New York; Pantheon Books, 2003).

32. Barry, *Rising Tide*, 158.

33. Barry Kohl, "Statement Presented to the Subcommittee on Environmental Pollution of the Senate Committee on Environment and Public Works," New Orleans, LA (June 27, 1977).

34. Professor Barry Kohl, in-person interview with author, Tulane University (Dec. 15, 2003).

35. U.S. Code Title 16, Chapter 59A, Sec. 3952–56 (2004). The Act created the Louisiana Coastal Wetlands Conservation and Restoration Task Force.

36. U.S. Army Corps of Engineers, New Orleans District, "Freshwater Diversion," brochure (n.d). For a critique of such projects, see Amanda Mascarelli, "Wetlands Not Aided by Mississippi Diversions," *Nature* (Aug. 5, 2011): doi:10.1038.

37. "Water Flows Through Davis Pond," at www.lacoast.gov/programs/Davis Pond/.

38. Fredine retired in 2006.

39. Team New Orleans, U.S. Army Corps of Engineers, "Louisiana Coastal Area Ecosystem Restoration Study," at www.mvn.usace.army.mil/environmental /lca.asp.

40. Water Resources and Development Act of 2007, Pub. L. 110-114 (2007).

41. Government Accountability Office, *Coastal Wetlands*, GAO-08-130 (Dec. 2007), 3.

42. Susan Milligan, "US Earlier Rebuffed Louisiana on Aid: Plan to Help Fund Coastline Project Was Cut from Bill," *Boston Globe*, Sept. 1, 2005.

43. Donna Cassata, "U.S. Faces Astonishing Post-Katrina Financial Toll," *Washington Post*, Sept. 11, 2005. Also see U.S. Army Corps of Engineers, New Orleans District, Mississippi Valley Division, *Louisiana Coastal Protection and Restoration Final Technical Report, Economics Appendix* (June 2009).

44. Michael Grunwald, "Louisiana Goes After Federal Billions," *Washington Post*, Sept. 26, 2005.

45. Douglas Brinkley, *The Great Deluge* (New York: HarperPerennial, 2006), 345–55.

46. Jenny Jarvie, "Coastal Buyout Talk Roils Lives in Mississippi," *Los Angeles Times*, Oct. 2, 2007.

47. Government Accountability Office, *Coastal Wetlands*, 5. Also see Coastal Protection and Restoration Authority of Louisiana, *Fiscal Year 2010 Annual Plan* (Apr. 27, 2009).

48. Amy Larson, "Administration Seeks to Drastically Alter Floodplain Management," *National Waterways Conference Newsletter* (July 2009): 1–2.

49. Mark Schleifstein, "Even the Mighty Mississippi's Sediment Won't Be Enough to Save Our Vanishing Coast," *New Orleans Metro Real-Time News*, June 29, 2009.

50. Andrew Humphreys, quoted in Barry, *Rising Tide*, 37.

51. "Rising Mississippi River Takes Aim at Delta Region," *USA Today*, May 13, 2011.

52. Melanie Eversley, "Mo. Farmers Return to Land Ruined by Blown Levee," *USA Today*, May 11, 2011.

53. Brian Vastag and Lisa Rein, "In Louisiana, Choice Between Two Floods," *Washington Post*, May 11, 2011; Peter Heller, "The Mississippi River Flood and the Katrina Risk," *Bloomberg Businessweek* (June 9, 2011); George Sorvalis, "Grave Decisions," *Water Protection Network* (May 3, 2011).

54. Paul Quinlan, "Once-Rare Mississippi River Flooding Now 'More Frequent and More Severe,'" *New York Times*, May 17, 2011; Ben Foster, "Mississippi Rising: Critics Say High Water Result of Man Not Mother Nature," *ABC News Online*, May 10, 2011, at http://abcnews.go.com/US/mississippi-rising-critics-high -water-result-man-mother/story?id=13570001#.TuL25bKLM24. For a technical critique of the Corps' flood data, see R. E. Criss and W. E. Winston, "Discharge Predictions of a Rainfall-Driven Theoretical Hydrograph Compared to Common Models and Observed Data," *Water Resources Research* 44 (2008), at http:// www.agu.org/pubs/crossref/2008/2007WR006415.shtml.

55. U.S. Army Corps of Engineers, Mississippi Valley Division, "2011 Flood Fight," at http://www.mvd.usace.army.mil/defaultex.php?plD=mission; U.S. Army Corps of Engineers, Team New Orleans, "2011 Flood Fight," at http://www.mvn .usace.army.mil/bcarre/floodfight.asp. Also see Tom Charlier, "Mississippi River Tried Changing Course During Flooding, Leaving Huge Bill," *Commercial Appeal* (Memphis, TN), Sept. 6, 2011.

56. See, for example, Douglas Clement, "The Failure of Flood Control," *Fedgazette* (Sept. 2001), at http://www.minneapolisfed.org/publications_papers/pub _display.cfm?id=2103.

57. U.S. Weather Service, "Flood Losses: Compilation of Flood Loss Statistics," at http://apps.weather.gov/hydro/hic/flood_stats/Flood_loss_time_series.shtml.

58. Gerald Galloway, "A Flood of Warnings," *Washington Post*, June 25, 2008.

59. Netherlands Water Partnership, http://www.waterland.net.

60. John Lee Hooker, "Tupelo," from *The Folklore of John Lee Hooker* (Vee Jay, 1961).

61. Bob Dylan, "Watching the River Flow," from *Bob Dylan's Greatest Hits, Vol. 2* (Big Sky, 1971).

8. DOWNSTREAM DILEMMA: WATER POLLUTION

1. Ellen Wohl, *Disconnected Rivers* (New Haven, CT: Yale University Press, 2004), 104.

2. Janice Petterchak, *Taming the Upper Mississippi: My Turn at Watch, 1935–1999* (Rochester, IL: Legacy Press, 2000), 67.

3. Shaun McKinnon, "Mines Still Threaten Colorado River, Foes Say," *Arizona Republic*, Aug. 11, 2008.

4. The official title of this law is the Federal Water Pollution Control Act Amendments, but nearly everyone called it the Clean Water Act (CWA). It was an amendment to the 1948 Act of the same name—an act that was toothless and ineffective. In a subsequent amendment the official name was changed to the popular name. See 33 U.S.C. 1251–1376.

5. Clean Water Act of 1977, Pub. L. 95-217; (1977); Water Quality Act of 1987, Pub. L. 100-4 (1987).

6. See, for example, National Research Council of the National Academies, Committee on the Mississippi River and the Clean Water Act, *Mississippi River Water Quality and the Clean Water Act* (Washington, DC: National Academies Press, 2008).

7. Pub. L. 93-523 (1974).

8. Pub. L. 104-182 (1996).

9. U.S. Environmental Protection Agency, Office of Wetlands, Oceans and Watersheds, "Options for Expressing Daily Loads in TMDLs," Draft (June 22, 2007), viii.

10. One of the most important cases in this controversy involved the Anacostia River, discussed later in this chapter. See *Friends of the Earth Inc. v. EPA*, No. 05-5015 (D.C. Cir. 2006).

11. Mary Cooper, "Water Quality," *CQ Researcher* 10.41 (Nov. 24, 2000): 955.

12. Environmental Protection Agency, *National Water Quality Inventory Report, Part 1: Water Quality Assessments* (2000).

13. Environmental Protection Agency, Office of Water, *National Water Quality Inventory: Report to Congress, 2004 Reporting Cycle*, EPA 841-R-08-001 (Jan. 2009).

14. Charles Duhigg, "Clean Water Laws Are Neglected, at a Cost in Suffering," *New York Times*, Sept. 13, 2009.

15. U.S. Geological Survey, U.S. Department of the Interior, *Water Quality in the Nation's Streams and Aquifers: Overview of Selected Findings, 1991–2001*, 4.

16. U.S. Geological Survey, U.S. Department of the Interior, "Pesticides in the Nation's Streams and Groundwater, 1992–2001," Circular 1291 (Feb. 15, 2007), 4, 8.

17. U.S. Government Accountability Office, *High-Risk Series: An Update*, GAO-09-271 (Jan. 2009), 22. Also see Charles Duhigg, "Debating How Much Weed Killer Is Safe in Your Water Glass," *New York Times*, Sept. 29, 2009.

18. David Fahrenthold, "Manure Becomes Pollutant as Its Volume Grows Unmanageable," *Washington Post*, Mar. 1, 2010; Seth Borenstein, "EPA Announces Rules to Cut Animal Waste in Waterways," Knight-Ridder News Service, Dec. 17, 2002; Eric Planin, "EPA Issues New Rules on Livestock Waste,"*Washington Post*, Dec. 17, 2002; Charles Duhigg, "Health Ills Abound as Farm Runoff Fouls Wells," *New York Times*, Sept. 17, 2009.

19. U.S. Geological Survey, *Sources of Nutrients Delivered to the Gulf of Mexico*, National Water-Quality Assessment (NAWQA) Program (2008); Kate Spinner, "Pollution Still Feeding Gulf Dead Zone," *Sarasota (FL) Herald Tribune*, July 28, 2009; "As Gulf Dead Zone Grows, GRN and Allies Petition EPA for Immediate Action," *Gulf Currents* (Fall 2008): 1.

20. U.S. Geological Survey, *Differences in Phosphorus and Nitrogen Delivery to the Gulf of Mexico From the Mississippi River Basin*, National Water Quality Assessment Program (n.d.); Committee on Environment and Natural Resources, Interagency Working Group on Harmful Algal Blooms, Hypoxia, and Human Health of the Joint Subcommittee on Ocean Science and Technology. *Scientific Assessment of Hypoxia in U.S. Coastal Waters*. Washington, DC, 2010); White House press release, "New Report Warns of Expanding Threat of Hypoxia in U.S. Coastal Waters" (Sept. 3, 2010); National Oceanic and Atmospheric Administration, "NOAA and Louisiana Scientists Predict Largest Gulf of Mexico 'Dead Zone' on Record," press release (July 15, 2008); Tom Philpott, "Clean Water Jacked," *Grist Mill* (Oct. 18, 2007), at http://gristmill.grist.org.

21. See Perry Beeman, "'Dead Zone' Strategy Rattles Farm Interests," *Des Moines Register*, July 31, 2009; Chris Kirkham, "Gulf's Dead Zone Growing, Despite Pledge to Control," *Times-Picayune* (New Orleans, LA), June 10, 2007. The dead zone may be enlarged due to the 2011 floods in the Mississippi River Valley. See Kenneth Weiss, "Mississippi River Floods May Bring Biggest Dead Zone," *Los Angeles Times*, June 14, 2011. On ethanol subsidies, see Taxpayers for Common Sense, "Big Oil, Big Corn: An In-Depth Look at the Volumetric Ethanol Excise Tax Credit" (June 2011), at http://taxpayer.net/user_uploads/file/Energy /Biofuels/2011/June2011_Ethanol_VEETC_Report.pdf; Environmental Working

Group, "Ethanol's Federal Subsidy Grab Leaves Little for Solar, Wind and Geothermal" (Jan. 8, 2009), at http://www.ewg.org/reports/Ethanols-Federal-Subsidy -Grab-Leaves-Little-For-SolarWind-And-Geothermal-Energy%20.

22. National Oceanic and Atmospheric Administration, U.S. Department of Commerce, *Links Between Gulf Hypoxia and the Oil Spill* (May 26, 2010); Sharon Begley, "What the Spill Will Kill," *Newsweek* (June 14, 2010): 25–28. At this writing it appears that the 112th Congress may finally kill the ethanol subsidy program.

23. Chesapeake Bay Program, *Bay Barometer*, CBP/TRS 293-09, EPA-903-R-09-001 (March 2009), at http://www.chesapeakebay.net/content/publications /cbp_50513.pdf; Felicity Barringer, "A Plan to Curb Farm-to-Watershed Pollution of Chesapeake Bay," *New York Times*, Apr. 13, 2007; "View from the Farm: Putting Best Management to Practice," *Chesapeake Quarterly* 9.3 (Sept. 2010), at http:// mdsg.umd.edu/CQ/V09N3/side1/.

24. Juliet Eilperin, "EPA to Scrutinize Permits for Mountain-Top Removal Mining," *Washington Post*, Mar. 25, 2009, A13; Ken Ward, "Mountaintop Removal Damage 'Irreversible,' Senate Hears DEP Official Only Witness to Defend Practice," *Charleston (WV) Gazette*, June 25, 2009; Jim Mann, "Coal Mine Poses Looming Threat to Water Quality," *Kalispell (MT) Daily Interlake*, Oct. 24, 2006.

25. Duhigg, "Clean Water Laws Are Neglected."

26. Ian Urbina, "Chemicals Were Injected Into Wells, Report Says," *New York Times*, Apr. 16, 2011; Juliet Eilperin, "EPA to Study Natural Gas Drilling's Effect on Water," *Washington Post*, Mar. 19, 2010; Ian Urbina, "Regulation Lax as Gas Wells' Tainted Water Hits Rivers," *New York Times*, Feb. 26, 2011.

27. See Matthew Brown, "EPA Raises Red Flag on Coal-Bed Methane Bill," *Missoulian*, (Missoula, MT), Apr. 20, 2007; Dale Rodebaugh, "Salazar Bill Addresses Reuse of Gas-Well Water," *Durango (CO) Herald*, Apr. 24, 2007; Dustin Bleizeffer, "CBM Task Force Struggles with Persistent Water Issues," *Casper (WY) Star-Tribune*, Apr. 26, 2007; Jim Robbins, "In the West, a Water Fight Over Quality, Not Quantity," *New York Times*, Sept. 10, 2006.

28. Sonja Lee, "Former Zortman and Landussky Mine Sites Short of Funding," *Great Falls (MT) Tribune*, May 31, 2006; "Idaho Mines Harm Yellowstone Fish, ISU Report States," *Idaho Statesman*, Jan. 12, 2007; Eve Byron, "Study Offers Mike Horse Tailings Options." *Helena Independent Record*, Sept. 22, 2011; Rob Chaney, "$116 Million Restoration Plan for Upper Clark Fork Set for Approval." *Missoulian*, Nov. 9, 2011; Matthew Brown, "Asbestos Leaking into Kootenai Spurs Call for Libby Mine Cleanup," *Ravalli Republic* (Montana), Nov. 26, 2011.

29. Trout Unlimited, Public Lands Institute, *Settled, Mined & Left Behind: The Legacy of Abandoned Hardrock Mines for the Rivers and Fish of the American West* (2004).

30. *Coeur Alaska v. Southeast Alaska Conservation Council,* 557 U.S. 07-984 (June 22, 2009). For an example, see Kim Murphy, "Alaska's Kensington Gold Mine Gets a Green Light," *Los Angeles Times,* Aug. 16, 2009.

31. U.S. Geological Survey, *Mercury in Aquatic Ecosystems: Recent Findings from the National Water Quality Assessment and Toxic Substances Hydrology Programs* (Aug. 21, 2009).

32. "Gold Diggers," editorial, *Salt Lake Tribune,* Apr. 21, 2010; Judy Fahys, "Tainted-Fish List Expected to Expand," *Salt Lake Tribune,* Oct. 4, 2006; Matt Christiansen, "Mercury Rising: It's Dangerous, It's Here, and It's Not Going Away," *Magic Valley Times* (Idaho), Nov. 5, 2006; Jeremy Meyer, "Rocky Mountain Lake Is High in Toxic Metal," *Denver Post,* Dec. 10, 2006; Dale Rodebaugh, "Vallecito Mercury Levels Up," *Durango Herald,* June 16, 2006; Traci Watson, "States Look Harder for Mercury," *USA Today,* Sept. 24, 2004.

33. Evan Pelligrino, "New UA Lab Researches Medicines in Water Supply," *Arizona Daily Star,* Apr. 13, 2009; Juliet Eilperin, "Pharmaceuticals in Waterways Raise Concern," *Washington Post,* June 23, 2005; Katy Human, "Effluent Tipping Scales on Fish Gender," *Denver Post,* Sept. 6, 2006; Michelle Wallar, "Flush with Chemicals," *Boulder (CO) Daily Camera,* Apr. 17, 2005.

34. It is not just the Potomac that is experiencing this problem; see U.S. Geological Survey, *Widespread Occurrence of Intersex Bass Found in U.S. Rivers* (Sept. 14, 2009); David Fahrenthold, "Study: Weedkiller in Waterways Can Change Frogs' Sex Traits," *Washington Post,* Mar. 2, 2010; "Pharmaceuticals Passing Unaltered from Humans Into Nation's Waterways," *U.S. Water News* (Nov. 2005): 24. Also see Bruce Finley, "Gender-Bending Bass Found in Yampa River," *Denver Post,* Sept. 15, 2009.

35. "Research Says American Lawns Play Important Role in Nutrient Cycles," *U.S. Water News* (March 2005): 10.

36. Ted Steinberg, *American Green: The Obsessive Quest for the Perfect Lawn* (New York: W. W. Norton, 2007).

37. Theo Stein, "For Some, River Not Worth Its Salt," *Denver Post,* Feb. 27, 2005.

38. Ralph Vartabedian, "Nuclear Scars: Tainted Water Runs Beneath Nevada Desert," *Los Angeles Times,* Nov. 13, 2009; U.S. Environmental Protection Agency, "Region 10 Superfund: Hanford, Washington," at http://yosemite.epa.gov/r10 /cleanup.nsf; U.S. Environmental Protection Agency, "Region 4 Superfund: U.S. DOE Oak Ridge Reservation," at www.epa.gov/region4/waste/npl/nplm/oakridtn .htm; Cindy Cole, "Radioactive Water near Hopi Springs," *Arizona Daily Sun,* Mar. 18, 2007.

39. Charles Duhugg, "That Tap Water Is Legal but May Be Unhealthy," *New York Times,* Dec. 17, 2009.

40. Brian Vastag, "EPA Reverses Bush-Era Water Safety Standards, Will Regulate Contaminants," *Washington Post*, Feb. 3, 2011; Garance Franks-Ruta, "Water and Space," *Washington Post*, May 6, 2008; Marla Cone, "Rocket-Fuel Chemical Found in Breast Milk," *Los Angles Times*, Feb. 23, 2005.

41. Lyndsey Layton, "Probable Carcinogen Hexavalent Chromium Found in Drinking Water of 31 U.S. Cities," *Washington Post*, Dec. 19, 2010.

42. Environmental Protection Agency, "National Listing of Fish Advisories, 2008," Fact Sheet (Sept. 2009), at http://water.epa.gov/scitech/swguidance/fish shellfish/fishadvisories/fs2008.cfm.

43. Eve Byron, "Cleanup Deal Expected from Arco, Asarco," *Helena (MT) Independent Record*, Apr. 25, 2008; Matthew Frank, "The Government Thinks It's Found the Best Spot to Bury Toxic Tailings," *Missoula Independent*, Mar. 24, 2011.

44. Water Infrastructure Network, *Water Infrastructure Now: Recommendations for Clean and Safe Water in the 21st Century* (2008).

45. *In Re Tri-State Water Rights Litigation*, Case No. 3:07-md-01 (PAM/JRK), Memorandum and Order of Judge Paul Magnuson (M.D. Fla., July 17, 2009), 4–10; Bill Rankin, "Dam Won Support, Not Finances," *Atlanta Journal-Constitution*, July 26, 2009.

46. *In Re Tri-State Water Rights Litigation*, 11.

47. This navigation project, consisting of three sets of locks and the dam and over 100 miles of dredged waterway, was a dismal failure, carrying a tiny portion of its projected traffic and seldom achieving the 9-foot authorized depth. Today it carries virtually no cargo. See Mobile District, U.S. Army Corps of Engineers, "Information Paper: Navigation on the Apalachicola," at www.sam.usace.army .mil/webdoc/apalachicola.pdf.

48. *In Re Tri-State Water Rights Litigation*.

49. Mobile District, U.S. Army Corps of Engineers, http://www.sam.usace .army.mil/lanier/.

50. *In Re Tri-State Water Rights Litigation*.

51. See Sally Bethea, "Cleaning Up Atlanta: Sewage Overhaul," *Waterkeeper Magazine* (Summer 2006): 32–36; U.S. Geological Survey, "The U.S. Geological Survey and City of Atlanta Water-Quality and Water-Quantity Monitoring Network," Fact Sheet 2005-3126 (Apr. 2006); City of Atlanta, Department of Watershed Management, *Performance Audit Response* (Apr. 2009).

52. Quoted in a handout provided by the Chattahoochee River National Recreation Area.

53. Dan Brown, Superintendent, Chattahoochee River National Recreation Area, in-person interview with author (July 25, 2009); Chattahoochee River National

Recreation Area, *Final General Management Plan/Environmental Impact Statement* (Sept. 2009).

54. Sally Bethea, Upper Chattahoochee River Riverkeeper, in-person interview with author, Atlanta, GA (July 21, 2009).

55. *United States and State of Georgia v. City of Atlanta*, Civil Action File No. 1:95-CV-2550-TWT (1998), and Civil Action File No. 1:98-CV-1956-TWT, "First Amended Consent Decree" (1999); Sally Mills, Deputy Commissioner, Bureau of Watershed Protection, City of Atlanta, in-person interview with author, Atlanta, GA (July 21, 2009).

56. Susan Rutherford, Watershed Manager, Bureau of Watershed Protection, City of Atlanta, in-person interview with author, Atlanta, GA (July 21, 2009).

57. See City of Atlanta, Department of Watershed Management, "Overview," at www.cleanwateratlanta.org/consentdecree/overview.htm; City of Atlanta, Department of Watershed Management, *Performance Audit Response* (Apr. 2009).

58. Shirley Franklin, "Good Government: Embracing Clean Water," *Waterkeeper Magazine* (Summer 2006): 34–35.

59. Bethea, "Cleaning Up Atlanta," 36; City of Atlanta, Department of Watershed Management, at www.cleanwateratlanta.gov/FAQ/.

60. Quoted in City of Atlanta, Department of Watershed Management, *Performance Audit Response*, 1.

61. American Rivers, "New Bill Makes Georgia a National Leader in Water Efficiency" (Mar. 11, 2010).

62. Sally Bethea, interview with author.

63. Sidney Lanier, *The Poems of Sidney Lanier* (Kindle edition).

64. Lee Cain, in-person interview with author, Bladensburg, MD (Oct. 10, 2008).

65. John Wennersten, *Anacostia: The Death and Life of an American River* (Baltimore: Chesapeake Bay Book Company, 2008), 164.

66. See Tom Turner, *Justice on Earth: Earthjustice and the People It Has Served* (White River Junction, VT: Chelsea Green Publishing, 2002), 149–63.

67. According to the Society's 2006 annual report, its volunteers have removed 842 tons of trash from the river. See Anacostia Watershed Society, *2006 Annual Report: Celebrating 18 Years of Service to the Anacostia River and Its Surrounding Communities*, 5. On Earth Day in 2009, volunteers removed 59 tons of trash, 289 tires, and 10,700 plastic bags from the river. *Voice of the River* 21.1 (Winter 2010).

68. Robert Boone, in-person interview with author, Bladensburg, MD (Dec. 5, 2008).

69. I did not use the name of this football team because I consider it a racist insult to American Indian peoples.

70. Jim Connolly, in-person interview with author, Bladensburg, MD (Dec. 5, 2008). At the time of our interview Jim was the executive director. He retired from that position in December 2009. The current president is Jim Foster.

71. See District of Columbia, Department of the Environment, "Washington Navy Yard," at http://ddoe.dc.gov/ddoe/lib/ddoe/information/pdf/hazwastenavy yard.pdf.

72. D'Vera Cohn, "Cleaning Up D.C.'s Dirty Little Secret," *Washington Post*, Apr. 22, 1991.

73. See the website for Ben Cardin, Senator from Maryland, at http://cardin .senate.gov/issues/record.cfm?id=27184&.

74. *Voice of the River* 21.1 (Winter 2010).

75. *Record* (Anacostia Watershed Society) 1.1 (Mar. 1989).

76. See Mike Sato, *The Price of Taming the River: The Decline of Puget Sound's Duwamish/Green Waterway* (Seattle: Mountaineers Books, 1997).

77. Jennie Goldberg and Judith Noble, in-person interviews with author on the Duwamish River (July 14, 2008).

78. See Wohl, *Disconnected Rivers*, 118–21.

79. See "Man Behind Hamm Creek Restoration Recognized," *Seattle Post-Intelligencer*, May 17, 2003; Robert McClure, "City Will Repair Wetlands Harmed by New Complex," *Seattle Post-Intelligencer*, Dec. 1, 2005.

80. David Bowermaster, "Naturalist, John Beal, 56, Was Hero of Hamm Creek," *Seattle Times*, June 27, 2006, at http://community.seattletimes.nwsource .com/archive/?date=20060627&slug=bealobit27m.

81. CERCLA is the 1980 Comprehensive Environmental Response, Compensation, and Liability Act (usually referred to as the superfund law), Pub. L. 96-510 (Dec. 11, 1980).

82. Elliott Bay/Duwamish Restoration Program Panel, *Elliott Bay/Duwamish Restoration Program* (July 2005).

83. See Green/Duwamish and Central Puget Sound Watershed, *Making Our Watershed Fit for a King*, Salmon Habitat Plan, Green/Duwamish and Central Puget Sound Watershed Water Resource Inventory Area 9 (Aug. 2005).

84. Lower Duwamish Waterway Group, at www.ldwg.org.

85. WRIA 9, "On-The Ground Actions for Salmon Habitat in the Green/ Duwamish and Central Puget Sound Watershed" (June 26, 2004), table adapted from similar tables in the WRIA 9 Near-Term Action Agenda (May 2002), produced by the Green-Duwamish and Central Puget Sound Forum of Local Governments.

86. Environmental Protection Agency, *The Clean Water and Drinking Water Infrastructure Gap Analysis*, Office of Water, EPA-816-02-020 (Feb. 2002).

87. Exodus 7:18 (King James Version).

88. Revelation 22:1 (King James Version).

9. RIVER CITY: URBAN RIVERSCAPES

1. Neil Young and Crazy Horse, "Down by the River," from *Everybody Knows This Is Nowhere* (Reprise, 1969); Tom Murphy quoted in John Dawes, "America's Three Rivers," at www.pittsburghgreenstory.org/html/3_rivers.html.

2. See Betsy Otto, Kathleen McCormick, and Michael Leccese, *Ecological Riverfront Design: Restoring Rivers, Connecting Communities*, Planning Advisory Service Report No. 518-519 (Chicago: American Planning Association, March 2004); Ann Riley, *Restoring Streams in Cities* (Washington, DC: Island Press, 1998).

3. See "Charter of the New Urbanism," at www.cnu.org/charter.

4. See Ann Breen and Dick Rigby, *The New Waterfront* (New York: McGraw-Hill, 1996); R. Marshall, *Waterfronts in Post-industrial Cities* (London: Taylor and Franklin, 2001).

5. See Friends of the James River Park, at: http://www.jamesriverpark.org/visit -the-park/about-the-park.php.

6. Councilmember Ed Reyes, in-person interview with author, Los Angeles City Hall (Jan. 27, 2006).

7. Fred Bernstein, "A Vision of a Park on a Restored Los Angeles River," *New York Times*, Sept. 28, 2010.

8. Ed P. Reyes, afterword to *Down by the Los Angeles River*, by Joe Linton (Berkeley, CA: Wilderness Press, 2005), 293.

9. Linton, *Down by the Los Angeles River*.

10. Friends of the Los Angeles River, "River Revitalization Corporation Take Their Seats," at http://folar.org/?page_id=6.

11. Jean Guerrero, "A River Really Runs Through It," *Wall Street Journal*, July 31, 2010; Alissa Walker, "As Los Angeles's River Runs Again, Designers Determine Its Course," *GOOD Magazine* (July 17, 2010), at http://www.good.is/post/as -los-angeles-s-river-runs-again-designers-determine-its-course/.

12. Renata von Tscharner, in-person interview with author, Boston, MA (Apr. 6, 2006).

13. For a history of this part of the river, see Karl Haglund, *Inventing the Charles River* (Cambridge, MA: MIT Press, 2002).

14. The Charles Eliot mentioned in chapter 2, the president of Harvard, was this man's father.

15. See Massachusetts Department of Conservation and Recreation, "Charles River Reservation," at www.mass.gov/dcr/parks/metroboston/charlesR.htm.

16. This park is in the planning stages now, having received some funding from the Stimulus Act, P. L. 111-5 (Feb. 17, 2009).

17. Kate Bowditch, in-person interview with author, Boston, MA (Apr. 7, 2006).

18. U.S. Environmental Protection Agency, *Clean Charles River Initiative*, at http://www.epa.gov/region1/charles/initiative.html.

19. Derrick Jackson, "A Trophy for the Charles," *Boston Globe*, July 2, 2011.

20. Charles River Watershed Association, at www.crwa.org.

21. The dam's name is sometimes spelled "Embrey."

22. Rusty Dennen, "Dam to Disappear," *Free Lance-Star* (Fredericksburg, VA), Mar. 16, 2003.

23. Senator Warner retired from the Senate in 2009.

24. Staff of the Free Lance-Star, *River Runs Free: Breaching the Rappahannock's Embrey Dam* (Fredericksburg, VA, 2004).

25. John Tippett, "Let the River Run," *Piedmont Virginian* (Autumn 2008): 43–44.

26. William Penn, quoted in John William Reps, *The Making of Urban America* (Princeton, NJ: Princeton University Press, 1965), 172.

27. Quoted in Sally Kalson, "Cartoonist Draws, Fires a Blank with Pittsburgh Joke," *Pittsburgh Post-Gazette*, Nov. 19, 2003.

28. "Mournful" was Jack Kerouac's word for the Susquehanna when he visited it in the 1950s. Jack Kerouac, *On the Road* (New York: Viking, 2007), 104.

29. Pennsylvania Department of Conservation and Natural Resources, "Point State Park," at http://www.dcnr.state.pa.us/stateparks/findapark/point/index .htm.

30. Mackenzie Carpenter, "Pittsburgh Ranked Tops in U.S. by the Economist," *Pittsburgh Post-Gazette*, June 10, 2009.

31. *The Riverfront Development Plan: A Comprehensive Plan for the Three Rivers*, City of Pittsburgh, Tom Murphy, Mayor.

32. The original plan was to direct the river underground into a culvert and pave the natural river corridor. See Lewis Fisher, *River Walk: The Epic Story of San Antonio's River* (San Antonio, TX: Maverick Publishing, 2006).

33. "Plan for Jordan River Rehabilitation Moves Forward," *Deseret News*, Apr. 29, 2010; Jeremiah Stettler, "Countywide Effort Needed for Jordan River, Mayors Say," *Salt Lake Tribune*, Aug. 28, 2009; Utah Reclamation and Mitigation Conservation Commission, U.S. Fish and Wildlife Service, Great Salt Lake Audubon

Society, and National Audubon Society, "Jordan River Natural Conservation Corridor," brochure (n.d.).

34. Tori Woods, "Cuyahoga Connections," *Land and People* (Fall/Winter 2010): 11–17.

35. "$3B Waterfront Plan in the Works for NYC," *USA Today*, Mar. 14, 2011.

36. See *River Reporter*, the newsletter of the Friends of the Chicago River; Michael Hawthorne, "State Pollution Board Orders Chicago River Cleanup," *Chicago Tribune*, June 2, 2011.

37. See, for example, Claudia Borchert, "Sustaining the Santa Fe River," *Southwest Hydrology* (Jan./Feb. 2010): 28–29; Amy Souers and Betsy Otto, "Restoring Rivers Within City Limits," *River Voices* 19.1 (2009).

38. William Least Heat-Moon, *River-Horse* (New York: Penguin Books, 1999), 48.

39. Deborah S. Lumia, Kristin S. Linsey, and Nancy L. Barber, U.S. Geological Survey, "Estimated Use of Water in the United States in 2000," Fact Sheet 2005-3051 (Washington, DC: U.S. Department of the Interior, 2005).

40. For a catalog of America's water-wasting tendencies, see Robert Glennon, *Unquenchable* (Washington, DC: Island Press, 2009).

41. Envision Utah, *Blueprint Jordan River* (Salt Lake City, UT, 2008), 2.

42. Laura A. Pratt, Debra J. Brody, and Qiuping Gu, "Antidepressant Use in Persons Aged 12 and Over: United States, 2005–2008," *NCHS Data Brief* No. 76 (Oct. 2011), at: http://www.cdc.gov/nchs/data/databriefs/db76.pdf. Also see Shankar Vedantam, "Antidepressant Use by U.S. Adults Soars," *Washington Post*, Dec. 3, 2004.

10. NET LOSSES: HABITAT AND ENDANGERED SPECIES

1. Alexander Whitaker, "Good News from Virginia," quoted in *Land of Rivers*, edited by Peter Mancall (Ithaca, NY: Cornell University Press, 1996), 30.

2. Quoted in John Bakeless, *America as Seen by Its First Explorers* (New York: Dover Publications, 1950), 222–23.

3. Quoted in Alice Outwater, *Water: A Natural History* (New York: Basic Books, 1996), 107.

4. Bakeless, *America*, 244.

5. Ibid., 300.

6. Ibid., 327.

7. *The Journals of Lewis and Clark*, abridged by Anthony Brandt (Washington, DC: National Geographic Adventure Classics, 2002), 188, 285.

8. Quoted in Timothy Egan, *The Good Rain* (New York: Vintage Books, 1990), 181.

9. President Obama recently poked fun at this: "The Interior Department is in charge of salmon while they're in fresh water, but the Commerce Department handles them when they're in saltwater. And I hear it gets even more complicated once they're smoked." William Yardley, "Along the Columbia River, Concerns for Salmon and Energy Production," *New York Times*, Feb. 19, 2011.

10. Institute for Water Resources, *Lands and Waters: Value to the Nation*, produced for the U.S. Army Corps of Engineers (n.d).

11. For an explanation of Maslow's "hierarchy of needs," see http://psychology .about.com/od/theoriesofpersonality/a/hierarchyneeds.htm.

12. International Union for Conservation of Nature, "Red List: Wildlife in a Changing World; An Analysis of the 2008 IUCN Red List of Threatened Species," 16, at http://data.iucn.org/dbtw-wpd.

13. U.S. Fish and Wildlife Service, *Species Reports: Environmental Conservation Online System*, "Summary of Listed Species Listed Population," at http://ecos .fws.gov/tess_public/TESSBoxscore.

14. U.S. Department of Commerce, NOAA Fisheries, "The Endangered Species Act—Protecting Marine Resources," at www.nmfs.noaa.gov/pr/pdfs/esa.factsheet .pdf.

15. See Krishna Gifford and Deborah Crouse, "Thirty-five Years of the Endangered Species Act," U.S. Fish and Wildlife Service, *Endangered Species Bulletin* 34.1 (Spring 2009): 4–7.

16. See Lisa Iams, "Agreement Reached to Protect Rio Grande Silvery Minnow," *Spillway* (Aug./Sept. 2000): 3; U.S. Department of the Interior, Office of the Secretary, "Landmark State-Federal Water Agreement Protects Endangered Fish and Water Users in New Mexico," news release (June 29, 2001); "Silvery Minnows Struggle as Rio Grande Dries," *Albuquerque Tribune*, June 21, 2006.

17. See Committee on Endangered and Threatened Species in the Platte River Basin, National Research Council, *Endangered and Threatened Species of the Platte River* (Washington, DC: National Academies Press, 2004).

18. See Matt Jenkins, "California Dreamin'," *High Country News* (Paonia, CO), Dec. 20, 2010, 10; CALFED Bay-Delta Program, www.calwater.ca.gov/delta /species/Fish.html; Colin Sullican, "Calif. Water Agency Changes Course on Delta Smelt," *New York Times*, May 12, 2009. CALFED is a consortium of twenty-five state and federal agencies working to solve water-related problems in the Delta. It is now part of the California State Water Plan.

19. See Todd Votteler, "The Edwards Aquifer: ESA-Driven Management," *Southwest Hydrology* 7 (July/Aug. 2008): 22–23.

20. Rocky Barker, "With Wolf Battle Raging, a Battle Over Bull Trout Looms," *Idaho Statesman*, Oct. 25, 2010; Jeff Barnard, "Feds Propose Expanding Bush's Bull Trout Habitat," ABC News/Associated Press, Jan. 29, 2010; Nicholas Geranios, "Fish and Wildlife Designates Bull Trout Habitat," *Casper (WY) Star-Tribune*, Sept. 26, 2005.

21. See Gary Pitzer, "Balancing the Colorado River's Ecosystem and Water Delivery Capability," *Colorado River Project River Report* (Winter 2009–10), published by the Water Education Foundation, Sacramento, CA; Robert Adler, *Restoring Colorado River Ecosystems* (Washington, DC: Island Press, 2007); Sue McClurg, "The Lower Colorado River MSCP," *Colorado River Project River Report* (Summer 2005); Juliet Eilperin, "Effort to Renew Colorado River Launched," *Washington Post*, Sept. 15, 2004.

22. See "Apalachicola River and Bay," *Nature Conservancy Magazine* (Winter 2003): 4.

23. Leslie Kaufman, "Seeing Trends, Coalition Works to Help River Adapt," *New York Times*, July 20, 2011; Rocky Barker, "Columbia River Basin Dam Managers to Adapt to Warmer Weather," *Idaho Statesman*, April 26, 2011; PEW Center on Global Climate Change, *Aquatic Ecosystems and Global Climate Change* (Philadelphia, Jan. 2002); Guy Gugliotta, "Warming May Threaten 37% of Species by 2050," *Washington Post*, Jan. 8, 2004.

24. See Rosamond Naylor et al., "Feeding Aquaculture in an Era of Finite Resources," *Proceedings of the National Academy of Sciences* 10.1073 (2009): 15103–10.

25. Ronald Hites et al., "Global Assessment of Organic Contaminants in Farmed Salmon," *Science* 303.5655 (Jan. 9, 2004): 226–29; *National Geographic News*, "Farmed Salmon: Decimating Wild Salmon," Feb. 12, 2008; Rebecca Claren, "Bracing Against the Tide," *High Country News* 35 (Mar. 17, 2003): 1, 8–11; Andrea Chang, "Target Drops All Farmed Salmon from Stores," *Los Angeles Times*, Jan. 26, 2010.

26. See Juliet Eilperin, "White House Seeks to Boost Fish Farms by Expanding Into Open Waters," *Washington Post*, June 8, 2005, A7; Randolph Schmid, "Proposal Would Allow Fish Farming off US Coasts," *Boston Globe*, June 8, 2005.

27. Jan Hoover, biologist, in-person interview with author, Corps of Engineers, Engineering Research and Development Center (Dec. 18, 2003); J. Hoover et al., *Suckermouth Catfishes: Threats to Aquatic Species*, Aquatic Nuisance Species Research Program Bulletin 04-1 (Vicksburg, MS, Feb. 2004).

28. Wayne Stroup, in-person interview with author, Corps of Engineers, Engineering Research and Development Center, Vicksburg, MS (Dec. 18, 2003).

29. Monica Davey, "U.S. Officials Plan $78.5 Million Effort to Keep Dangerous Carp out of Great Lakes," *New York Times*, Feb. 9, 2010; Eartha Jane Melzer, "Invasive Asian Carp Close to Entering Great Lakes," *Michigan Messenger*, Aug. 11, 2009; Dan Egan, "Army Corps to Activate Carp Barrier Wednesday," *Milwaukee Journal-Sentinel*, Apr. 7, 2009.

30. See Lee Blumenthal, "Mussel Invasion Closes In on Northwest Waters," *Seattle Times*, Aug. 31, 2009.

31. U.S. Army Corps of Engineers, Jacksonville District, Operations Division.

32. Felicity Barringer, "Fewer Marshes + More Man-Made Ponds = Increased Wetlands," *New York Times*, Mar. 31, 2006.

33. *Rapanos v. United States*, 126 S. Ct. 2208 (2006). See Charles Duhigg and Janet Roberts, "Rulings Restrict Clean Water Act," *New York Times*, Feb. 28, 2010.

34. "Clean Water Act Jurisdiction Following the U.S. Supreme Court's Decision in *Rapanos v. U.S. & Carabell v. U.S.*," memorandum signed by the Environmental Protection Agency and the Department of the Army, Civil Works (Dec. 12, 2008). Also see "Stephen Power, "Administration to Expand Federal Waterways Oversight," *Wall Street Journal*, Apr. 26, 2011.

35. Clean Water Restoration Act of 2009, S. 787 (2009). See "Threats to U.S. Waters," *National Wildlife* (April 14, 2010), at http://www.nwf.org/News-and -Magazines/National-Wildlife/News-and-Views/Archives/2010/Action-Report -June-July-2010.aspx.

36. Clean Water Cooperative Federalism Act, H.R. 2018, 112th Cong., 1st Sess. (May 26, 2011).

37. Ransom Myers et al., "Hatcheries and Endangered Salmon," *Science* 303 (Mar. 26, 2004): 1980. Also see Glen Martin, "Salmon Jeopardized by Method Used in Run Count, Scientists Say," *San Francisco Chronicle*, Mar. 26, 2004; Craig Welch, "Salmon Panel Goes Public in Dispute Over Hatchery Fish," *Seattle Times*, Mar. 26, 2004; "Under Plan, Fall Chinook Could Be Delisted," *Idaho Statesman*, May 2, 2004.

38. Federal Energy Regulatory Commission, *Final Environmental Impact Statement* (FEIS-0103), Condit Hydroelectric Project, FERC No. 2342-055 (Oct. 1996).

39. David Becker, "The Challenges of Dam Removal: The History and Lessons of the Condit Dam and Potential Threats from the 2005 Federal Power Act Amendments," *Environmental Law* 36 (Summer 2006): 811–826.

40. Jesse Burkhardt, "Condit Dam Hearing: Hundreds Turn Out, Offer Views to FERC," *Enterprise* (Salmon, WA), Mar. 21, 2002; Opinion, "Trouble in a Song," *Enterprise*, Mar. 21, 2002.

41. Condit Hydroelectric Project, FERC No. 2342, "Removal Plan Summary," 7, affixed to the Condit Hydroelectric Project Settlement Agreement, signed Sept. 17, 2002.

42. Jim Anderson, letter to the editor, *Enterprise*, Mar. 21, 2002; Jerry Smith, letter to the editor, *Enterprise*, Jan. 17, 2002.

43. Pat Baumgarden, "Condit Dam Removal Is an Insane Idea," *Enterprise*, Feb. 21, 2002.

44. L. Vanda Bruggen, letter to the editor, *Enterprise*, Nov. 22, 2001.

45. Letter to Magalie Salas, Secretary, FERC, from John Whittaker of Winston and Strawn, representing Klickitat and Skamania Counties (Dec. 9, 2002); Jesse Burkhardt, "County Blasts DOE Over Dam," *Enterprise*, Feb. 6, 2003.

46. Condit Hydroelectric Project, "Removal Plan Summary," 2–4; Washington Department of Ecology, *Condit Dam Removal: SEPA Supplemental Environmental Impact Statement* (Sept. 30, 2005), 1-5 to 1-6.

47. See Letter from Brett Swift of American Rivers to the Washington Department of Ecology, re "Proposed Condit Dam Removal Project, FERC No. 2342; Draft Supplemental EIS," Washington Department of Ecology Publication 05-06-022 (Seattle: URS Corporation, 2005), 1-6 to 1-9; National Marine Fisheries Service, "Biological Opinion for Interim Operation, Decommissioning, and Removal of the Condit Hydroelectric Project (FERC 2342)," NOAA Fisheries Consultation No. 2002/00977 (Oct. 12, 2006).

48. Gail Miller and Terry Flores, in-person interviews with author, Portland, OR (June 16, 2003).

49. Don Struck, in-person interview with author, Klickitat County Courthouse (June 18, 2003).

50. Ibid. Former Commissioner Struck currently serves as chair of the county's Solid Waste Task Force.

51. In-person interview with Phyllis Clausen, at Condit Dam, June 19, 2003. In 2011 Clausen was still active in issues affecting the White Salmon and the Columbia River Gorge.

52. Washington Department of Ecology, *Condit Dam Removal Draft Second Supplemental Environmental Impact Statement*, Publication no. 09-12-008 (June 2009).

53. *Order Accepting Surrender of License, Authorizing Removal of Project Facilities, and Dismissing Application for New License*, Federal Energy Regulatory Commission. Project Nos. 2342–005 and 2342–011 (Dec. 16, 2010).

54. This dramatic event can be seen on YouTube: http://www.youtube.com/watch?v=J9cudp1eCdc. Also see Lynda Mapes, "Condit Dam to Be Demolished

Wednesday," *Seattle Times*, Oct. 25, 2011; PacifiCorp, "Condit Dam Decommissioning Schedule of Activities," at http://www.pacificorp.com/condit#.

55. John Banks, in-person interview with author, Penobscot Nation, Maine (May 19, 2004).

56. Statement of Barry Dana, Chief of the Penobscot Nation, Office of the Governor and Council, Indian Island, Maine (n.d., c. 2003).

57. Penobscot Rivers Restoration Trust, "Project Details," at www.penob scotriver.org/Project%20Details.htm; Atlantic Salmon Federation, "The Sea-Run Fisheries of the Penobscot River," brochure (n.d).

58. The Lower Penobscot Multiparty Settlement Agreement, signed June 5, 2005. See Pam Belluck, "Agreement in Maine Will Remove Dams for Salmon's Sake," *New York Times*, Oct. 7, 2003; Mary Wiltenburg, "Historic Maine Bargain Opens Way for Return of Atlantic Salmon Runs," *Christian Science Monitor*, Oct. 7, 2003; Douglas Watts, "Penobscot Reborn," *Atlantic Salmon Journal* 52.4 (Winter 2003): 27–31.

59. Biologists recently counted 2,276 salmon at the dam, evidence that restoration efforts are already having an impact. John Holyoke, "Penobscot Salmon Total Tops 2,200," *Bangor (ME) Daily News*, June 24, 2011.

60. "Penobscot River Restoration Project Takes Major Strides Forward," *Penobscot Passages* (Jan. 2009); U.S. Federal Energy Regulatory Commission, Penobscot River Restoration Trust, Project Nos. 2403-056, 23120019, 2721-020, *Draft Environmental Assessment: Application for Surrender of License* (Aug. 2009).

61. Penobscot River Restoration Trust, "Vision of the Penobscot River Once Again Teeming with Life While Continuing to Generate Energy Is Closer to Becoming a Reality," press release (May 31, 2006).

62. Penobscot River Restoration Trust, "Timeline" (updated June 2011), at http://penobscotriver/org/content/4031/Timeline/; Kevin Miller, "Penobscot River Dam Removal, Fish Restoration Project Approved," *Bangor Daily News*, July 21, 2011; Kevin Miller, "Penobscot River Dam Removal, Fish Restoration Project Approved," *Bangor Daily News*, July 21, 2010; Penobscot River Restoration Trust, "Penobscot River Restoration Project to Boost Maine Economy, Restore Fish," press release (June 30, 2009). Also see U.S. Federal Energy Regulatory Commission, "Notice of Availability of Final Environmental Assessment," Project Nos. 2403-056, 2312-019, and 2721-020 (May 18, 2010).

63. Gale Courey Toesing, "Restoring Fish, Preserving Culture," *Indian Country Today* (Rapid City, SD), May 19, 2008.

64. Friends of the Presumpscot River, Portland, Maine, "The Presumpscot River," at http://www.presumpscotriver.org/about-the-river.html.

65. Gail Wippelhauser et al., "Draft Fishery Management Plan for the Presumpscot River Drainage," unpublished paper.

66. Natural Resources Council of Maine, "A Citizen's Guide to Dams, Hydropower, and River Restoration in Maine" (Augusta, ME, 2002); Presumpscot River Management Plan Steering Committee, "A Plan for the Future of the Presumpscot River" (Aug. 18, 2003).

67. Dusti Faucher, in-person interview with author, Portland, ME (May 27, 2004).

68. "Making Ripples, Making Waves," *Presumpscot River News* 13.1 (Jan. 2005).

69. "The Presumpscot River—Background," at http://presumpscotcoalition .org/docs/presumpscot_river_fact_sheet.pdf.

70. Presumpscot River Plan Steering Committee, "Fisheries Conditions, Issues and Options for the Presumpscot River," background paper for the development of the plan for the future of the Presumpscot River, prepared for the Casco Bay Estuary Project (May 28, 2002).

71. Casco Bay Estuary Partnership, at: http://www.cascobay.us.maine.edu /overview.html.

72. *S. D. Warren v. Maine Board of Environmental Protection*, 547 U.S. 370 (May 15, 2006). S. D. Warren is the same company as SAPPI.

73. Roland Martin, Commissioner, Maine Department of Inland Fisheries and Wildlife, *In the Matter of Cumberland Mills Dam Fishway Proceeding*, Findings of Fact and Decision (2009).

74. Ken Fletcher, in-person interview, Winslow, ME (May 19, 2004).

75. *Edwards Mfg. Co.*, 84 FERC 61227, 62092–93, 62097 (1998). This was part of the settlement that led to the removal of Edwards Dam in the same watershed but on the Kennebec River.

76. *Save Our Sebasticook v. FERC*, No. 04-2005 (D.C. Cir. Dec. 9, 2005).

77. Representative Fletcher is now serving his fourth term in the legislature.

78. Colin Hickery, "Fort Halifax Dam Breached," *Kennebec (ME) Journal*, July 18, 2008.

79. U.S. Department of the Interior, *The Department of the Interior's Economic Contributions* (June 21, 2011), 16–17.

80. Recreational Boating and Fishing Foundation, *Participation in Boating and Fishing: A Literature Review* (Alexandria, VA, Sept. 2000).

81. National Survey on Recreation and the Environment, "Recreation Statistics Update," *Update Report* No. 1 (Aug. 2004).

82. U.S. Department of the Interior, U.S. Fish and Wildlife Service, U.S. Department of Commerce, Economics and Statistics Administration, and U.S.

Census Bureau, *2006 National Survey of Fishing, Hunting, and Wildlife-Associated Recreation*, FHW/06-NAT (2006).

83. American Sportfishing Association, *Sportfishing in America: An Economic Engine and Conservation Powerhouse* (revised Jan. 2008).

84. U.S. Census Bureau, *The 2009 Statistical Abstract: The National Data Book*, table 853, "Fisheries—Quantity and Value of Domestic Catch: 1980 to 2006."

85. Congressional Sportsmen's Foundation, "Take a Closer Look," *American Sportsman*, at http://sportsmenslink.org.

86. See EPA, "Highlights of National Trends," *EPA's Report on the Environment* (2008), 17.

87. To understand this concept, see Edella Schlager and William Blomquist, *Embracing Watershed Politics* (Boulder: University Press of Colorado, 2008).

88. Quoted in Chris Stetkiewicz, "U.S. Rule Change Could Weaken Salmon Protections," Reuters News Service, May 4, 2004.

11. PLAYGROUND ON THE MOVE:
RIVER RECREATION

1. Travel Industry Association of America, "Power of Travel," at www.tia.org/marketing/toolkit_tourism_facts.html.

2. Data are from 2007, quoted in American Recreation Coalition, "Outdoor Recreation in America, Research and Statistics," 8, at www.funoutdoors.com/research.

3. California Ocean and Coastal Environmental Access Network, "Category: Tourism and Recreation," at http://ceres/ca.gov/ocean/theme/tourism_background.htm.

4. President's Commission on Americans Outdoors, *Americans Outdoors: The Legacy, the Challenge, with Case Studies; The Report of the President's Commission* (Washington, DC: Island Press, 1987).

5. H. Ken Cordell, Sporting Goods Manufacturers Association (U.S.), *Emerging Markets for Outdoor Recreation in the United States*, a Report to the Sporting Goods Manufacturing Association and the Outdoor Products Council (1995).

6. Ecotourism Society, "USA Ecotourism Statistical Factsheet" (North Bennington, VT, 1999), at www.ecotourism.org; International Ecotourism Society, "Fact Sheet: Ecotourism in the U.S." (Washington, DC, 2006).

7. David Mahood, "Tourism and Biophilia: Protecting the World's Remaining Natural Habitats," *EcoCurrents* (Aug. 17, 2009), at http://www.yourtravelchoice.org/2009/08/tourism-and-biophilia/.

8. U.S. Army Corps of Engineers, "National Recreation Lakes Act of 2001" (March 6, 2001), at http://corpslakes.usace.army.mil/employees/cecwon/pdf.

9. National Marine Manufacturers Association, "2003 Recreational Boating Statistical Abstract" (2004), at www.nmma.org/facts/boatingstats/2003/files /Abstract.pdf.

10. National Survey on Recreation and the Environment, "Recreation Statistics Update," *Update Report* No. 1 (Aug. 2004).

11. National Research Council, *Restoration of Aquatic Ecosystems: Science, Technology, and Public Policy* (Washington, DC: National Academy Press, 1992), 22.

12. National Recreation Lakes Study Commission, *Reservoirs of Opportunity*, Final Report (June 1999).

13. S. V. Smith et al., "Distribution and Significance of Small Artificial Water Bodies," *Science Total Environment* 21 (2002): 24; U.S. Department of the Interior, National Biological Service, *Our Living Resources* (Washington, DC, 1995), 418.

14. U.S. Environmental Protection Agency, *Technical Document: EPA's Draft Report on the Environment* (2003), 5–42.

15. In recent reports on the proposal to remove four dams on the lower Snake River, the Corps has finally begun to recognize that moving-water recreation is a benefit to dam removal. For examples of the abuse of flat-water estimations in cost-benefit analyses, see Jimmy Carter's hit list, described in chapter 2.

16. National Survey on Recreation and the Environment, "Recreation Statistics Update."

17. Water Infrastructure Network, "Water Infrastructure Now: Recommendations for Clean and Safe Water in the 21st Century," at www.win-water.org/reports /winow.pdf.

18. U.S. Department of the Interior, U.S. Fish and Wildlife Service, U.S. Department of Commerce, Economics and Statistics Administration, and U.S. Census Bureau, *2006 National Survey of Fishing, Hunting, and Wildlife-Associated Recreation*, FHW/06-NAT (2006).

19. International Association of Fish and Wildlife Agencies, *The Economic Importance of Hunting in America* (2002), at http://www.fishwildlife.org/files /Hunting_Economic_Impact.pdf.

20. See National Park Service, Rivers, Trails, and Conservation Assistance Program, *Economic Benefits of Conserved Rivers: An Annotated Bibliography* (June 2001).

21. American Hiking Society, "The Economic Benefits of Trails" (Feb. 2004), at www.americanhiking.org/news/pdf/econ_ben.pdf.

22. Outdoor Industry Foundation, *Outdoor Recreation Participation and Spending Study* (2003).

23. The data are from the U.S. Fish and Wildlife Service, as reported in Francis Clines, "As Their Numbers Soar, Birders Seek Political Influence to Match," *New York Times*, Feb. 4, 2001.

24. U.S. Census Bureau, *The 2009 Statistical Abstract: The National Data Book*, tables 1213 and 1217.

25. Daniel Stynes, *National Park Visitor Spending and Payroll Impacts, 2007*, National Park Service, Social Science Program (Sept. 2008). For even larger estimates, see National Parks and Conservation Association, *America's Great Outdoors* (2010), at http://www.npca.org/protecting-our-parks/policy-legislation /americas-great-outdoors/.

26. "USDA National Visitor Use Monitoring Results: 2005–2009," as reported in *America's Great Outdoors: A Promise to Future Generations* (Feb. 2011), 54, at http://americasgreatoutdoors.gov/.

27. U.S. Department of the Interior, *The Department of the Interior's Economic Contributions* (June 21, 2011). Also see Department of the Interior, *America's Great Outdoors, Fifty-State Report* (Nov. 2011).

28. Outdoor Industry Foundation, *National Economic Impact Reports: The Active Outdoor Recreation Economy Report* (2011); Peter Metcalf, "Outdoor Recreation a Long-Term Economic Boon," *Salt Lake Tribune*, Aug. 4, 2010.

29. Mike Gorrell, "Outdoor Industry Sales Up 4 Percent in 2010," *Salt Lake Tribune*, Feb. 14, 2011.

30. Frank Hugelmeyer, "Herbert, Utah Leaders Betraying Recreation Industry," *Salt Lake Tribune*, June 18, 2011.

31. *America's Great Outdoors: A Promise to Future Generations*, 8.

32. See Sonoran Institute, Theodore Roosevelt Conservation Partnership, *Backcountry Bounty: Hunters, Anglers and Prosperity in the American West* (June 2006); Western States Tourism Policy Council, "Tourism Statistics: Economic Impact of Tourism in the West" (n.d.), at www.wstpc.org/Publications/Economic Impact.htm.

33. "Recreation Wealth," op-ed, *Salt Lake Tribune*, July 4, 2011. Also see Protect the Flows, a consortium of businesses dependent on instream flow in the Colorado River, at http://protecttheflow.com.

34. Utah Tourism Industry Coalition, *State of the Utah Tourism Industry* (2009). For more on the importance of water to parks in that region, see Center for Park Research, National Parks and Conservation Association, *National Parks of the Colorado River* (April 2011).

35. "Getting Outdoors," *Salt Lake Tribune*, Aug. 10, 2010; "Tourism Gold," *Salt Lake Tribune*, Sept. 26, 2011.

36. Outdoor Industry Foundation, *National Economic Impact Reports: The Active Outdoor Recreation Economy Report*.

37. Headwaters Economics, "Summary: The Economic Importance of National Monuments to Local Communities" (2011), at http://headwaterseconomics .org/wphw/wp-content/uploads/NatlMon_Summary.pdf.

38. Wilderness Society, "Natural Dividends: Wildland Protection and the Changing Economy of the Rocky Mountain West" (2009), at www.wilderness .org/Library/Documents/NaturalDividends.cfm.

39. Pub. L. 90-542 (1968).

40. Pub. L. 111-11-Mar. 30, 2009, 123 Stat. 94. See "Huge Victory for Rivers," *American Rivers* (Fall 2009): 6–9.

41. For recent examples, see Staci Matlock, "Governor Joins Forces to Protect Waterways," *Santa Fe New Mexican*, April 22, 2008; Associated Press, "Study Examines Gallatin River," *Billings (MT) Gazette*, Sept. 12, 2006.

42. National Park Service, National Center for Recreation and Conservation, *Nationwide Rivers Inventory* (n.d.).

43. For examples, see American Rivers, "Congaree River Blue Trail Honored with National Designation," press release (June 7, 2008); Jennifer McMillan, "By Land, Water, or Wheel," *River Management Society News* 18.2 (Summer 2005): 1–2; North American Water Trails, Silver Spring, MD, www.watertrails.org; National Park Service, *Economic Impacts of Protecting Rivers, Trails, and Greenway Corridors*" 4th ed., revised (1995).

44. *America's Great Outdoors: A Promise to Future Generations*.

45. See Colleen Slevin, "Whitewater Parks Roil Debate," *Rocky Mountain News*, May 30, 2005; Dave Curtin, "Rivers' Balance Imperiled," *Denver Post*, July 17, 2006; Natalie Bartley, "Bartley: Cascade, Riggins and Boise Work on Water Parks," *Idaho Statesman*, Oct. 29, 2007; Scott Sonner, "Kayakers Flock to Truckee River for Reno River Festival," *Las Vegas Sun*, May 11, 2006; Shane Benjamin, "City Pursues Recreation Water Rights," *Durango (CO) Herald*, Oct. 5, 2005.

46. America Outdoors Association, at: http://www.facebook.com/pages/America -Outdoors-Association/326605924178.

47. See Brett Prettyman, "Cataract Classroom," *Salt Lake Tribune*, June 23, 2005.

48. Two dam sites in Cataract Canyon were considered; one at the confluence of the Green and the Colorado rivers, and the other at Dark Canyon. The confluence site is now in Canyonlands National Park, and the mouth of Dark Canyon is covered by Lake Powell except in low-water years, of which there have been many recently. See Robert Webb, Jayne Belknap, and John Weisheit, *Cataract Canyon* (Salt Lake City: University of Utah Press, 2004).

49. You can watch Norm and Kent at work on YouTube at www.youtube.com /watch?v=9_VYNyT3w.

50. Craig Childs, *Stone Desert* (Englewood, CO: Westcliffe Publishers, 1995), 193.

51. See Kent Frost, *My Canyonlands* (New York: Abelard-Schuman, 1971).

52. See Ray Grass, "Adventurers Can Expect a Wet and Wild Season," *Deseret News*, May 25, 2006.

53. In the summer of 2011 the reservoir increased to 76 percent of full pool after a record-breaking year of precipitation.

54. See Niagara County Department of Planning, Development and Tourism, Samuel M. Ferraro, Commissioner, *Niagara County Town of Newfane Eighteenmile Creek Restoration Project* (Sanborn, NY, c. 2004); U.S. Environmental Protection Agency, "Eighteenmile Creek Area of Concern," at www.epa.gov/glnpo /aoc/eighteenmile.html.

55. This story was told by several of the speakers at the workshop.

12. THE RIVER COMMONS

1. Donald Worster, *A River Running West* (New York: Oxford University Press, 2001), 194.

2. J. W. Powell, *The Exploration of the Colorado River and Its Canyons* (1895) (New York: Dover, 1961), 285.

3. A similar experience can be had by boating the Gates of Lodore on the Green River or the lower Yampa River. Just below their confluence, at Echo Park, is evidence of a massive dam that would have inundated both rivers and parts of Dinosaur National Monument, as described in chapter 3. See Judy Fahys, "A River Saved," *Salt Lake Tribune*, June 16, 2003.

4. Richard Westwood, *Rough-Water Man* (Reno: University of Nevada Press, 1992), 231–32.

5. The exciting drama of the proposed dams in the Grand Canyon has been told in several excellent books: Russell Martin, *A Story That Stands like a Dam* (New York: Henry Holt, 1989), chap. 10; Steven Pyne, *How the Canyon Became Grand* (New York: Viking, 1998), 150–58; James Lawrence Powell, *Dead Pool* (Berkeley: University of California Press, 2008), 134–42.

6. This list became so notorious among river runners that the Grand Canyon Private Boaters Association called its newsletter the *Waiting List*. Before the waiting list was abandoned in 2006, it had 7,000 people on it. Shannon McKinnon, "Feds Will Let More People Raft Canyon," *Arizona Republic*, Nov. 11, 2005.

7. U.S. Department of the Interior, National Park Service, Grand Canyon National Park, *Colorado River Management Plan, Record of Decision* (Feb. 17, 2006).

8. The saga of the "motor fight" is told succinctly in *River Runners for Wilderness v. Martin*, No. 08-15112, D.C. No. CV-06-00894-DGC, Opinion (9th Cir. July 21, 2009).

9. "Thanks for Sharing," *Outside Magazine* (Nov. 27, 2010): 124.

10. Ives had every intention of returning to the Grand Canyon, but the Civil War broke out, and he joined the Confederate army and was killed in action. See Worster, *River Running West*, 130; Frank Waters, *The Colorado* (New York: Holt, Rinehart and Winston, 1946), 234–35.

11. See G. T. Pederson et al., "The Unusual Nature of Recent Snowpack Declines in the North American Cordillera," *Science Express* (June 9, 2011); "Global Climate Change Impacts in the United States: Regional Highlights, Southwest" (April 2011), at www.globalchange.gov/usimpacts; Craig Welch, "Study of 800-Year Old Tree Rings Backs Global Warming," *Seattle Times*, June 9, 2011; Felicity Barringer, "Southwestern Water: Going, Going, Gone?" *New York Times*, Feb. 11, 2011; Felicity Barringer, "Water Use in Southwest Heads for a Day of Reckoning," *New York Times*, Sept. 27, 2010; Jean Reid Norman, "Lake Mead Braces for Lowest Level since 1965," *Las Vegas Sun*, Apr. 24, 2009; Tim Barnett and David Pierce, "Sustainable Water Deliveries from the Colorado River in a Changing Climate," *Proceedings of the National Academy of Sciences* 108.50 (Mar. 6, 2009): 7334–38; Martin Hoerling et al., "Reconciling Projections of Colorado River Streamflow," *Southwest Hydrology* (May/June 2009): 20–31; Balaji Rajagopalan et al., "Water Supply Risk on the Colorado River: Can Management Mitigate?" *Water Resources Research* 45 (2009, W08201), doi:10.1029/2008WR007652; National Research Council, *Colorado River Basin Water Management: Evaluating and Adjusting to Hydroclimatic Variability* (Washington, DC: National Academies Press, 2007).

12. Amy Joi O'Donoghue, "Critics Blast Las Vegas Pipeline Proposal," *Deseret News*, Nov. 19, 2011; Josh Loftin, "Vegas Water Pipeline Could Cost $15B, Study Shows." *Bloomberg Businessweek* (Aug. 24, 2011); Jim Carlton, "Wet Winter Can't Slake West's Thirst," *Wall Street Journal*, Mar. 31, 2011; Patty Henetz, "Official: Water Deal Critical," *Salt Lake Tribune*, Apr. 26, 2010; Henry Brean, "Critics of Pipeline Project to Flood Crucial Meeting," *Las Vegas Review-Journal*, Aug. 17, 2009.

13. Mulroy shared this proposal with several people over lunch at the Spring Run-Off Conference, Utah State University, Logan, April 3, 2009. For background, see Pat Mulroy, "Climate Change and the Law of the River: A Southern Nevada Perspective," *West-Northwest Journal of Environmental Law and Policy*

14.2 (Summer 2008): 1–2; Sonoran Institute, *Growth and Sustainability in the Las Vegas Valley* (Jan. 2010).

    14. Lynn Arave, "Pipeline to Bring Water to Wasatch Front?" *Deseret News,* Oct. 18, 2009; David Ovard, "The Bear River Project: Utah's Last Water Hole," *Policy Perspectives* (Feb. 28, 2006): 1.

    15. "Cheaper Alternatives to Dams and Diversions on the Bear River," *Waterlines,* Utah Rivers Council News (Fall 2006): 1–2; Joe Baird, "Water Projects Moving Forward," *Salt Lake Tribune,* Nov. 10, 2005; "Conserving Water," *Utah Daily Herald,* Sept. 1, 2009.

    16. "Down the Drain," editorial, *Salt Lake Tribune,* Dec. 13, 2011; Brandon Loomis, "Task Force OKs State Tax Revenues for Lake Powell Pipeline," *Salt Lake Tribune,* Nov. 15, 2011; Patty Henetz, "Desert Mirage," *Salt Lake Tribune,* Jan. 6, 2010.

    17. See U.S. Global Change Research Program, *Global Climate Change Impacts* (New York: Cambridge University Press, 2009), 129–30; Barnett and Pierce, "Sustainable Water Deliveries from the Colorado River in a Changing Climate"; "Lake Mead Could Be Dry by 2021," *Scripps Institution of Oceanography News* (Feb. 12, 2008), at http://scrippsnews.ucsd.edu/Releases/?releaseID=876; U.S. Climate Change Science Program, "Synthesis and Assessment Product" (Jan. 16, 2009), at www.climatescience.gov/Library/sap/sap4–3/default.php; Tony Davis, "Research: SW Must Reduce Its Water Use," *Arizona Daily Star,* Dec. 14, 2010; Tom Yulsman, "Climate Change Tempers Good News About Colorado River Basin Water Supply," *Climate Central* (July 13, 2011).

    18. Associated Press, "Wyoming-Colorado Water Pipeline Proposal Draws Criticism," *Billings Gazette,* Dec. 13, 2011; Catherine Tsaj, "Army Corps Says It Terminated Flaming Gorge Study," *Casper (WY) Star-Tribune,* July 18, 2011; Bret Prettyman, "Pipeline Controversy: Tapping the Green River," *Salt Lake Tribune,* Sept. 18, 2010; Ben Neary, "Feds: Not Enough Green River Water for Pipeline," *Deseret News,* Sept. 20, 2009.

    19. See, for example, Colorado River Governance Initiative, "Rethinking the Future of the Colorado River," draft report (Dec. 2010), at www.waterpolicy.info; Rajagoplan et al., "Water Supply Risk on the Colorado River," 7652.

    20. "World's Largest Aquifer Going Dry," *U.S. Water News Online* (Feb. 2006), at www.uswaternews.com/archives/arcsupply/6worllarg2.html; Elizabeth Brooks, "Depletion and Restoration of the Ogallala Aquifer," in Andrew S. Goudie and David J. Cuff, eds., *Encyclopedia of Global Change* (Oxford: Oxford University Press, 2002), 144–48; Jane Braxton Little, "The Ogallala Aquifer: Saving a Vital U.S. Water Source," *Scientific American* (Mar. 30, 2009), at http://www.scientific american.com/article.cfm?id=the-ogallala-aquifer.

21. California Department of Water Resources, "Where Rivers Meet—The Sacramento–San Joaquin Delta" (n.d.), at www.water.ca.gov/swp/delta.cfm; S. E. Ingebritsen et al., *Delta Subsidence in California: The Sinking Heart of the State*, U.S. Geological Survey, FS-005-00 (Apr. 2000).

22. Jeffrey Mount and Robert Twiss, "Subsidence, Sea Level Rise, and Seismicity in the Sacramento-San Joaquin Delta," *San Francisco Estuary and Watershed Science* (March 2005): 1–18.

23. David Whitney, "Doolittle, Auburn Dam Win Key Ally," *Sacramento Bee*, Mar. 11, 2006; "Editorial: Natomas Delay? No, It's Too Risky," *Sacramento Bee*, Mar. 31, 2009; "Editorial: Natomas Levees Can't Be Delayed," *Sacramento Bee*, Aug. 21, 2008; Sacramento Area Flood Control Agency, *Natomas Levee Improvement Program Update* (July 31, 2009).

24. See Steven Hawley, *Recovering a Lost River* (Boston: Beacon Books, 2011), chap. 7; *Stormwater System Improvements Project*, funded by the Federal Emergency Management Agency (FEMA) and the city of Lewiston, Idaho (n.d.).

25. Charles Duhigg, "Cleansing the Air at the Expense of Waterways," *New York Times*, Oct. 13, 2009; Gary Pitzer, "The Water-Energy Nexus in the Colorado River Basin," Colorado River Project, *River Report* (Summer 2009): 1.

26. See U.S. Government Accountability Office, *Energy-Water Nexus*, GAO-11-35 (Oct. 2010); Keith Schneider, "A High-Risk Energy Boom Sweeps Across America," *Yale Environment 360* (2010), at http://e360.yale.edu/content/feature.msp?id=2324; Environmental Defense Fund and Western Resource Advocates, "Protecting the Lifeline of the West" (Boulder, CO, 2010); Western Resource Advocates, *Water on the Rocks: Oil Shale Water Rights in Colorado* (Boulder, CO, 2009); Sally Adee and Samuel Moore, "In the American Southwest, the Energy Problem Is Water," *IEEE Spectrum*, "Water vs. Energy Special Report" (June 20, 2010); Kirk Siegler, "Yampa River Runs with Possibility and Protest," National Public Radio, July 1, 2009.

27. Simon Donner and Christopher Kucharik, "Corn-Based Ethanol Production Compromises Goal of Reducing Nitrogen Export by the Mississippi River," *Proceedings of the National Academy of Sciences* 105.11 (Mar. 18, 2008): 4513–18; Tom Philpott, "More Corn for Meat and Ethanol, Less Habitat for Gulf Fish," *Grist* (Aug. 2, 2010), at http://www.grist.org/article/2010-08-02-more-corn-for-meat-and-ethanol-less-habitat-for-gulf-fish.

28. Chelsea Krotzer, "Oil Cleanup Continues on the Yellowstone River," *Billings (MT) Gazette*, July 11, 2011; Matthew Brown, "Yellowstone River: Rising Waters Could Push Crude Into Undamaged Areas," *Missoulian*, July 5, 2011.

29. Bevan Griffiths-Sattenspiel and Wendy Wilson, "The Carbon Footprint of Water," a River Network Report (May 2009), at www.rivernetowrk.org.

30. See J. C. Neff et al., "Recent Increases in Eolian Dust Deposition Due to Human Activity in the Western United States," *Nature Geosciences* 10.1038 (Feb. 24, 2008): 189–95; Charles Trentelman, "Study: Global Warming Would Hit Utah Hard," *Ogden (UT) Standard-Examiner*, Sept. 8, 2009; Allen Best, "Why Development in the Desert Means Lower Rivers and Less Snowpack in the Rockies," NewWest.Net, Nov. 30, 2010, at http://www.newwest.net/topic/article/why _development_in_the_desert_means_lower_rivers_and_less_snowpack_in_the _r/C35/L35/?utm_source=feedburner&utm_medium=feed&utm _campaign=Feed:+newwest/main+New+West+Network+Front+Page.

31. See Mary Battiata, "Silent Streams," *Washington Post*, Nov. 27, 2005; Hal Herring, "Wild Rivers and Riprap: The Case of the Yellowstone," NewWest.Net, Oct. 2, 2006; at http://www.newwest.net/topic/article/wild_rivers_and_riprap _the_case_of_the_yellowstone2/C505/L33/; Scott McMillion, "Army Corps to Monitor Impacts of Projects on Yellowstone River," *Bozeman (MT) Daily Chronicle*, Feb. 22, 2007.

32. Julia Pyper, "World's Dams Unprepared for Climate Change Conditions," *Scientific American* (Sept. 16, 2011): 30–31; Daren Carlisle et al., "Alteration of Streamflow Magnitudes and Political Ecological Consequences," *Frontiers in Ecology and the Environment*, doi:10.1890/100053 (2010); Rebecca Carter and Susan Culp, *Planning for Climate Change in the West*, Policy Focus Report (Cambridge, MA: Lincoln Institute of Land Policy, 2010); Doyle Rice and William Welch, "The Flooding The Tornadoes The Wildfires," *USA Today*, June 7, 2011.

33. Fred Pearce, "U.S. Plans Its First Megadam in 40 Years," *New Scientist* (Dec. 8, 2011); Cory Hatch, "Commission Against Green Dam," *Jackson Hole (WY) News & Guide*, Nov. 10, 2010; Monte Whaley, "200 Rally for Northern Front Range Reservoir Project," *Denver Post*, July 16, 2010; Rocky Barker, "New Dams on the Boise River?" *Idaho Statesman*, June 29, 2010; David Lester, "Study of Water Storage Options in Yakima Basin Gets $2 Million," *Yakima (WA) Herald-Republic*, Apr. 29, 2009; Joey Burch, "Reservoir Plan Resurfaces with New Pitch from Senator," *Denver Post*, May 29, 2005; Michael Booth, "Divide Develops Over Dam," *Denver Post*, April 23, 2008; "The Last of a Dying Breed," *Waste Basket* (Taxpayers for Common Sense) 2.12 (April 15, 1997): 1–2; Shaun McKinnon, "New Yuma-Area Reservoir to Guard Against Drought," *Arizona Republic*, Jan. 12, 2008; Randal Archibold, "An Arid West No Longer Waits for Rain," *New York Times*, Apr. 4, 2007; "FOR Opposes New Dams at Congressional Hearing," *Headwaters* (Friends of the River) (Summer 2003): 10; Nicholas Geranios, "Dam-Building Era May Not Be Over in West," *Deseret Morning News*, Mar. 3, 2008; Nate Poppino, "High-Desert Hydro Developers Envision More Hydropower near China Mountain,"

Magic Valley Times (Twin Falls, ID), Mar. 5, 2009; "California Contemplates a New Era of Dam-Building," *U.S. Water News* 19.8 (Aug. 2002).

34. Bureau of Reclamation, *Framework for a Special Study on Water Supply on the Henrys Fork of the Snake River Basin* (Washington, DC: Department of the Interior, n.d.); Rocky Barker, "Barker: Could the Teton Dam Be Rebuilt?" *Idaho Statesman*, Feb. 7, 2011; Cory Hatch, "Group Puts Teton River on 'Endangered' List," *Jackson Hole News*, June 3, 2010; "Study Mulls Rebuilding Dam," *Idaho State Journal*, June 11, 2009; Kim Trotter, "Arguments for Rebuilding Teton Dam Don't Hold Water," *Idaho Statesman*, July 15, 2009; David Whitney, "House Committee Advances Revival of an Auburn Dam," *Contra Costa (CA) Times*, May 12, 2006.

35. Robert Black, "Wyoming Aims for More Dams," *Casper Star-Tribune*, Aug. 25, 2005; Jeff Gearino, "Dam Projects Face Hurdles," *Casper Star-Tribune*, Aug. 28, 2005.

36. U.S. Army Corps of Engineers, Headquarters, "President's Fiscal Year 2012 Budget for U.S. Army Corps of Engineers Released" (Feb. 14, 2011), at http://www .usace.army.mil/CEPA/NewsReleases/archive/2011/02/14/presidents-fiscal-year -2012-budget-for-u-s-army-corps-of-engineers%E2%80%99-civil-works-released .aspx.

37. Mark Twain, *Life on the Mississippi* (New York: Collier and Son, 1874), 18.

38. George Byron Merrick, *Old Times on the Upper Mississippi* (1909) (Minneapolis: University of Minnesota Press, 2001), 14. Merrick quit his job as a riverboat pilot to fight in the Civil War. He later became a newspaper editor. A state park in Wisconsin is named for him.

39. Ibid.

40. Walter Havighurst, *Voices on the River: The Story of the Mississippi Waterways* (1964) (Minneapolis: University of Minnesota Press, 2003), 139–40.

41. See, for example, Holly Jones and Oswald Schmitz, "Rapid Recovery of Damaged Ecosystems," *PLoS ONE* 4.5 (2009): e5653; Federal Interagency Stream Restoration Working Group, *Stream Corridor Restoration* (National Oceanic and Atmospheric Administration, NOAA Restoration Center, n.d.); Marcus Hall, *Greening History: The Presence of the Past in Environmental Restoration* (London: Routledge, 2010); Peter Friederici, *Nature's Restoration* (Washington, DC: Island Press, 2006).

42. See, for example, Economics of Ecosystems and Biodiversity, "Mainstreaming the Economics of Nature," The Economics of Ecosystems and Biodiversity (TEEB) (2010), http://www.teebweb.org/; Storm Cunningham, *Rewealth* (New York: McGraw-Hill, 2008); Joel Makower, *Strategies for the Green Economy* (New York: McGraw-Hill, 2008); Paul Hawken, *Natural Capitalism* (New

York: Back Bay Books, 2008); Western Progress and Progressive States Network, *Building a Restoration Economy: Legislation and Practices at the State Level* (New York, n.d.); Robert Costanza et al., *An Introduction to Ecological Economics* (Boca Raton, FL: CRC Press, 1997).

43. Restoration projects are, quite literally, too numerous to mention.

44. See, for example, Karen Filipovich, "A Watershed Approach: There's a New Homegrown Democratic Process at Work in Montana and the West," *Headwaters News*, Jan. 4, 2007, at http://www.headwatersnews.org/; Jerd Smith, "Water Coalition in the Making," *Rocky Mountain News*, Nov. 8, 2004; Jerd Smith, "Plan Would Create Water Roundtables," *Rocky Mountain News*, Jan. 21, 2005.

45. American Rivers, *River Budget: National Priorities for Local River Conservation Fiscal Year 2007* (Washington, DC, 2007).

46. Interior Alliance, *The Treaty Initiative to Share and Protect the Global Water Commons* (Kamloops, British Columbia, June 14, 2001); Peter Gleick, "Global Freshwater Resources: Soft-Path Solutions for the 21st Century," *Science* 302 (Nov. 28, 2003): 1524–28; Rocky Mountain Institute, "A Paradigm Shift for Water Management," at www.rmi.org/sitepages/pid279.php; John Leshy, "Notes on a National Water Policy," *Harvard Law and Policy Review* 3 (2009): 133–59; Edella Schlager and William Blomquist, *Embracing Watershed Politics* (Boulder: University Press of Colorado, 2008); Mark Lubell et al. "Watershed Partnerships and the Emergence of Collaborative Action Institutions," *American Journal of Political Science* 46.1 (2002): 148–63. For an in-depth analysis of collaborative watershed partnerships, see Paul Sabatier et al., eds., *Swimming Upstream: Collaborative Approaches to Watershed Management* (Cambridge, MA: MIT Press, 2005).

47. See, for example, Mark Jaffe, "Unusual Alliances Give Dams Upgrades," *Denver Post*, Jan. 11, 2009.

48. There are many voluble critics of water privatization. See Jeffrey Rothfeder, *Every Drop for Sale* (New York: Tarcher/Putnam, 2001); Maude Barlow and Tony Clarke, *Blue Gold* (New York: New Press, 2002); Vandana Shiva, *Water Wars: Privatization, Pollution, and Profit* (Cambridge, MA: South End Press, 2002); Jon Luoma, "Water for Profit," *Mother Jones Magazine* (Nov./Dec. 2002): 34; Peter Gleick et al., *The New Economy of Water* (Oakland, CA: Pacific Institute for Studies in Development, Environment, and Security, 2002).

49. Peter Carrels, *Uphill Against Water* (Lincoln: University of Nebraska Press, 1999), 208.

50. Wallace Stegner, *The Sound of Mountain Water* (New York: E. P. Dutton, 1946), 36.

51. Aldo Leopold, *A Sand County Almanac* (New York: Oxford University Press, 1949), 204.

52. This quote is attributed to Gandhi, but is quite similar to a speech given by a labor organizer to the Amalgamated Clothing Workers of America in 1914.

53. Chad Pregracke, *From the Bottom Up: One Man's Crusade to Clean America's Rivers* (Washington, DC: National Geographic, 2007), 283.

54. James Madison, "The Federalist No. 45," in *The Federalist Papers* (New York: Washington Square Press, 1964), 98.

55. "Race Discrimination: Online Information on Race Discrimination," at http://www.racediscrimination.org/2011/12/11/jackie-robinson-civil-rights-pioneer -note-the-epitaph-and-brief-biography.html.

56. For a fascinating discussion of such challenges, see Brock Bahler, "A Catholic Bishop and Two Atheists Bringing Justice to the World," *Fides Quarens Intellectum* 4.1 (Fall 2007): 77.

57. Marjory Stoneman Douglas and John Rothchild, *Marjory Stoneman Douglas: Voice of the River* (Sarasota, FL: Pineapple Press, 1990), 211.

58. Twain, *Life on the Mississippi*, 317.

59. Quoted in Edmund Morris, *Theodore Rex* (New York: Random House, 2001), 516.

60. Amos 5:24 (Revised Standard Edition).

61. Musqueam Territory, *Indigenous Declaration on Water* (Vancouver, British Columbia, July 8, 2001).

62. Edward O. Wilson, *The Diversity of Life* (Cambridge, MA: Belknap Press of Harvard University Press, 1992), 350–51.

63. Not to be confused with the medical term.

64. Gretel Ehrlich, "On Water," in *The Solace of Open Spaces* (New York: Penguin, 1985), 77.

65. Barry Lopez, *Crossing Open Ground* (New York: Vintage Books, 1978), 65.

66. Annie Dillard, *Pilgrim at Tinker Creek* (New York: Bantam Books, 1975), 10.

67. John McPhee, *Encounters with the Archdruid* (New York: Farrar, Straus, Giroux, 1971), xx.

68. Henry David Thoreau, *Walking* (Boston: Beacon Press, 1994), 122.

INDEX